Official Languages
English
Spanish
French
Dutch
ATLANTIC OCEAN
ma Islands
co
era
San Salvador
Rum Cay
Turks & Caicos
Haiti
Dominican Republic
Puerto Rico
Virgin Islands
Anguilla
St. Maarten/St. Martin
Saba
Antigua
Guadeloupe
Dominica
Martinique
BBEAN SEA
St. Lucia
St. Vincent & the Grenadines
Barbados
Grenada
Aruba
Curacao
Bonaire
Los Roques
Margarita Is.
Cubagua
Tobago
Trinidad
Venezuela
Guyana
Colombia
AF719922

Nudibranch & Sea Slug Identification

FLORIDA CARIBBEAN BAHAMAS

LES WILK

NEW WORLD PUBLICATIONS, INC.

Jacksonville, Florida USA

CREDITS

Cover Photo: Les Wilk
Title Page Photo: Keri Wilk
Back Cover Photos: Les Wilk (Tl), Kris Wilk (Tr), Keri Wilk (Bl & Br)

Photography Editor: Les Wilk
Format and Layout: Any Wilk
Editor: Any Wilk
Prepress Consultant: Eric Riesch
Print Consultant: D'Print Pte Ltd., Singapore
Publisher: New World Publications, Inc.

ISBN 978-1-878348-72-2

Distributed by New World Publications, Inc., 1861 Cornell Road, Jacksonville, FL 32207
Phone (904) 737-6558, fishid.com, orders@fishid.com

Acknowledgements

This guide rests upon the foundation of information made available by numerous individuals within a wide-ranging community, including:

- Professional marine scientists whose books and peer-reviewed scientific publications underpin our current knowledge of sea slugs.
- Highly knowledgeable citizen scientists who conduct targeted research, document unusual behaviors and external morphologies with videos and macro photographs, and occasionally discover species new to science. The results of their dedication have often been published in journals, marine-related websites and image archives. Some have published their own books.
- The large community of enthusiastic underwater macro photographers who contribute by recording and sharing data that may be significant to the advancement of science.

I especially thank the following individuals for their helpful and generous responses to my requests and inquiries:

Ángel Valdés
Jeff Hamann
David Behrens
Anne DuPont
Terence Gosliner
Bernard Picton
Patrick Krug
Hans Bertsch
Michael Middlebrooks
Phil Allman
Nestor Ardila
Jesús Ortea
Yan Buske
Manuel Caballer
Alvaro Esteves Migotto
Stéphane Hourdez
Stephen C. Kempf
Karen Osborn
Alicia Hermosillo McCowen
Gustav Paulay (FMNH)
Sofía Paz Sedano
Athila Bertoncini
Andrea Bernecker
Senem Çağlar & Bilal Ozturk
Mike G. Rutherford
Matthew J. Scripter
Nancy Sheridan (Florida FWC)
Robert Aguilar (Smithsonian)
Joshua Stone
Jesús Souza Troncoso
Deneb Ortigosa
Carlos Sigmaringa
Dean Janiak
Vinicius Padula & Joao Oliveira

Other Resources

Websites:
iNaturalist: inaturalist.org
Nudibranch Domain: nudibranchdomain.org
OPK opistobranquis: opistobranquis.info/en
Sea Slug Forum: seaslugforum.net

Facebook:
Cayman Society of Underwater Photographers (CAYSOUP)
Cozumel Diving
Nudibase
Nudibranch Central
Nudibranch Lovers
Underwater Macro in the Americas (formerly Nudibranch Hunters of Florida)

Online galleries:
Ray Haberman, St. Vincent: raymondhaberman.smugmug.com
Melodie Caussat, Guadeloupe: seaslugs-guadeloupe.com
Alain Goyeau, Guadeloupe: alaingoyeau.fr/fr/limaces-de-guadeloupe-et-de-la-caraibe
Julie Mellinger, Guadeloupe: juliemwildphotography.com
Ellen Muller, Bonaire: pbase.com/imagine
Suzan Meldonian, Florida: niteflightphoto.com

Authors:
Valdés et al (2006):
Caribbean Sea Slugs
Colin Redfern:
Bahamian Seashells: 1161 Species from Abaco, Bahamas
Everett M. Turner Jr., Essi Evans & Cynthia Abgarian:
Cayman Has Nudibranchs and Other Sea Slugs
Mickey Charteris:
Caribbean Reef Life
Linda Ianniello & Susan Mears:
Blackwater Creatures: A Guide to Southeast Florida Blackwater Diving

Personal Support

No long term project of this magnitude can be completed without the sustaining support of family. I extend heartfelt thanks to my immediate family members for their technical support, proofreading, helpful suggestions, and encouragement throughout this endeavor: **Kris, Keri, Daniela, Any**.

I reserve my deepest gratitude for my wife Any whose meticulous and tireless work was instrumental in shaping the final quality of this book: page layouts, page formatting, proofreading, creation of indexes and photo credits pages, and for many suggestions that improved both the text and the structure.

Introduction

Naming Conventions

The universal convention currently being used for assigning unique scientific names for species of organisms was established by Carl Linnaeus in the 18th century. A name consists of two separate parts: a **generic name** (genus) followed by a **specific epithet** (species), each being typically a Latin (or Latinized) form of a descriptive or otherwise meaningful term applicable to the organism. When written, both parts are in lower case **italics** except for the first letter of the generic name, which is always capitalized.

In this book, each species description begins with a header having Left and Right components:

Left: Its scientific name followed by the name(s) of the individual(s) who first described the species and the year in which the description was published. For example: ***Elysia crispata*** **Mörch, 1863**

Right: Its **common name**, which is a non-scientific name that has become associated with the species through common usage. Such names often do not exist. Moreover, they can be problematic for species that have acquired different common names in different geographical regions (some very far apart). In such cases the least confusing way of referring to them is by their scientific names.

When a species is reclassified into another genus after its original description, the generic name is modified accordingly and the scientist's name (or names) and the date of the original publication are enclosed in parentheses. For example: in 2011 ***Chelidonura berolina*** **Er. Marcus & Ev. Marcus, 1970** was reclassified into the genus *Camachoaglaja*, thus becoming ***Camachoaglaja berolina*** **(Er. Marcus & Ev. Marcus, 1970).**

An animal that does not yet have a formal scientific name is called an undescribed species. If it can be assigned confidently to a well known genus then it can be referred to by the informal (incomplete) scientific name consisting of the known genus followed by the specific epithet **sp.**, which is shorthand for saying "a species that belongs to the indicated genus". For example: ***Tyrinna*** **sp.** If there are several undescribed species within the same genus they can be designated individually with a unique sequence number. For example: ***Cuthona*** **sp. 1**, ***Cuthona*** **sp. 2**, etc. The plural of **sp.** is **spp.**, which represents a group of undescribed species within the same genus. For example: ***Doto*** **spp**.

Animals that are *probably* a described species but warrant some additional scrutiny for confirmation because of doubts are indicated by inserting **cf.** between the genus name and the specific epithet. The contraction **cf** is short for the Latin word "confer", meaning "compare with". For example: ***Costasiella*** **cf.** ***patricki***.

An animal with an external appearance that is nearly identical to that of a described species has usually been considered to be that species, with minor differences assumed to be the products of intraspecies variability. However, the increased use of molecular analyses has uncovered numerous **cryptic species** within the bounds of such variability, i.e. organisms nearly identical in appearance and anatomy but different enough genetically to be considered unique species. In many cases, such studies have pin-pointed subtle external features that can now be used to distinguish between these species. Accordingly, increased attention must now be given to undescribed differences.

Presentation

The basic hierarchy of ranks within the Linnaean taxonomy of organisms is, from top down: Domain, Kingdom, Phylum, Class, Order, Family, Genus, Species. Intermediate ranks have also been defined as necessary, such as Subclass, Infraclass, Subterclass, Superorder, Superfamily, etc. In the pages with slug descriptions the taxonomy is limited to Order, Family, Genus, and Species. If an Order has not been defined, a suitable upper taxon is selected. For example, since the hierarchy for the family Acteonidae does not include an Order, it is nested under the **Superfamily Acteonoidea** instead.

Families, Genera, and Species are presented alphabetically, with minor exceptions. The left header on each page indicates the Order (or an alternate if it does not exist) and the right header indicates the Family. The Quick Reference near the front, and also on the inside of the back cover, alphabetically lists the Orders and their Families for the species included in this book. Each Family section begins with a brief family description, followed by images and descriptions for the species within it.

For convenience the page header background for each order is given a unique color.

Additional Content

To simplify the interpretation of species descriptions, annotated drawings of slug body parts are shown after the next page.

For those interested in more information or details there is also a list of primary references at the back of the book, followed by a Glossary of Terms used in the descriptions.

Indexes

At the back of the book there are two indexes for species names. One is for scientific names and the other is for common names. Note that most species do not have common names. Photo credits immediately follow them, organized by both page number and photographer name.

Nomenclature and the Pace of Discovery

The scientific naming of marine organisms, particularly sea slugs, is currently undergoing a period of unprecedented change, driven largely by modern molecular phylogenetic analyses. This process frequently results in some species being moved to new genera, the splitting of an existing species into a complex of new species, the resurrection of old names previously classified as *nomina dubia*, and even the partitioning of higher taxa. For example, 13 families from the order Nudibranchia were recently removed from it and placed in the resurrected order Doridida.

While this book utilizes the most current and verified names available at the time of publication, paper-based field guides such as this face a continuing inherent challenge: they are static snapshots of a continuously evolving scientific consensus. Unlike electronic databases or online resources, a printed guide cannot update its taxonomy in real time. If any name presented in this book differs from that in the World Register of Marine Species (WoRMS) it **could** be an error, but more likely it is evidence of a scientifically accepted nomenclatural change that needs to be incorporated into the next book printing.

We encourage users to cross-reference with up-to-date online resources for the very latest taxonomic placements.

About the Author

Les Wilk holds a Ph.D. in theoretical physics and spent his professional career applying his knowledge in academia and in the nuclear energy industry. However, his "second career" – and his true passion – is underwater. He has spent nearly five decades photographing, researching, and writing about marine life.

Travelling and diving had always been activities that Les' family enjoyed – spending weeks each year exploring the Caribbean. In the 1990s, when one of his sons founded ReefNet (a diving technology company), these interests deepened. He initially applied his scientific background to explain underwater physics and optics phenomena in articles published at **scubageek.com**. His curiosity soon expanded into marine biology as the family began working toward publishing an electronic field guide to Caribbean fishes. Casual dive trips evolved into serious pursuits of underwater photography and identification, with Les taking the role of lead researcher.

Applying the same scientific rigour he used as a physicist, Les has spent nearly every day since then distilling scientific literature, consulting with experts, and collecting and cataloging photographs to build thousands of comprehensive species profiles. This work is evident in ReefNet's interactive field guide **Reef Fish Identification**, and more recently in his collaboration with Paul Humann and Ned DeLoach to expand their definitive book **Reef Creature Identification: Florida Caribbean Bahamas**.

Despite having no formal education in biology, Les has become widely recognized as an authority on Caribbean marine life. His meticulous research bridges the gap between marine scientists and recreational naturalists. This book reflects a specialized focus of that ongoing work.

Les lives near Toronto, Canada with his wife and longtime collaborator, Any.

--- Kris Wilk & Keri Wilk ---

External Morphology Terms

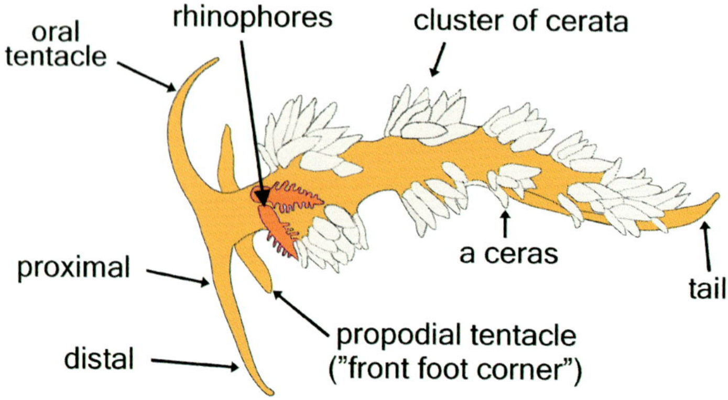

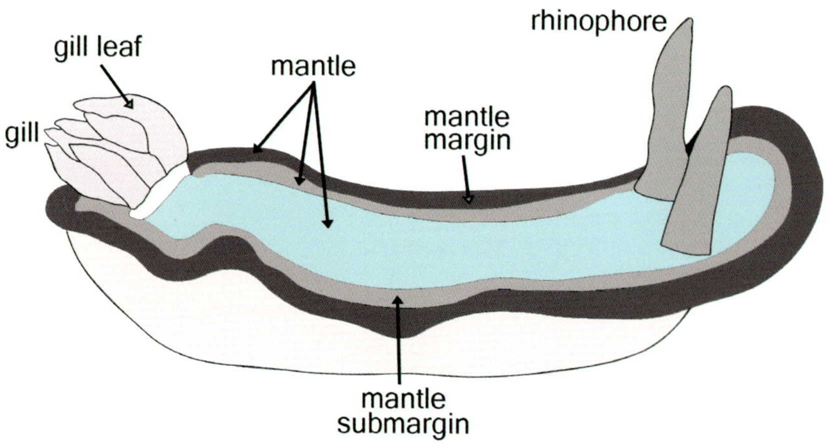

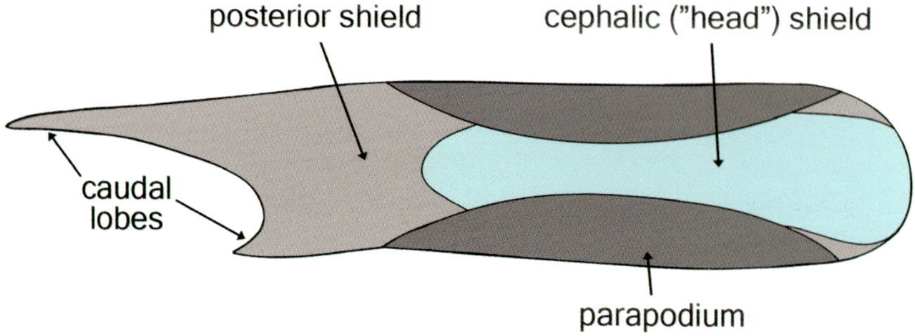

Common Rhinophore Types

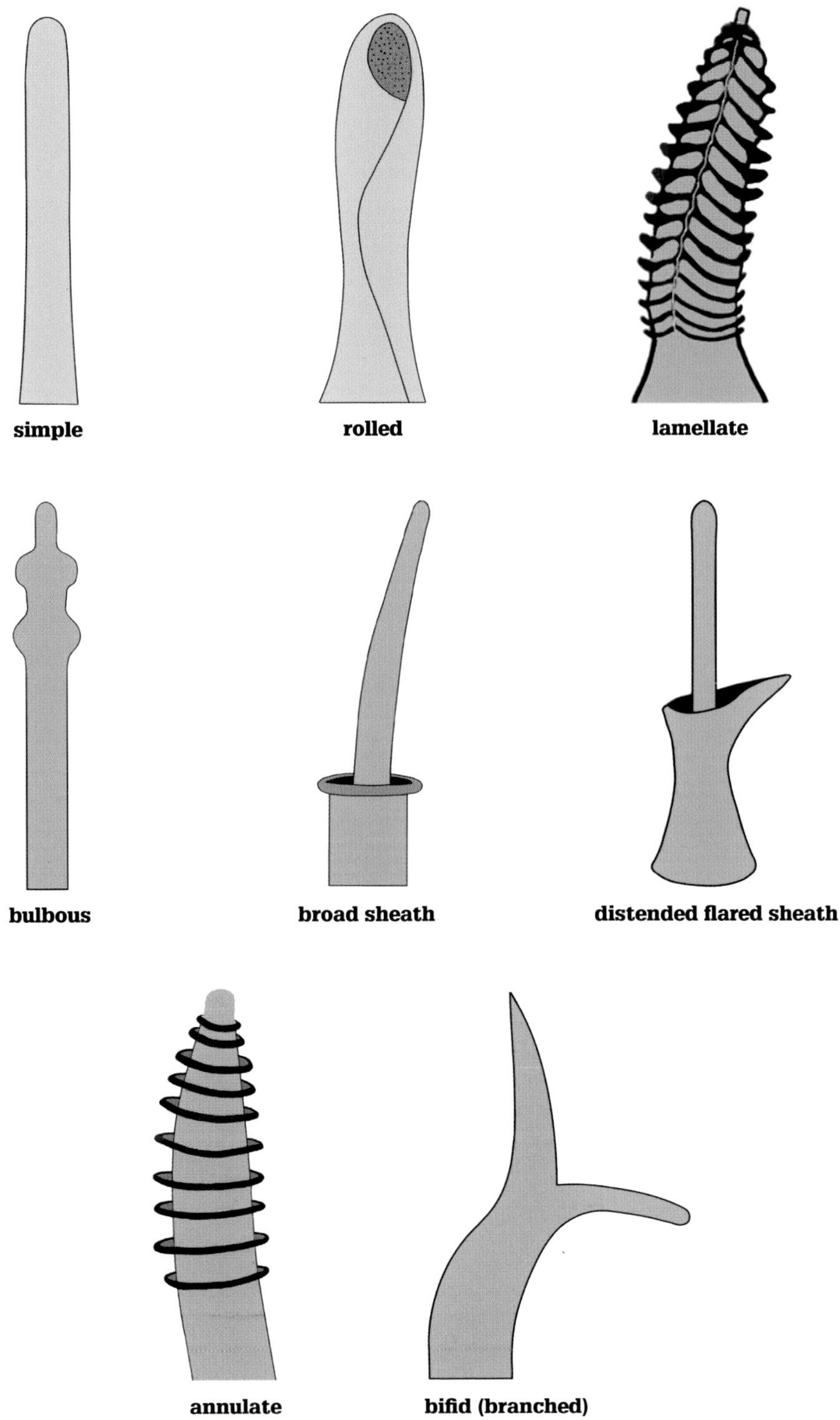

Common Cerata Types & Example Genera

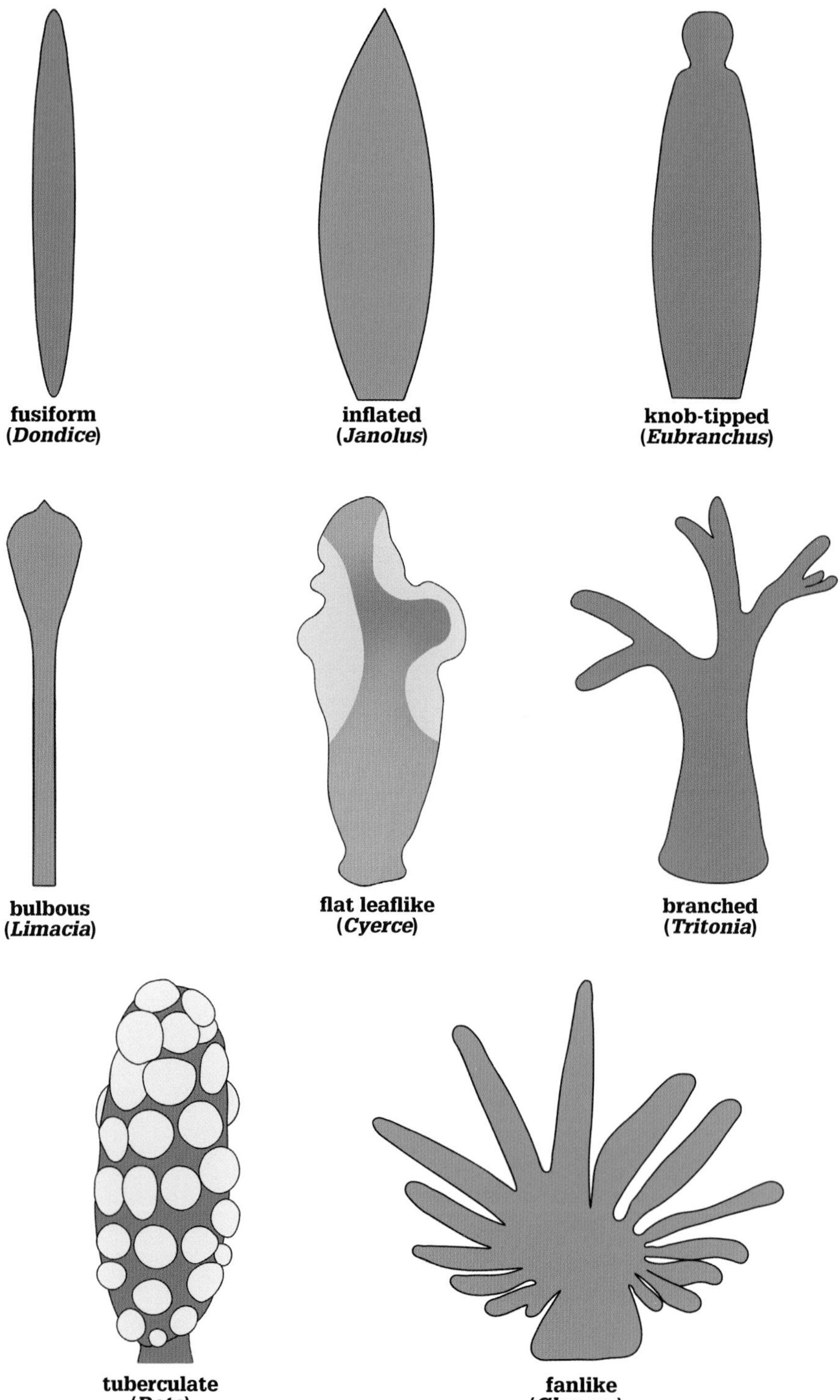

Quick Reference (Order & Family)

Do not be afraid of nature's complexity. Embrace it. In every small organism, in every plankton, is a vast library of genetic intricacies waiting to be read and to have their secrets revealed.

This book is dedicated to my favorite
pair of budding naturalists:

My Grandchildren

Hugo & Coral

Family ACTEONIDAE A. d'Orbigny, 1842

Shell external, ovoid to fusiform, stocky; spire moderately elevated, usually conical, up to 8 whorls and a prominent protoconch; aperture narrow posteriorly, flares forward to an oval bounded by a long gently curved outer lip; columella with one to several strong folds; whorls rounded, sutures pronounced, body whorl sculptured with numerous punctate spiral grooves; periostracum thin; shell length usually less than 25 mm.

Body short and wide, fully retractable into shell, usually supplied with a chitinous operculum to seal the aperture following a defensive retraction. Cephalic shield with a pair of anterior lobes and a larger pair of anterior lobes that cover the front of the shell. No parapodia. Foot usually short, broad, front corners angular or tentacular, rear end shorter than shell.

Inhabit both shallow and deep waters (to at least 10,000 feet). Prey on polychaete worms.

***Japonacteon punctostriatus* (C. B. Adams, 1840)** **pitted baby-bubble**
Size: to 7.5 mm Acteonidae
ID: Shell solid, elongate to moderately globose; spire conical, 2-3 whorls; suture weakly channeled; protoconch globose, 1.5 whorls; no umbilicus; the aperture broadens quickly to form a rounded anterior and protruding lip; many spiral grooves with small oval pits. Shell with animal inside brown-blotched. No other information for animal. Intertidal to at least 60 ft. Florida, Gulf of Mexico, Antilles to Costa Rica.

***Acteon candens* Rehder, 1939** **Rehder's baby-bubble**
Size: to 10 mm Acteonidae
ID: Shell large, solid; spire large, conical; animal's cephalic shield with frontal notch & two small rear lobes. Shell opaque glossy white, reddish bands at edges of early whorls. Animal white. Intertidal to 70 ft. Florida, Gulf of Mexico, Cuba, Costa Rica, Brazil.

***Acteon danaida* Dall, 1881**
Size: to 11 mm Acteonidae
ID: Shell opaque glossy white, elongate, moderately pointed; ~6 rounded whorls on the spire, 6 spiral grooves per whorl; body whorl with ~25 spiral grooves crossed by growth lines; aperture with thin outer lip. Animal unknown. Beach drift to 2000 ft. South Florida, Bahamas, northern Brazil.

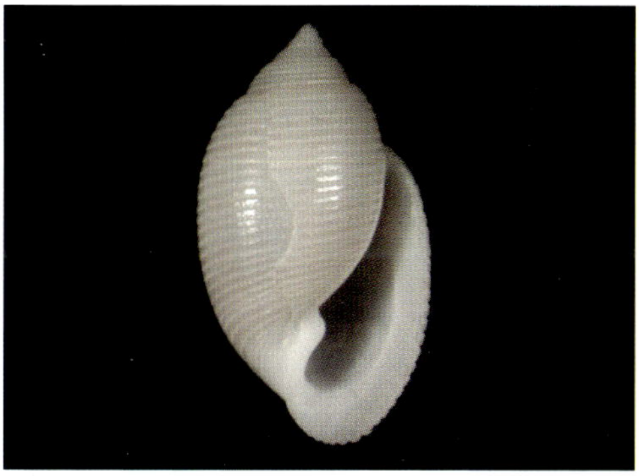

***Acteon finlayi* McGinty, 1955**
Size: to 15 mm Acteonidae
ID: Shell pinkish, ovate, spire conical, ~6 slightly rounded whorls, sutures impressed; umbilicus almost closed; body whorl with spiral rounded threads and thinner longitudinal ones. Animal details unknown. To at least 240 feet. East Florida, Cuba, Virgin Islands, Guadeloupe, Grenadines, Brazil.

***Mysouffa cumingii* (A. Adams, 1855)**
Size: to 20 mm Acteonidae
ID: Shell large, solid, globose; spire short, conical; surface covered with punctate grooves. Animal with cephalic shield bilobed anteriorly, 2 large rear lobes that cover front of shell. 30-230 feet. Southeast Florida, Guadeloupe, Antigua, St. Vincent and the Grenadines, Panama, Brazil.

Family APLUSTRIDAE Gray, 1847

Circumglobal in temperate to tropical seas. Only two genera occur in the tropical West Atlantic – *Hydatina* and *Micromelo*. External shell globose to ovoid, thin-walled, fragile, exposed sunken spire with 2-3 whorls, apical whorl a tiny central knob, body whorl very large, auricular aperture narrow posteriorly but broadly flared toward anterior. Animal can protrude substantially from the aperture but cannot retract completely into the shell, so there is no operculum.

Body oval to elongate. Headshield large, with three pairs of lobes – a moderately large semi-rolled frontal pair, followed by a smaller pair of rolled funnel-like siphonal lobes that direct water toward the Hancock's chemosensory organ beneath them, and then a large rounded flap-like posterior pair that extends back and sideways to partially cover the front of the shell. A rear extension of the mantle forms an infra-pallial lobe that folds up and over the shell's apex. Foot oval, broad, flat, delicate, extends beyond mantle with wavy sides often raised like parapodia. Small close-set eyespots framed by opening in headshield. Large plicate gill sometimes visible extending from the right side of its large mantle cavity. Inhabit shallow subtidal and intertidal waters where they crawl in the open on reefs, sand, and seagrass flats in search of their main prey – polychaete worms. They seem adapted for this specialized feeding by having a long muscular proboscis-like oral tube that is well suited for probing worm tubes. For defensive purposes they rely on swimming, camouflage, and the extraction and deployment of worm toxins.

Florida

Bonaire

Hydatina physis **(Linnaeus, 1758)**

brown-line paper bubble

Size: to 50 mm — Aplustridae

ID: Shell globose, translucent white with narrow black to brown spiral lines, usually with two broad blackish ones bracketing a narrower brown one. Parapodia and headshield pink to purplish-red with white or iridescent blue margins. Depth: usually < 20 feet but up to 120 feet. Bermuda, Florida, Bahamas, Antilles, French Guiana, Brazil, Venezuela, Mexico, Jamaica.

St. Vincent

Micromelo undatus **(Bruguiere, 1792)** **miniature melo**

Size: to 30 mm — Aplustridae

ID: Shell glossy white, 3 transverse reddish-brown spiral lines joined by axial arcs. Animal translucent gray with large white oval spots; margins of mantle, foot, and rhinophoral lobes with thin brown edge, followed progressively by bands of yellow/orange to whitish to greenish/bluish. To 17 feet. Bermuda, Florida, Bahamas, Caribbean, and Brazil.

with spawn; Grand Cayman

Puerto Rico

Family AGLAJIDAE Pilsbry, 1895

A family of 16 genera and almost 100 species of benthic carnivores occurring worldwide, most commonly in tropical and subtropical latitudes in depths from shallows to almost 700 feet. Sometimes prominently colored. Lengths up to 60 mm.

Body usually elongate and narrowly cylindrical; posterior visceral hump with one or more blunt to tail-like caudal lobes, a pair usually asymmetrical with one larger/longer than the other; cephalic shield large, somewhat triangular with broad anterior base (often with lobed corners) and posterior narrowed portion often raised vertically or lying flat over the anterior part of the visceral hump; front edge of shield with clusters of sensory bristles flanking the mouth for tracking prey; a parapodial flap extends upwards from each side of the foot, either covering the sides of the body or continuing up to the dorsal midline where they sometimes meet. Gill concealed within the laterally-opening mantle cavity; radula generally absent, sometimes vestigial. Shell internal, reduced to a small translucent plate; either membranous or wholly/partially calcified.

Usually found crawling on shallow sandy bottoms among rocks, corals, algae, and seagrasses, where they are active feeders on mobile prey such as polychaete and nemertean worms, acoel flatworms, small crustaceans, ctenophores, bubble snails, and even other slugs (including their own species). Lacking a radula, they evert their buccal bulb (very large in *Navanax*) and use it as a suction pump to trap the prey and ingest it whole. Some species are known to sequester alarm pheromones from their prey for signaling the presence of danger to their conspecifics.

Recent studies have indicated that color patterns and external morphologies are not reliable features for species identification. For example, molecular analyses of morphologically identical populations of *Navanax aenigmaticus* from geographically isolated populations uncovered a complex of three distinct lineages. This finding casts some doubt on the circumtropical distribution of other species, such as *Chelidonura hirundinina*, which has long been considered to have a presence in both the Atlantic and the Indo-Pacific.

Species are broadly variable in appearance and identical looking animals may be distinct species based on internal anatomy differences. In the following, their identifications are both historically-based and necessarily tentative pending further studies.

Bahamas

Riviera Beach, Florida

Bermuda

Roatan

St. Vincent

Camachoaglaja berolina **(Er. Marcus & Ev. Marcus, 1970)** **berolina aglaja**

Size: to 12 mm Aglajidae

ID: Highly variable. Left tail lobe long, sharply pointed. Ground color reddish-brown, pale/dark brown, black; dorsum may have white or yellow patches/spots/speckles. Parapodia and front of headshield often with yellow/orange submarginal band, sometimes broken or bordered by blue spots. Posteriorly, the headshield and body may have a white or yellow band or transverse blotch. Depth: < 10 feet. Bermuda, south Florida, most of Caribbean Sea.

↓Bahamas↓

↓Roatan↑

Camachoaglaja mariagordae **(Ortea Espinosa & Moro, 2004)** **Norman's aglaja**

Size: to 12 mm Aglajidae

ID: The external morphologies of this species and *Camachoaglaja berolina* (previous page) are very similar and their variable color patterns also overlap considerably, making external appearance alone an unreliable tool for distinguishing between them. The type locality for *C. mariagordae* is Maria la Gorda, Cuba. The top left animal was initially described as the genetically distinct species *Chelidonura normani* Ornelas-Gatdula, Dupont & Á. Valdés, 2011, which is currently synonymized with *C. mariagordae* in WoRMS (but that may change). The other two animals above are assumed to be variants; in Cuba the bottom left one is locally referred to as the "blue sand" variant. The black one is considered by some to be a variant of *C. berolina*. Clarifications are needed. So far, the presence of this species has been confirmed only in Cuba, the Bahamas, and Roatan, but more locations will likely be discovered. Depth and habitat as for *C. berolina*.

Riviera Beach, Florida

Camachoaglaja **sp. 1**

Size: 10 mm Aglajidae

ID: Body somewhat broad; front of headshield bilobed, with two large round unpigmented areas, each covering a small black eye at its inner edge; parapodia small; left tail lobe longer and sharply pointed, right lobe short and rounded. The dorsum is white with a coarse reddish-brown net-like pattern. About 20 feet deep. Riviera Beach, Florida.

Bonaire

Camachoaglaja **sp. 2**

Size: 10 mm Aglajidae

ID: Very similar to *Camachoaglaja* sp. 1, but the net-like pattern is considerably finer, especially at the rear of the body. It could be a variant of that species. Depth: ~20 feet.

Riviera Beach, Florida

Camachoaglaja **sp. 3**

Size: 10 mm Aglajidae

ID: Headshield broad, front bilobed, each lobe with a large round unpigmented area containing a tiny black eye. Parapodia small, partly cover sides of shield and visceral hump. Left tail lobe long and sharply pointed, right tail lobe very short. White with dark brown mottling, vague traces of net-like pattern. Depth: ~10 feet. Riviera Beach, Florida.

Grand Cayman

Chelidonura **sp.**

Size: 7 mm Aglajidae

ID: Blackish. Covered with many irregular and often broken whitish longitudinal lines. Parapodial margins with some blue spots and orange streaks; four orange spots near front of headshield. Tail lobes pointed, longer left one with orange margins and median blue stripe near tip. Might be a color variant of *C. cubana* (next page). Depth: 10 feet. Grand Cayman.

feeding on acoel flatworms

↓Florida Guadeloupe↑

Riviera Beach, Florida

Chelidonura cubana **Ortea & Martinez, 1998**

Cuban aglaja

Size: 9 mm — Aglajidae

ID: Reddish-brown to blackish. Headshield with white patch at rear, 3 white/bluish front lobes (middle widest). Widely-spaced blue/green spots along edges of parapodia, tails, and one behind each small frontal lobe looped by long broken orange stripes. Scattered orange spots, e.g. on parapodial outer faces. To 20 feet. Florida, Caribbean. Rare.

St. Vincent

mating aggregation; Riviera Beach, Florida

Chelidonura hirundina **(Quoy & Gaimard, 1833)**

swallowtail aglaja

Size: 25 mm — Aglajidae

ID: Ground colors red, orange, yellowish, brown, blackish. Headshield has black margins, median black-edged blue/green "T", posterior rear white crescent. Parapodial margins and tails black with median blue/green stripe; tail pattern loops onto visceral hump. Tails may be equally long. Depth: to 30 feet. Florida, Bahamas, Caribbean. Also referred to as the "leech aglaja".

laying eggs; St. Vincent

St. Vincent

Riviera Beach, Florida

Guanaja, Honduras

Nakamigawaia felis **(Er. Marcus & Ev. Marcus, 1970)**

black aglaja

Size: to 12 mm Aglajidae

ID: Usually solid black, but occasional white-patterned black morphs have small spots or crowded broken thin stripes and a crossband of larger spots at the rear of the headshield. A striped red variant is a rare find (Guanaja, Bonaire). Tail lobes are usually short and rounded. Typically < 40 feet deep. Common in Florida, Bahamas, and the Caribbean.

Roatan

Roatan

Nakamigawaia **sp.** **lava aglaja**

Size: 10 mm Aglajidae

ID: Reddish-orange with variable black spotting and patching. The external morphology being so similar to that of *N. felis* suggests that these animals may be newly discovered variants of that species. However, identifying them as such is premature in the absence of molecular comparisons or observed mating behavior with *N. felis*. Photographed off Roatan at 80 feet.

↓St. Vincent↑

Cuba

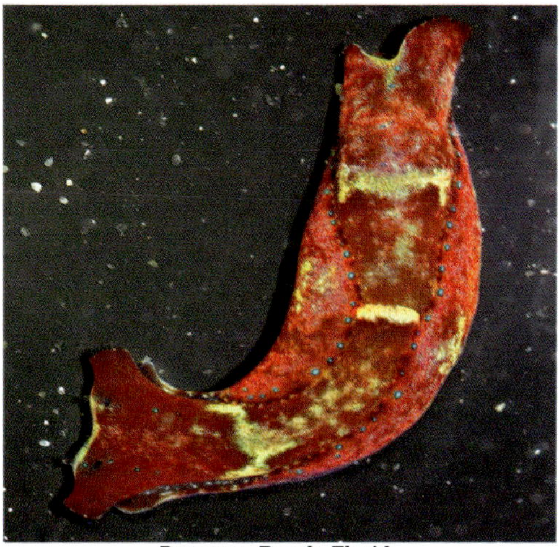

Navanax gemmatus **(Mörch, 1863)** **striated aglaja**
Size: to 50 mm Aglajidae
ID: Two rounded lobes on head; tail lobes short, broad, rounded, left one larger. Cream, yellowish, olive, gray, brown, purplish. Back with crowded dark and pale longitudinal lines, often 3-4 white to pinkish crossbands, sometimes pale/dark mottling. Parapodial edges with brilliant blue or greenish black-edged spots. To 30 feet. Common in Florida, Bahamas, Caribbean.

Pompano Beach, Florida

↓Bahamas Bonaire↑

Philinopsis depicta **(Renier, 1807)** **painted aglaja**
Size: to 25 mm Aglajidae
ID: Headshield reaches midbody, ends with erect triangular lobe; tail lobes short and rounded, equal in length. Dark brown to black; may have white spots on or behind shield and on parapodial faces; parapodial edges bright blue, margins with orange spots or dashes; shield's front has elongate orange dashes, lobe's base has orange spot on each side and orange markings near apex. Depth: to 15 feet. Animals with this coloration have been identified as either *P. depicta* (left photo) or *P. bagaensis* (right photo). Both could be either. Reliable identification by external appearance alone may not be possible.

Riviera Beach, Florida

St. Vincent

Roatan

Philinopsis pusa **(Ev. Marcus and Er. Marcus, 1966)** **pusa aglaja**
Size: to 45 mm Aglajidae
ID: Front with two short squarish lobes; cephalic shield extends to midbody, ends in an erect triangular lobe, slightly overlaps visceral hump; 2 short rounded equal-length tail lobes. Color patterns extremely variable: whitish with dark spots, brown with white and/or yellow/orange spots, uniformly pale orange or brown, black, sometimes with yellow patches at front or rear. Parapodial margins usually neon blue, often interrupted by black spots. Shield apex of dark individuals also neon blue. Has been observed feeding on *Gastropteron chacmol*, *Aglaja felis*, and its own species. To 15 feet. Florida, Bahamas, Caribbean, Mexico, Brazil.

St. Vincent

Riviera Beach, Florida

Spinoaglaja aeci **(Ortea & Espinosa, 2001)** **red-ring aglaja**
Size: to 22 mm Aglajidae
ID: Cylindrical body narrows behind a truncate head, widens to a midbody swelling that crosses the rear of the head shield and continues onto the parapodia. Many short conical papillae. Tail lobes equally short, triangular to rounded. Dark to yellowish-green, orange-green; specked with yellow/white, many yellowish papillae, some with **red rings basally**. Specks coalesce into diffuse crossbands on head, swelling, and tail lobe margins. Depth: 6-50 feet. Bermuda, Florida, Bahamas, much of Caribbean, Brazil.

Riviera Beach, Florida

Dominica

Guadeloupe

Bahamas

with egg mass; Guadeloupe

Key Largo, Florida

Riviera Beach, Florida

Spinoaglaja petra **(Ev. Marcus, 1976)** **rock aglaja**
Size: to 22 mm Aglajidae
ID: Body cylindrical, smooth to rough-looking due to many small papillae, noticeable lateral swellings at midbody; headshield extends to beyond midbody, slightly overlaps visceral hump; two short, triangularly rounded "tails" at rear; parapodia extend along entire length of body, fold up to cover only its sides; head blunt. Yellowish-orange, shades of brown, blackish, peppered with white spots/dots; white patches on head, rear of headshield, rear of back, tail margins, and parapodial edges at swellings. Depth: to 50 feet. Bermuda, Florida, Bahamas, much of Caribbean, Brazil. The four animals in the right column were originally identified as *Philinopsis anneae* Ornelas-Gatdula & Á. Valdés, 2012. The bright orange one looks much like the Mediterranean *S. wildpretii* (possibly a synonym of *S. aeci*). The bottom left animal with the egg mass could also be *S. aeci*. Identifications are necessarily somewhat tentative without more conclusive external diagnostic features.

Family NOTODIAPHANIDAE Thiele, 1931

Currently a family with one genus and one species, with both family and genus having uncertain status (nomina dubia) until more research is done. The family description is necessarily that of the single species, *Notodiaphana atlantica* Ortea, Moro & Espinosa, 2013. Occurs in the tropical west and east Atlantic and the Mediterranean Sea.

Shell external, small, fragile, translucent; ovoid to cylindrical to pyriform, length about twice width; spire deeply sunken, with a narrow apical umbilicus, almost engulfed by the body whorl; aperture elongate, anterior rounded and wider than posterior; outer lip sharp, straight to slightly concave, rises smoothly from the apertural side of a deep apical pit and projects above the apex; columella smooth and thick; surface of body whorl appears faintly reticulated – thin, crowded, irregular spiral grooves crossed by thinner sinuous axial lines; axial sculpture more prominent inside the apical depression.

Body elongate oval; cephalic shield with two triangular lobes, front edge ciliated, lobes merge medianly in front of eyes; foot bilobed in front, with thickened short extension of mantle at rear to enable crawling; foot forms small parapodial lobes that fold up to partially cover front of shell.

Live specimens have been found among seaweeds such as *Caulerpa racemosa*, *Halimeda discoidea*, *Cladophora* spp., attached *Sargassum* species, and other algae in relatively shallow water.

Riviera Beach, Florida

Bermuda

Notodiaphana atlantica **Ortea, Moro & Espinosa, 2013** **wing bubble**

Size: to 4 mm Notodiaphanidae
ID: Shell translucent white. Body white with translucent areas but appears pale yellow with a mid-dorsal red dot and a ventral black spot (digestive gland). Depth: to 8 feet. Amphiatlantic. Bermuda, Florida, Bahamas, Cuba, Venezuela.

Family CYLICHNIDAE H. Adams & A. Adams, 1854

A circumglobal family of just over 100 species in 10 genera found at all latitudes. In the tropical West Atlantic the most commonly reported species are in the genera *Cylichna* and *Cylichnella*, the former being by far the most speciose genus in the family. Most records are of empty shells either found in beach drift or obtained by dredging.

Shell external, strongly calcified, cylindrical to ovate to pyriform; spire short or sunken; surface grooves and striae usually absent or inconspicuous; aperture as long as shell, narrow posteriorly, wide and rounded anteriorly; lengths 3-40 mm.

Body oval to elongate; considerably shorter than shell but can withdraw completely into it; usually lacks an operculum; cephalic shield short, with a pair of flat posterior lobes that extend back to protect the mantle cavity; foot short, without parapodia.

Usually buried just below the surface of muddy substrates where they probably feed on benthic foraminiferans, segmented worms, small crustaceans, and other mollusks. Depths range from intertidal to abyssal.

Cylichnella bidentata **(d'Orbigny, 1841)** **two-toothed barrel-bubble**

Size: to 4 mm Cylichnidae
ID: Shell elongate-cylindrical; smooth except for irregular axial growth lines and spiral threads near base; aperture's outer lip thin, lower lip rounded; columella with prominent plicae and a pointed tubercle. Body and shell translucent white. Depth: 0-1200 feet. Throughout the tropical West Atlantic.

Family COLINATYDIDAE

Oskars, Bouchet & Malaquias, 2015

Currently a family with one genus (*Colinatys*) created following recent molecular studies of cephalaspideans. Its type species is the one initially described as *Atys alayoi* Espinosa & Ortea, 2004 based on specimens from Cuba. Currently, as *Colinatys alayoi*, it is the only described species in this family.

The shell is small, very thin and fragile, almost transparent, quadrangular, wider anteriorly, with slightly convex sides above the midpoint; spire involute, apical depression wide; umbilicus narrow, partially covered by a projection of the thin columellar callus; columella with no visible folds; aperture long, narrowed posteriorly and broadened anteriorly, with a thin and sharp outer lip that protrudes slightly above the spire. Although the shell's outer surface is smooth, there are deeper layers forming a conspicuous internal reticulate pattern – a white net-like mesh of fine spiral lines crossed by axial threads with small nodules at their points of intersection. Shells are up to 2 mm long.

The body is small enough to fit completely into the shell, but it does not have an operculum; cephalic shield bilobed in front, its tentacles widely spaced, short, and rounded; foot shorter than shell; eye spots prominent.

Rare. Depth: to at least 60 feet. Has been found on *Sargassum* but its ecology is not well known.

Palm Beach Inlet, Florida

Colinatys alayoi **(Espinosa & Ortea, 2004)**

checkerboard barrel-bubble

Size: to 2 mm — Colinatydidae

ID: See family description at left. Has been seen on attached *Sargassum*, under balled *Cladophora* & under a coral slab in very shallow water, protruding from its shell cautiously as far as its eyespots. To 60 feet. Cuba, Florida, Bahamas, Guadeloupe.

Family RETUSIDAE Thiele, 1925

A taxon with four genera and over one hundred species of small infaunal marine gastropods having external shells that are rarely longer than 10 mm. They are found worldwide on/in sandy or muddy bottoms in depths ranging from intertidal shallows (shores, bays, estuaries) to abyssal plains deeper than 16,500 feet.

Body elongate; cephalic shield with a shallow median notch anteriorly, bilobed posteriorly, usually not overlapping the shell; foot broad and shorter than the shell, does not develop parapodia; gill not visible externally; indistinct eye spots on headshield near base of posterior lobes; radula and jaws both absent, food usually rasped by a gizzard with three oval plates; a few species have a small operculum.

Shell cylindrical or pyriform; posterior whorls typically sunken but the protoconch is usually visible as a short rounded or pointed projection; rear of shell typically truncate; aperture as long as shell, broadens anteriorly; surface usually smooth, sometimes with spiral grooves and/or growth lines that are especially noticeable at the ends of the body whorl; accommodates almost the entire retracted animal.

They are normally buried in sediment where they feed selectively on diatoms, foraminiferans, and other small mollusks. In turn, they are preyed upon by aglajid sea slugs.

Brazil

Pyrunculus caelatus **(Bush, 1885)**

engraved barrel-bubble

Size: to 4 mm — Retusidae

ID: Shell whitish, fragile, elongate, wider anteriorly, apex umbilicated. Conspicuous growth lines at rear. Depth: >50 feet. Florida, Gulf of Mexico, parts of Caribbean, Brazil.

Family RHIZORIDAE Dell, 1952

A family of less than 50 species of marine gastropods that are known worldwide primarily from their small external shells (less than 9 mm long) being either dredged from soft substrates or occasionally washed up onto beaches.

Shells translucent, tapering, ovate or cylindrical, extended at both ends; apex recessed; aperture narrow, elongate, anteriorly rounded, lip's posterior margin bends toward the columellar axis and terminates at an open spine of variable shape and length, aperture extends as high as tip of spine; columella smooth, somewhat twisted; shell imperforate anteriorly; surface of most species sculpted with thin, incised, widely spaced spiral lines at each end of shell, and even finer more crowded wavy lines incised over the middle of the body whorl.

There is little ecological information available. Some species apparently feed on rhizoids or vesicles of *Caulerpa racemosa* alga that is usually anchored on vertical rock faces and mangrove roots in areas of strong water circulation. Live animals have been found as deep as 150 feet.

Indian Shores, Florida

Volvulella persimilis **(Mörch, 1875)** — **southern spindle-bubble**

Size: to 4 mm — Rhizoridae

ID: Shell translucent grayish-white, glossy, terete, fragile, apex pointed; aperture long, rear two-thirds very narrow; umbilicus small; very fine spiral lines at each end, microscopic striae in between. No information is available about the animal. Depth: 10-150 feet. Throughout the tropical West Atlantic.

Family BULLIDAE Gray, 1847

A small family of primarily circumtropical "bubble shells" consisting of a single genus (*Bulla*) with seventeen species, two of which occur in the tropical West Atlantic.

Shell external, solid, ovoid to slender, smooth or with some spiral striations; columella smooth and thick; parietal callus thin; spire deeply and narrowly involute, has concentric striae, may have conspicuous sutures; visible top of protoconch convex and smooth; aperture long, narrow posteriorly but flares considerably toward the anterior, resembling an inverted comma when shell posterior is positioned at the top; outer lip thin; surface mottled in shades of brown.

Body elongate; anterior of cephalic shield with a lateral pair of short inrolled siphon-like tentacles, beneath it on each side of the head there is a lamellate and ciliated Hancock's sensory organ; posterior of shield with a pair of long and broad triangular lobes that overlap the front of the animal's external shell; sides of foot fold up to form a pair of small parapodia that also partially cover the shell; body small enough to retract completely into the shell, but lacks an operculum to seal the aperture.

They inhabit primarily intertidal to shallowly subtidal areas, including lagoons and estuaries, preferably with sandy or muddy bottoms, green algae, and seagrasses, but have also been found as deep as 200 feet. Their distinctive empty shells are commonly washed up onto beaches. By day they usually remain hidden in the substrate, emerging at night to graze on algae.

***Bulla clausa* Dall, 1889** **imperforate bubble**

Size: to 11.5 mm Bullidae

ID: Shell solid, somewhat globose, smooth; well-marked growth lines and extremely small wavy spiral striae; apex imperforate; aperture as long as shell, widens considerably toward base, callus strongly arched. Pale yellowish-brown to salmon. No information is available for the animal. Depth: 0-900 feet. South Florida and northwest Gulf of Mexico.

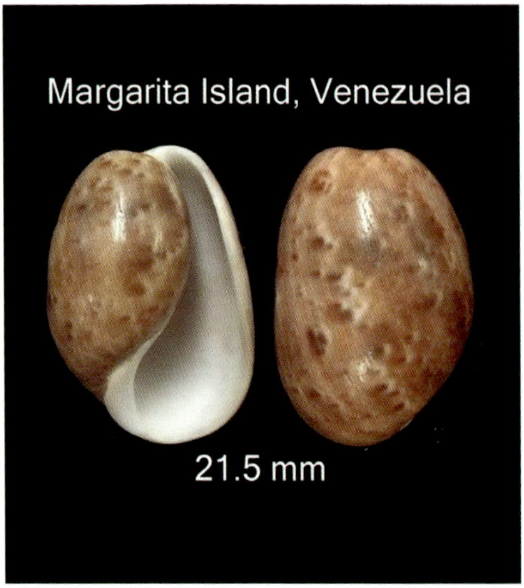

***Bulla solida* Gmelin, 1791** **solid bubble**

Size: to 51 mm Bullidae

ID: Shell thick, somewhat globose; faint growth lines, no spiral grooves; aperture widens greatly toward base; parietal callus thin; columella concave, smooth; inner lip thickish, rounded. No information is available for the animal. Depth: 0-130 feet. Egg masses have been found on plants and rocks in shallow brackish lagoons. Uncommon Florida, Gulf of Mexico, Caribbean.

***Bulla occidentalis* A. Adams, 1889** **western striate bubble**

Size: to 50 mm Bullidae

ID: Shell ovate, smooth; spire sunken; very thin; tightly packed wavy spiral lines. The cephalic shield's front corners are short rolled tentacles. Shell color is variable – whitish, shades of brown, greenish, mottled with brown dots that may coalesce into small blotches. Body translucent gray; head and parapodia white-spotted with some brown to reddish pigment. Spawn is a tangled gelatinous string of yellow eggs on a sandy bottom. Depth: typically to 10 feet. Throughout the tropical West Atlantic.

continued on next page

Guadeloupe

Guadeloupe

Bermuda

juvenile; Bonaire

aggregation laying eggs; Florida

Guadeloupe

Florida

Family HAMINOEIDAE Pilsbry, 1895

A family of bubble snails with possibly more than 200 species in the size range 3-30 mm. Many are undescribed and known only from their shells. They are distributed worldwide in tropical to temperate seas, and only a few of them have been recorded in the tropical West Atlantic.

Shell external, fragile, translucent, ovoid to cylindrical to globose, sometimes flattened, widest at midpoint; aperture narrow posteriorly, broadens anteriorly, its thin outer lip usually projects above the apex; spire usually sunken and umbilicate or entirely resorbed; columella smooth, may also be resorbed; body whorl smooth or with fine spiral striae; periostracum very thin; calcification light to moderate.

Body oval to elongate; cephalic shield broad, anterior rounded with corners folded down, posterior often with a pair of lobes; foot broad, sides fold up as parapodia that often meet or overlap at midback to partially or entirely cover the shell; rear of mantle may form a protruding tail; entire body can retract completely into the shell, especially if the spire and columella were resorbed.

They inhabit sand and mud flats in shallow subtidal and intertidal regions, including bays, estuaries, and even tidepools. Most species are infaunal and some are epifaunal, but all are herbivores. The infaunal ones feed on diatoms and filamentous algae; epifaunal ones on various species of green algae.

← Guadeloupe

8 mm, St. Vincent →

← Bocas del Drago, Panama →

Atys caribaeus **(d'Orbigny, 1841)** **Caribbean glassy-bubble**

Size: to 20 mm Haminoeidae

ID: Body very elongate; cephalic shield deeply notched at rear; parapodia short, thin, flap-like, cover only front end of shell. Shell elongate, ovate, outer lip rises above anterior and twists down sharply; close-set spiral grooves everywhere but most prominent at front and rear. Animal usually translucent white with irregular opaque white spotting and black dots, sometimes densely covered with brown to reddish-brown dots. Shell smooth, glossy, translucent whitish. Depth: 0-600 feet. North Carolina to Florida, Bahamas, Antilles, Venezuela to Yucatan, Suriname, southern Brazil

Yucatan

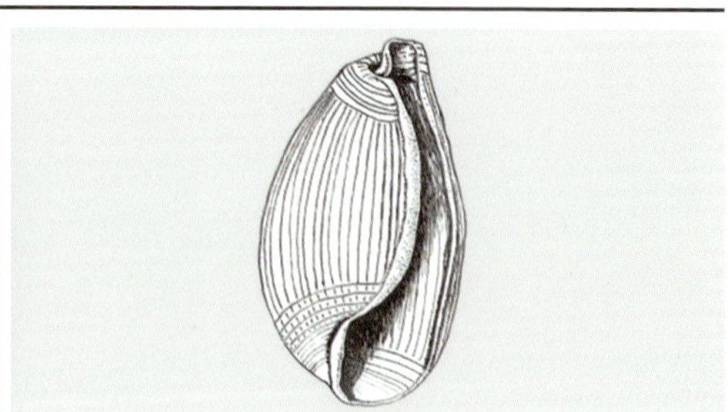

Brazil

Atys riiseanus **Mörch, 1875**

Size: to 20 mm Haminoeidae

ID: Shell fragile, ovate, outer lip high, variably spaced spiral grooves prominent at front and rear, axial growth lines most distinct in middle. Animal's cephalic shield notched at front, two pointed flaps at rear; parapodia cover only front of shell. Animal translucent white; head and parapodia dotted with black and/or white; brownish-orange patches of digestive gland visible through shell. Shell translucent whitish, glossy. Depth: 0-300 feet. Florida, Bahamas, much of Caribbean and Gulf of Mexico, southern Brazil.

Yucatan

Atys sandersoni **Dall, 1881**

Size: to 15 mm Haminoeidae

ID: Body very elongate; posterior shield with notch; parapodia short, cover front of shell. Shell thin, elongate, slenderer anteriorly; aperture long, oval; transverse growth lines and spiral striae indistinct; apex a sunken pit. Body translucent white, densely black on head and parapodia. Shell glossy translucent white. Often under mangrove roots. Depth: 3-300 feet. Florida, Bahamas, Antilles, Panama to Yucatan. Brazil.

Jupiter Sound, Florida

Atys sharpi **Vanatta, 1901** **slender glassy-bubble**

Size: to 8 mm Haminoeidae

ID: Body elongate, parapodia cover front of shell. Shell solid, smooth, fine spiral striations strongest at ends; apex sunken, with a very small perforation. Body translucent white; head and parapodia with many small white spots (some spots orange-brown); spots are larger and on gray patches within the shell. The shell is translucent pale orange. Depth: to at least 50 feet. Florida, Bahamas, Antilles, Venezuela.

Riviera Beach, Florida

Atys **sp.**

Size: 12 mm Haminoeidae

ID: Body ground color translucent gray. Cephalic shield with many clusters of white dots. Back and sides with patches of white dots and randomly oriented orange scrawls and lines that are clearly visible through the shell, as well as (sometimes) a row of black spots along the lower sides. Close-set black eyes in large unpigmented regions. Shell opaque to semi-translucent white. In areas of sand and algae, Depth: to 20 feet. This is *Atys* sp. on page 264 of *Bahamian Seashells* (Redfern, 2013). Florida, Bahamas.

Weinkauffia macandrewii **(E. A. Smith, 1872)** **banded glassy-bubble**

Size: typically 5 mm, but as long as 10 mm Haminoeidae

ID: Body very elongate; cephalic shield large, bilobed at rear; parapodia barely cover front of shell. Shell elongate ovoid; apex sunken, outer lip rises slightly above spire; middle third smooth, other two with 7-12 equally spaced spiral grooves. Body translucent white with opaque white dots/spots; cephalic shield and parapodia may have some black spots; viscera brownish to greenish. Shell usually with narrow white spiral bands between the spiral grooves; smooth central region usually white, sometimes also banded. Depth: 0-90 feet. Amphiatlantic. Bermuda, Louisiana, Bahamas, Antilles, Panama, Costa Rica, Brazil. Rare.

Cozumel

Panama

Florida

Martinique

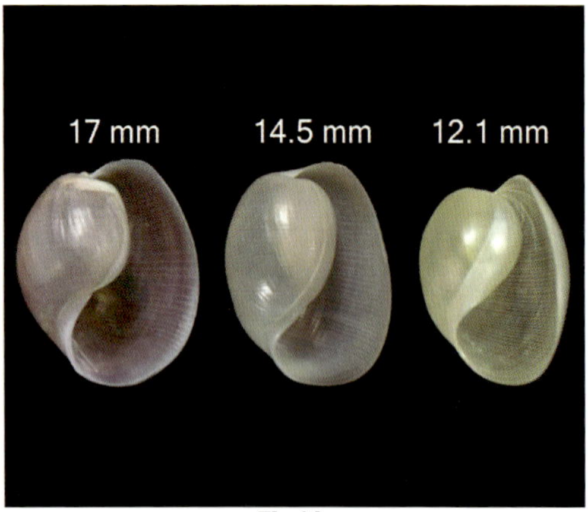

Florida

***Haminoea elegans* (Gray, 1825)** **elegant glassy-bubble**

Size: to 35 mm Haminoeidae

ID: Body elongate, broad; headshield extends to midbody, **deeply notched at front with small lateral tentacles, not bilobed at rear**; parapodia large, extend back almost to rear of body, overlap slightly at midback, cover sides of body and front of shell; **tail flared, rear edge crenulated**; eyes close-set, centered in large unpigmented circles. Shell ovoid, partly covers the body but can accommodate the entire retracted animal. Animal translucent gray to yellowish-tan with many **blackish/brown spots**, and often with many **small white blotches** (clusters of tiny white dots). Shell translucent white, grayish-white, yellowish, pale lavender, to reddish-orange. Primarily a nocturnal feeder. Depth: to 10 feet. Amphiatlantic, occurs throughout the tropical West Atlantic.

Guadeloupe

Florida

Grand Cayman

Panama

with egg mass; Riviera Beach, Florida

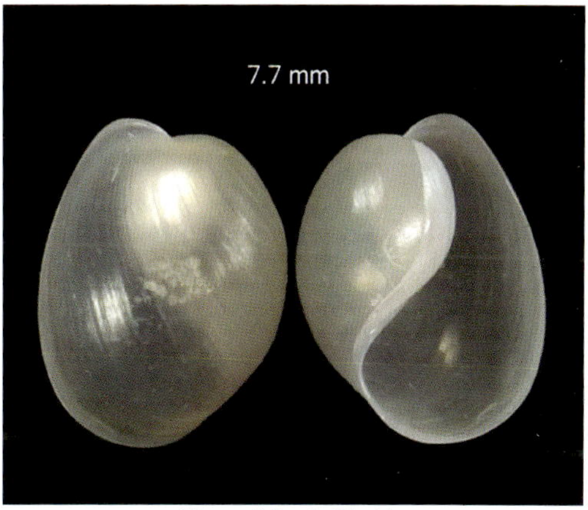

Marco Island, Florida

Haminoea antillarum (d'Orbigny, 1841) **Antilles glassy-bubble**

Size: to 40 mm Haminoeidae

ID: Body elongate and broad; headshield short and broad, does not reach midbody, **shallowly bilobed at front with well-defined lateral tentacles, rear bilobed with median cleft**; parapodia extend back almost to rear of body, cover front of shell but leave most of it exposed; eyes well separated, centered in large unpigmented circles; **tail broad, flat, rounded**. Shell globose, very thin, glossy, transparent, partly covers the body but can accommodate the entire retracted animal. Body translucent gray to brown, often densely covered with white and blackish spots and blotches; **dark brown and orange spots** on viscera are clearly visible through the shell. Depth: to 25 feet. Common throughout the tropical West Atlantic.

Roatan

Bahamas

Florida

Grand Cayman

Florida

Cedar Key, Florida

Haminoea succinea **(Conrad, 1846)** **amber glassy-bubble**

Size: to 20 mm Haminoeidae

ID: Headshield short with broad rounded front having a short wide tentacle on each side; rear with deep midline cleft. Parapodia do not reach dorsal midline, cover lower front sides of shell leaving rest exposed; eyes close-set, beneath epidermis in unpigmented circles; tail broad, flat, rounded. Shell posterior, somewhat cylindrical, extremely fragile, partially covers body but can accommodate entire retracted animal. Body ground color translucent gray to amber. Headshield, foot, and parapodia with a dense distribution of dark brown to black spots and a scattering of opaque white dots and blotches, which are visible through the shell. Shell translucent white. Depth: typically less than 10 feet but has been found as deep as 180 feet. Bermuda, Chesapeake Bay to Florida, Gulf of Mexico, Bahamas, Antilles, Venezuela to Mexico, Brazil.

Haminoea petitii **(d'Orbigny, 1841)** **straight glassy-bubble**
Size: to 12 mm Haminoeidae
ID: Headshield short, front notched, rear bilobed. Parapodia small, cover only front of shell. Shell fragile, ovate-cylindrical, inflated looking, apex broad and imperforate, spiral striae extremely fine to absent. Animal has orange-spotted viscera. Shell glossy, greenish-yellow. Depth: 3-150 feet. Bermuda, Florida to Louisiana, Bahamas, Antilles, Venezuela to Mexico, Brazil.

Haminoea **sp. 1**
Size: to 20 mm Haminoeidae
ID: Body elongate; headshield with a pair of broadly rounded lobes at front, appears deeply incised at rear; parapodia low, extend back almost to rear of body, cover sides of shell but leave most of it exposed. Shell posterior, external, inflated, half as long as body. Body green. Exposed region of shell mottled with small and large dark green to black blotches. Observed on algae attached to boulders. Depth: 23-33 feet. Martinique.

Haminoea **sp. 2**
Size: to 20 mm Haminoeidae
ID: Headshield short, front indented, lobes broadly rounded, lateral tentacles broadly auriculate; parapodia low, extend almost to rear of body, cover sides of shell leaving most of it exposed. Shell posterior, external, inflated. Pale yellowish-green with scattered olive-green small spots. Shell with clusters of black and white spots. About 26 feet deep near algae. Martinique.

Haminoea **sp. 3**
Size: to 15 mm Haminoeidae
ID: Headshield slightly bilobed at front and smoothly rounded at rear, without apparent incisions. Parapodia higher than in *Haminoea* sp. 1 and 2, covering more of the shell. Shield and parapodia densely covered with minute green specks and large white blotches. Body inside shell has many black spots. Depth: ~10 feet. Cayman Brac.

Family GASTROPTERIDAE Swainson, 1840

A family of 5 genera and over 40 described species (and many undescribed ones) of mostly strikingly colored slugs that are distributed mainly throughout the tropical and sub-tropical regions of the Indo-Pacific. It is represented in the tropical West Atlantic by only the most widespread genus – *Gastropteron*.

Body oval to elongate, length usually less than 10 mm; cephalic shield short, usually triangular with a narrowed rear end that is raised, curved forward, and folded into a median inhalant siphon; visceral hump large, elongate, with a ridge on its right side corresponding to the mantle cavity's top free edge; hump overlain by an elongate posterior shield; a flagellum near the rear of the visceral hump; mantle cavity small or absent; wide foot longer than visceral hump, its sides forming large thin parapodia that fold up to either meet on the dorsum or partially cover the visceral hump; gill plicate, sometimes visible between the body and the right parapodium; penial orifice on right side behind mouth. Shell internal and reduced, or absent.

When disturbed, most species can swim for long periods of time by flapping their large parapodia, which is why they are referred to as "bat-winged slugs". Little is known about their ecology, but some species have been found on seagrasses and macroalgae, and some appear to feed on non-spiculate sponges and the flatworms that live on them. The siphon may function as a chemosensor, like rhinophores in other slugs.

Roatan

Gastropteron **sp. 1**

Size: 10 mm Gastropteridae

ID: Ground color is cream, densely mottled with small stellate clusters of brownish, reddish, and purplish dots. Parapodia with an orange wash, white blotches, yellow on the rear margin. Posterior of siphon white-blotched; foot/tail purplish. Resembles the Eastern Pacific species *G. pacificum*. Depth: ~20 feet.

Grand Cayman

Gastropteron **sp. 2**

Size: 10 mm Gastropteridae

ID: The ground color is yellowish, densely mottled with clusters of dark purple dots. The parapodia and the rear of the siphon have white blotches, spots, and dots; the parapodial edges are bright yellow. Flagellum white; tail white with purple spots. Depth: ~20 feet.

Bahamas

Gastropteron **sp. 3**

Size: much longer than 10 mm Gastropteridae

ID: Body ovate, high, laterally compressed; front of head rounded downwards; parapodia long, very broad, flat, rounded, fold up from foot to overlap dorsally at midbody; short flagellum on right posterior, projects only slightly; tail short, broadly triangular. Uniformly blackish. Depth: ~6 feet. Bahamas. The closest West Atlantic match is *G. vespertilium* Gosliner & Armes, 1984, which has been found only in Tampa Bay, Florida. It is charcoal-gray to purplish-black, but also has yellow parapodial and siphonal margins, a longer and white flagellum, a narrower and longer tail, and reaches a maximum length of only 8 mm.

Cozumel

Grand Cayman

Grand Cayman

San Andres

Florida

San Andres

Gastropteron chacmol **Gosliner, 1989** **flapping dingbat**

Size: to 10 mm Gastropteridae

ID: Body oval; front of cephalic shield rounded, bilobed with a deep median cleft; shield's posterior folds up, curls forward to form a simple involute upright long siphon that usually leans over the front of the head; parapodia long, high, wrap upwards to overlap above the back; a long pointed flagellum protrudes well above the parapodia from the lower right rear of the visceral hump, usually pointing obliquely backwards; tail short, triangularly pointed. Shell internal, cashew-shaped, about one-sixth of the body length. Ground color variable – deep red to brownish-purple to black, sometimes with a scattering of white, yellow, or bluish spots and blotches. Parapodia with narrow white, bright yellow, or bluish margins. Siphon and flagellum usually ground color but sometimes appear almost white due to heavy spotting. Usually found on shallow sand and grass bottoms, either crawling on or burying themselves in the sand, sometimes trailing one another closely. Depth: to 30 feet. Uncommon. Florida, Bahamas, Cuba, Grand Cayman, Texas, Sint Maarten, Guadeloupe, Venezuela, San Andres, Roatan, Cozumel, Belize, Bay of Campeche, Brazil.

Family PHILINIDAE J. E. Gray, 1850

Marine gastropods in the order Cephalaspidea (headshield slugs) with over a hundred species distributed worldwide across all latitudes. Most are infaunal, inhabiting soft substrates in both shallow and deep water.

Body relatively large, sturdy, cylindrical or depressed and wedge shaped; cephalic shield long (> half body length), broad anteriorly, tapers quickly or is truncate posteriorly, often with a median groove; the visceral hump behind the shield is longer, posteriorly squared off or multi-lobed; the foot is short and broad, its thick side flaps folding up as parapodia to partially wrap the body; gill not visible externally, eyes rarely visible, Hancock's organ inconspicuous.

Shell usually internal, completely enclosed within rear half of body; ovoid to elongate to almost quadrate, 1-40 mm long, fragile; spire low to sunken; aperture broad; surface can be smooth or have chains of small pits that sometimes fuse to form spiral grooves or create a reticulated appearance from the crossing of raised spiral and axial lines.

Although some are epifaunal, most spend their adult lives within a thin layer of sand or mud substrate where they move around using their cephalic shield as a plow or shovel. Most are carnivores that feed on foraminifera, small gastropods and bivalves, echinoids, polychaetes, crustaceans, fish remains, even each other.

Venezuela

Philine buchensis **Caballer & Ortea, 2015**

Venezuelan paper-bubble

Size: to 4 mm — Philinidae

ID: Body elongate; headshield triangular at rear; parapodia short, reach but do not cover headshield. Shell internal, one third of the body length, very thin and fragile, rear edge of aperture does not extend over the truncated apex, many transverse growth striae, and 6-18 prominent paired spiral bands that occupy up to slightly more than half the shell's length, bands composed of perforations. Translucent yellowish with irregular opaque white dots everywhere, including sole of foot. Visceral mass visible as pinkish-brown mass. Shell translucent. Inhabits mangrove lagoons. Depth: to 3 feet.

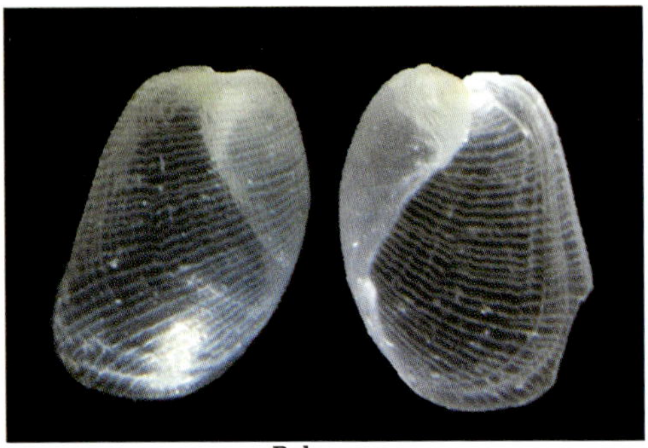

Bahamas

Philine caballeri **Ortea, Espinosa & Moro, 2001**

Manuel's paper-bubble

Size: to 4 mm — Philinidae

ID: Body elongate, headshield narrows toward the rear; parapodia narrow. Shell internal; half to two-thirds body length; in front of posterior shield; thin and fragile; spire sunken; aperture large; body whorl with closely spaced irregularly-edged spiral cords and several transverse growth striae. Animal bright orange, paler on caudal lobe, sometimes with white dots. Shell white, covered with semi-transparent periostracum. Has been found on *Cladophoropsis*, *Cladophora*, anchored *Sargassum*, mixed algae, and on rocky substrates. Depth: to 30 feet. Cuba, Guadeloupe, Bahamas, Costa Rica. Rare.

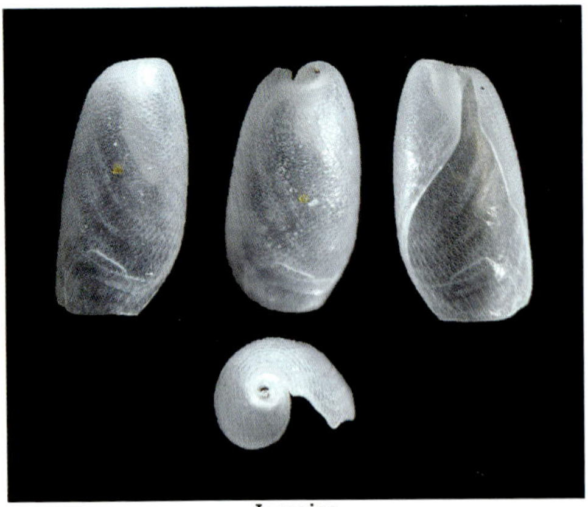

Jamaica

Philine sagra **(d'Orbigny, 1841)**

crenulate paper-bubble

Size: to 6 mm — Philinidae

ID: Shell oblong, thin, fragile, slightly widened and truncated anteriorly; aperture broad anteriorly, narrow posteriorly; apex of body whorl open, not depressed; surface with closely spaced oblique "chains" of oblong rings linked end-to-end, the chains separated by wavy striae; shell edges finely crenulated. Uniformly white. No information about the animal. Depth: 3-260 feet. Throughout most of the tropical West Atlantic.

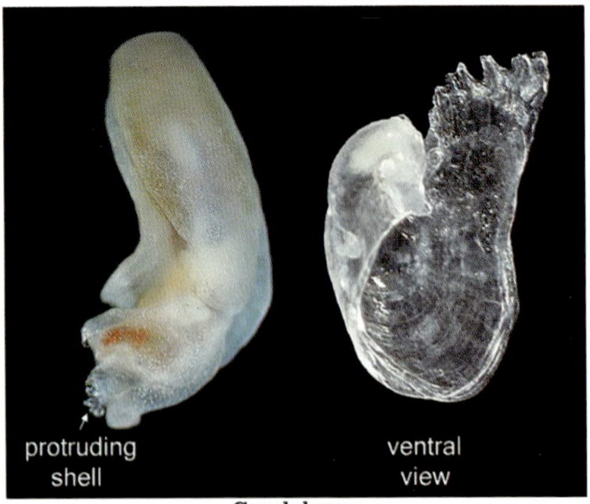

Guadeloupe

Spiniphiline persei **Caballer & Ortea, 2015**

poet's paper-bubble

Size: to 3 mm — Philinidae

ID: Body elongate; head twice as long as the posterior shield and overlaps it; shield with two downward pointing lobes posteriorly, right one larger; parapodia short, narrow, barely overlap the two shields. Shell one-third body length, internal but projects from rear with six reinforced spines. Body white-dotted; viscera yellowish; shell translucent whitish. Depth: to at least 96 feet off Guadeloupe. Rare.

Family TORNATINIDAE P. Fischer, 1883

The familial placement of the barrel-bubble genus *Acteocina* J. E. Gray, 1847 has been varied and contentious. Family names have included Scaphandridae, Cylichnidae, and Acteocinidae. Currently it is Tornatinidae, consisting of one genus – *Acteocina*.

Shell external, solid, elongate cylindrical, ovoid, globose; larger ones tend toward pyriform with greatest width posteriorly; spire height variable but usually less than one-fifth of the shell length and with a conspicuously tilted heterostrophic protoconch; teleoconch smooth, with up to three whorls; aperture 10-20% shorter than shell, narrow posteriorly, flares anteriorly; outer lip thin, arched in the middle, retracted anteriorly; surface of body whorl can be smooth, ornamented with spiral striae or fine axial lines; columella short, concave, with a moderately strong fold separating it from a thin parietal callus. Shell length 2-9 mm.

Body elongate; anterior of cephalic shield with median notch and ciliated margin, posterior with rounded flaps; foot broad, front notched; parapodia small and round; mantle opening matches size of aperture; gizzard plates heavily calcified.

It has a worldwide distribution in high to low latitudes, in both frigid and temperate/tropical marine and estuarine waters. Depth: intertidal to at least 150 feet. All species are infaunal, living and feeding in the upper 10 mm of a soft substrate, where their prey includes sand-living invertebrates, worms, and other mollusks.

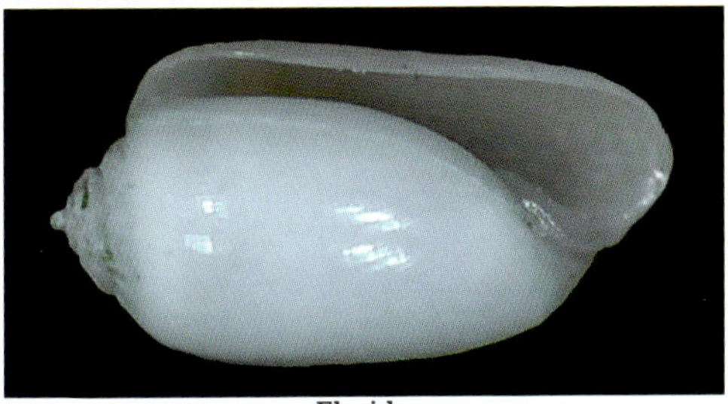

Florida

Acteocina atrata **P. S. Mikkelsen & P. M. Mikkelsen, 1984** — **blackback barrel-bubble**

Size: to 6 mm — Tornatidae

ID: Shell dextral, thin, smooth; protoconch globose. No parapodia, headshield with 2 large rear lobes. Shell translucent off-white. Animal translucent gray with irregular white spots, tissues of internal organs black. Egg mass ovoid, white, tethered to top of sediment. Depth: to 6 feet. Florida, Bahamas.

Bermuda

Acteocina candei **(d'Orbigny, 1841)**

Size: to 5 mm — Tornatidae

ID: Shell solid, smooth, elongate cylindrical, protoconch highly protruding. No parapodia, headshield has 2 large rear lobes. Animal translucent whitish; shell whitish to yellowish to green; viscera visible as an irregular white pattern on a pale reddish background. Depth: to 6 feet. Throughout the tropical West Atlantic.

Bermuda

Acteocina lepta **Woodring, 1928** — **Woodring's barrel-bubble**

Size: to 5 mm — Tornatidae

ID: Shell ovoid to elongate, spire short, almost flat, no umbilicus, body whorl tapers gently at base. No parapodia, headshield with two large rear lobes. Shell and body translucent whitish, headshield white-dotted, body inside shell has white spots and scrawls, viscera orange to reddish. Depth: 20-700 feet. Bermuda, North Carolina to Mexico, Bahamas, Antilles, south to Brazil.

Acteocina canaliculata **(Say, 1826)** — **channeled barrel-bubble**

Size: to 6 mm — Tornatidae

ID: Shell cylindrical to slightly pyriform; spire conical; protoconch partially sunken, shoulder keeled (juveniles) to rounded (adults). No parapodia, headshield with two large rear lobes. Shell porcelain white, partly translucent. Animal translucent white to grayish, head with some opaque white spots, internal organs pale orange to pink. Depth: intertidal to 130 feet. Throughout much of the tropical West Atlantic.

Family RUNCINIDAE H. Adams & A. Adams, 1854

A family of small headshield slugs consisting of nearly 80 species in 12 genera, distributed globally in temperate to tropical waters.

Body narrow to elongate-oval to pyriform, flattened; dorsal shield (notum) undivided transversely, rear end rounded to bilobed, overlaps edges of body; foot undivided, without lateral extensions, posterior protrudes as a broadly triangular to rounded tail; no tentacles; external gills either absent or small on the right posterior side under the mantle to the right of the anus, sometimes exposed; internal shell usually absent, but when present is small, fragile, abalone-shaped, and posterior; jaws and radula are present, usually four grinding gizzard plates.

They are bottom dwellers in usually shallow areas, to about 3 feet, where they are assumed to feed on diatoms and filamentous algae. Body lengths to 8 mm.

← Cuba Guadeloupe →

Edmundsina prieta **(Ortea, Moro & Espinosa, 2007)**
Size: 2 mm Runcinidae
ID: Black, elongate, somewhat rounded; front of head rounded, notched medianly; foot slightly wider than body; protrudes as a broadly triangular rounded tail. A sand dweller in depths of 12-33 feet. Cuba.

Karukerina antola **Ortea, 2013**
Size: 10 mm Runcinidae
ID: Head narrow, squarish; body widens to rear; mantle's rear is asymmetrically bilobed, above a 2-leaved gill. Mantle green, whi6910te-spotted, front of head with white patch on each side. Gill leaves yellowish-green. Found at 50 feet off Guadeloupe.

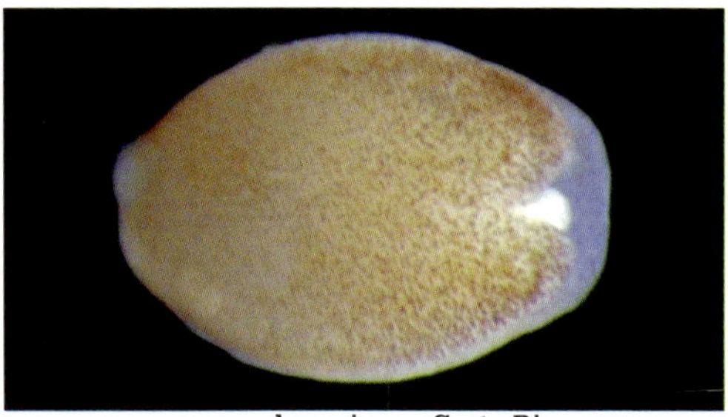

preserved specimen; Costa Rica

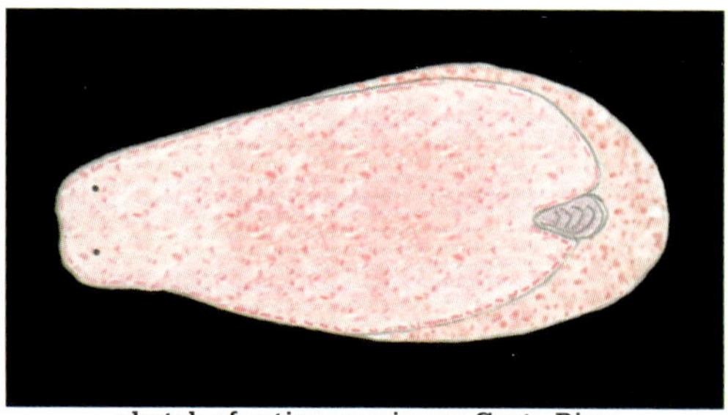

sketch of active specimen; Costa Rica

Lapinura aestus **Ortea, Moro & Espinosa, 2017**
Size: to 2.5 mm Runcinidae
ID: Body pyriform, broadening toward rear; front of head straight to slightly concave, small eyes slightly behind; mantle bilobed posteriorly; left lobe somewhat triangular, right lobe wider and more rounded. Tail rounded, protrudes broadly behind mantle lobes. Shell perforated, transparent, contains a lamellate gill; visible in posterior cleft between mantle lobes. An external gill (not visible) is present under the right lobe. Mantle and tail reddish-orange to orange-brown. Mantle peppered with red dots, edge prominent with line of bright red dots and dashes. Dark viscera sometimes visible. Found on rocks with grass cover. Very slow moving. Depth: tidal zone to 3 feet. Costa Rica.

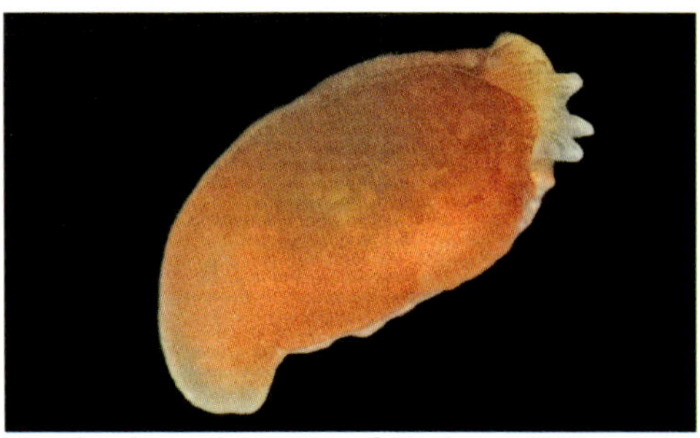

← Cuba Bocas del Toro, Panama →

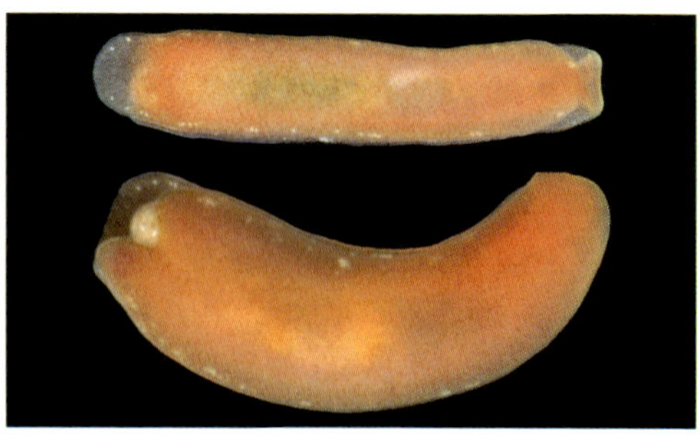

Runcina dorcae **Ortea, Moro & Espinosa, 2007**
Size: to 8 mm Runcinidae
ID: Body pyriform, broadens to rear, narrows and elongates when moving forward; front of head slightly concave; mantle's rear edge forms a wide gill cavity; gill large, usually protrudes as a 5-leaved fan behind a rounded tail; eyes not visible. Uniformly bright red, gill leaves white with reddish stalks. Forms a red ball when alarmed. Depth: 20-60 feet. Cuba, Martinique, Panama. Rare.

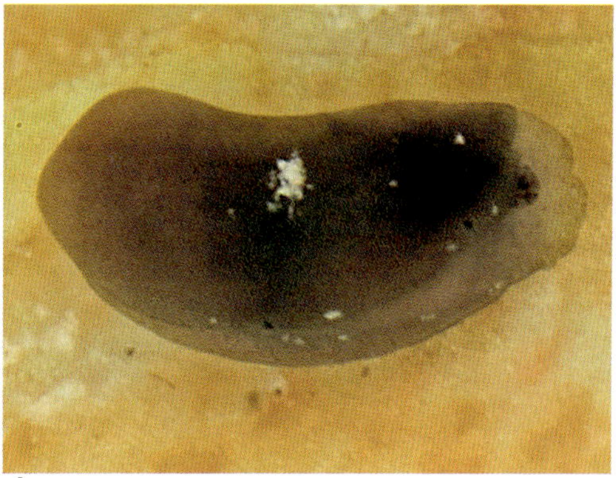

Bermuda

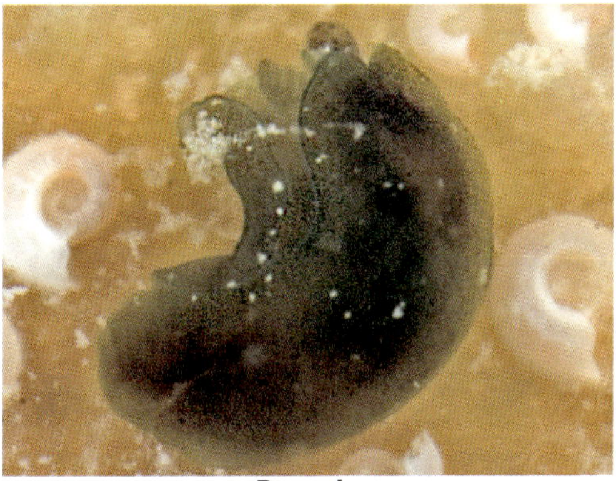

Bermuda

Cuba

Lapinura divae (Ev. Marcus & Er. Marcus, 1963) **Diva's lapinura**

Size: to 3 mm Runcinidae

ID: Body small, elongate, flattish; rear of mantle bilobed and broader; mantle separated from foot by a narrow lateral groove; tail short, broad, truncate; gill beneath right posterior lobe of mantle; large eyes inside the lateral groove near the front. Shell external, vestigial (~0.2 mm), cup-like, slightly more than one whorl with a broad aperture, between posterior lobes of mantle. **Animal brown to reddish to scarlet to green**, often with a scattering of white specks and spots; the spots usually larger along the mantle edges. The internal organs are brownish. Found on sand and rock bottoms with some algae cover. Probably feeds on algae. Depth: to 7 feet. Has been reported from Bermuda, Southeast Florida, Bahamas, Cuba, Guadeloupe, Martinique, Bonaire, Curacao, Venezuela, Costa Rica, Yucatan Peninsula, Jamaica, Brazil.

Bermuda

Lapinura sp.

Size: to 3 mm Runcinidae

ID: Body small, elongate, flattish; rear of mantle bilobed and broader. Animal brown. A white patch at each rounded front corner of the head and a black eye visible behind each patch. The fact that the eyes are visible from above makes it unlikely for this animal to be a variant of *Lapinura divae*. The eyes of that species (see above) are concealed within a lateral groove and not apparent from a dorsal view.

Order Pteropoda Cuvier, 1804

The Pteropoda are holoplanktonic marine snails/slugs divided into two main groups, the **thecosomes (sea butterflies)** and the **gymnosomes (sea angels)**. The thecosomes consist of 8 families with about 70 species in the **suborder Euthecosomata** and 3 families with about 20 species in the **suborder Pseudothecosomata**. The 55 gymnosomes are distributed among 6 families in the suborder Gymnosomata. The term Pteropoda derives from the Greek *ptero* (wing) and *poda* (footed), in reference to the species' shared characteristic of a foot that has been modified to create a pair of swimming "wings", a feature that is helpful for animals that spend their entire lives in the water column.

Sea butterflies share the following features: body small (1-30 mm long); shell present for animal to retract into; some species with an operculum on the foot; shell shapes include straight and pointed, bottle-shaped, globose, and pyramidal; most have features for buoyancy control; they spend most of their time floating as opposed to swimming, often seeming stationary; some move hundreds of meters up the water column to feed nocturnally near the surface and return to the depths when daylight approaches; they feed by deploying sticky mucous webs for capturing small planktonic organisms; they are protandrous hermaphrodites, most releasing egg masses into the water column for the development of veliger larvae.

Suborder Euthecosomata (sea butterflies)

Circumglobal in all seas, including the Antarctic and central Arctic Oceans, with population densities highest in frigid waters but species diversity greatest in warm areas. The shell is external, calcareous, fragile, sometimes sharply pointed. When the animal is in its normal swimming/feeding orientation, it is upside-down and the following are partially exposed – the head, two small median foot lobes that flank the depressed mouth, and a pair of foot-derived, broad, muscular, dorso-lateral wings from the front of the shell's aperture, as well as (in some species) smaller ventrally (upward) protruding mantle lobes. Although the latter two features can both assist with buoyancy regulation, the wings provide the major swimming capability. Swimming upwards is a jerky dragging of the shell forward, consisting of dorsal-ventral sculling of the wings with the body inclined upwards, speeds up to 40 cm/s; swimming downwards is basically "sinking". Their daily vertical migrations are typically about 200 m each way, but some species go much deeper.

These animals' feeding behaviors are exemplified by that of *Cavolinea inflexa*. It remains at depths of 300-400 feet during daylight hours and migrates vertically into the epipelagic zone between dusk and dawn for intermittent feeding purposes. Migration speeds are 20-70 mm/s. The feeding process is as follows: it hangs upside-down and motionless as it drifts with currents with mantle lobes extended and wings expanded; it creates a large globular mucous web in less than three minutes and deploys it above itself, the web often much larger than the animal (web diameters for the Euthecosomata are (10-200 mm); particles of food and detritus stick to the web, which is drawn toward the mouth by the foot lobe cilia; dorsal and median mantle lobe cilia direct non-food particles trapped in mucous (pseudofeces) away from the mouth; the web plus attached food particles are drawn into the stomach as four separate strings within about 3 minutes. Actual feces are released from a lateral slit, away from both the mouth and the web, often seen as a dense cloud beneath the animal. The deployment of the web also assists with the maintenance of neutral buoyancy during feeding by providing drag resistance to sinking. Prey consists of both phytoplankton and microzooplankton (tintinnids, larvae of crustaceans and gastropods), but mostly the latter. The animal is often seen with a pair of very long tentacles (central appendages of the median mantle lobe) but their function is not understood. One possibility is to act as visual lures for swimming prey. When disturbed while feeding it breaks loose from the web and swims horizontally (sometimes only 20 cm) and then reassumes its motionless feeding posture. Predators include fishes and sea angels.

Suborder Pseudothecosomata (sea butterflies)

Circumglobal primarily in temperate to subtropical seas. These animals differ from the Euthecosomata by having: a firm, gelatinous, transparent, internal pseudoconch that replaces the external calcareous shell; two swimming wings fused into a single plate; two symmetrical cephalic tentacles; a proboscis developed from foot lobe elongations; usually flat or funnel-shaped feeding webs; a usually deeper habitat.

The presence of a pseudoconch reduces body weight, thereby reducing the effort required for maintaining neutral buoyancy and swimming. Some species can rapidly swim away from danger. When swimming, the animal keeps the rear end of its pseudoconch (apex) forward while **pushing** it from behind with dorsal-ventral flexions of its wingplate (the Euthecosomata **pull** the shell). Swimming speed is controlled by rotating the body to reposition the wingplate – above the pseudoconch for slow, below it for fast.

Their differently shaped feeding webs can be quite large, some flat ones being up to 2 meters across. Feeding begins with the animal inverted just below the mucous sheet (or funnel) and touching it with its extended proboscis. As food particles become entangled in the mucous, the mixture is drawn by ciliary action toward the proboscis and then into the mouth, while some of it is discarded as pseudofeces. Prey consists of juvenile snails, foraminifera, and even juvenile thecosomes.

Suborder Gymnosomata (sea angels)

The rarely encountered epipelagic or mesopelagic sea angels are found in all seas, including the Antarctic and Arctic Oceans. The name "gymnosome" derives from Greek for their lack of a shell – *gymno* = naked, *some* = body. Their other morphological features are as follows: head well-defined, neck narrowed; a pair of retractile tactile tentacles at the front of the head and a pair of smaller light-sensitive tentacles in dorsal neck depressions just behind the head; body elongate to bag-like, surface tough and elastic, bilaterally symmetric, transparent to opaque, length 2-85 mm; a pair of lateral foot-derived muscular swimming wings, smaller than in thecosomes (relatively shorter and broader); simple thin gills at the rear of the trunk, either encircling it, lateral, or at its rear tip; buccal mass eversible, usually with a radula, hook sacs, and a jaw, sometimes at the tip of a long proboscis; all species except one have a radula, usually with a serrate rachidian tooth flanked by up to 15 long curved laterals; in most species the radula is flanked by a pair of hook sacs with numerous curved hooks that help to grip and pull prey into the buccal cavity; some have sucker-bearing tentacles inside the buccal cavity.

Because gymnosomes are rarely seen their behavior is not well known. It is probable that many of them migrate vertically for feeding like the thecosomes. Their swimming ability is very good – quick movements, good maneuverability, maximum speeds in the range 350-1000 mm/s. They scull obliquely with their wings, getting upward/forward thrust on a downstroke, downward/forward thrust on an upstroke, rarely seeming stationary except during mating or spawning. Since they are known to be carnivores that feed primarily on thecosomes as well as on other large zooplankton, their feeding structures and swimming capabilities seem to have evolved for the pursuit, capture, and ingestion of such prey. Some have also been observed resting on top of a mucous web and waiting to be drawn toward its thecosome prey as the web is ingested.

adult

juvenile

adult

Cavolinia inflexa **(Leseur, 1813)** **inflexed cavoline**
Size: Shell length to 7 mm Cavoliniidae
ID: Shell bilaterally symmetric; relatively long and depressed compared to other cavoliniids, with a flat dorsal and slightly convex ventral face; anterior broad and rounded, curving back on each side to its greatest width at a short and stout postero-laterally directed spine at mid-length or considerably forward of it; the rest tapering posteriorly as a broad-based long and conical median caudal spine that terminates with a slightly dorsally-curved pointed apex; dorsum protrudes forward to considerably overhang the ventrum, its front third with two longitudinal ribs; main aperture is anterior, with smaller slits near lateral spines; tip of caudal spine (the embryonic shell) may be absent in adults.

The animal's foot has lobes and two larger ciliated parapodia that can squeeze through each side of the aperture, the parapodia able to unfold as broad flat "wings" to facilitate drifting/swimming. It's external mantle, which covers the shell, has four sections: (i) a front lobe that protrudes through the main aperture and forms a collar around the body behind the wings and foot lobes (ii) two strongly arched dorsal lobes that protrude through the side slits and create a large cavity between them and the shell that is filled with a gelatinous "pseudoconch"(iii) two ventral lobes that protrude from the side slits; (iv) two heavily ciliated median lobes between the dorsal and ventral lobes that also protrude through the side slits; each lobe has three appendages: a flat anterior one, a long central one, a more prominent rear one. The shell is translucent white, sometimes with brown and pinkish patches and amber highlights. Orange visceral mass clearly visible. Circumglobal in warm to temperate waters. Atlantic latitudinal range is approximately from 55° N to 45° S. Depth: 1-2200 feet. Reported from the Carolinas, Bermuda, Florida, Texas, Mexico, Cuba, Puerto Rico, St. Croix, Guadeloupe, Martinique, Venezuela, and south to Brazil. Not reported from north coast of South America. Common.

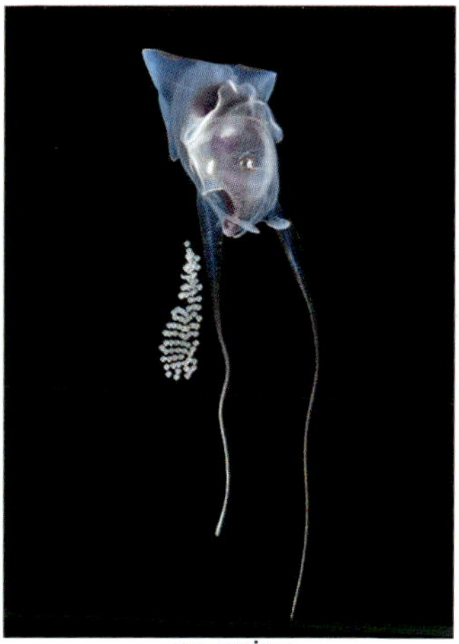
spawning

with dinoflagellate parasites

Cavolinia uncinata **(d'Orbigny, 1835)** **uncinate cavoline**
Size: Shell length to 11 mm Cavoliniidae
ID: Shell bilaterally symmetric, dorsally flattish, ventrally highly convex; anterior broad and rounded; each side nearly straight longitudinally, extending far back to a short and sharply pointed postero-laterally directed spine that is slightly flexed dorsally, the pair flanking a longer dorsally-curved caudal spine; the shell margin between the median spine and each lateral spine is concave; the aperture from the front to each lateral spine is slit-like, wider at the front and near the spines; the dorsal side has 5 low longitudinal ribs except on its hood-like front that protrudes well beyond the aperture and curves down over it; the front of the ventral side has ~10 close-set transverse curved ribs; the tip of the caudal spine is usually absent in adults. The visceral mass is compact and central. The body morphology is similar to that of *Cavolinia inflexa*. Shell: translucent whitish, rib area of ventral lip brownish. The protruding body parts are translucent whitish, the visceral mass orange. Depth: 0-2300 feet. Reported from Massachusetts to the Gulf of Mexico, Bermuda, Bahamas, Greater Antilles, Venezuela, Brazil, Argentina.

ventral aspect

egg string

Cavolinia tridentata **(Forsskål, 1775)** **three-tooth cavoline**

Size: Shell length to 20 mm Cavoliniidae

ID: A sea butterfly. The shell is bilaterally symmetric, thin, fragile, elongate, ovoid, with a rounded front; dorsally flattish; ventrally moderately convex, most convex anteriorly, prominently striated; there is a short pointed spine ("tooth") at the rear on each side, the pair flanking a longer straight spine that is usually blunt because of a broken tip; the shell margin between the median spine and each lateral spine is concave; the aperture is at the front, very broad transversely, narrow vertically, and triangular; at the side of each lateral spine there is a short, narrow, linear slit; the dorsal side has a hood-like front that protrudes well beyond the aperture and curves down over it.

The animal has foot lobes and two larger ciliated parapodia that can squeeze through each side of the aperture, the parapodia unfolding as broad flat "wings" to facilitate drifting and swimming. When fully deployed, each wing is about twice as wide as the shell, roughly triangular, its front edge the longest, the side edge trilobate, and the two front edges aligning to form a nearly straight line with a small median notch. In the process of retracting, the wings contract to one-fourth their maximum size. A very large leaf-like trilobate structure sometimes protrudes from each postero-lateral slit. The animal also has a gelatinous pseudoconch like that of *Cavolinia inflexa*. The shell is translucent yellowish-brown. The wings and small mantle appendages have blue, white, and pink highlights; the large postero-dorsal leaf-like appendages are green. The orange-brown visceral mass is clearly visible.

Depth: 0-400 feet, but has been recorded as deep as 16,000 feet. Occurs along the east coast of North America from Newfoundland to Florida, Bermuda, Bahamas, Gulf of Mexico, and northeast South America.

with mucous web

Corolla spectabilis **Dall, 1871** **spectacular corolla**

Size: Pseudoconch to 40 mm; wing span to 80 mm. Cymbuliidae

ID: A sea butterfly without a calcareous shell. Parapodia fused into a single, broad, oval to quadrangular wing plate with conspicuous mucous glands along its periphery; a broad flattened pair of fused foot lobes form a ventrally extendable proboscis with ciliated lateral grooves; proboscis with distal mouth opening; two small knob-like tentacles near base of proboscis; the animal's structure and the protection for its vital organs are provided by a pseudoconch (false shell)"; pseudoconch broadly oval, stiffly gelatinous, warty, slipper-like (with pocket), half as long as the wing span, perpendicular to the wing plate's long axis, extending forward from near the plate's middle and considerably overhanging its anterior edge. Transparent. Viscera visible as a compact reddish-orange mass inside the pseudoconch. Depth: 0 to >730 feet. Range is 43°N-5°S latitude in the East Pacific and West Atlantic; recently reported from the eastern Mediterranean Sea. Bermuda, eastern Gulf of Mexico, east Florida, Bahamas, Venezuela, Brazil.

***Diacavolinia* sp.**
Size: Shell length to 10 mm. Cavoliniidae
ID: Shell calcareous, roughly triangular, slightly longer than wide, widest posteriorly; dorsally flattish to slightly convex with several longitudinal ribs, sometimes with transverse growth lines; a broad front projection with a large lip curves strongly down over the front aperture; ventrally strongly convex, 20-40 transverse ribs; each rear corner has a postero-laterally directed spine that is short, stout, sharp, hooked, sometimes arched dorsally; medianly at the rear there is a short broadly truncate projection without a septum, as if a broad-based caudal spine had detached and the dorsal and ventral sides had fused; a large to very narrow aperture extends from the front to each lateral spine. The protruding (but retractable) soft body parts are very similar to those of *Cavolinia uncinata* (q.v.). The shell is transparent to translucent whitish with pink tints and may have brownish pigmentation along the margin of the ventral lip and midside on the dorsum and ventrum. Body parts are transparent to translucent whitish, wings yellow centrally, visceral mass compact and yellow. It's a diurnal vertical migrator, a mucous web feeder on zooplankton and phytoplankton, and a good swimmer, usually seen with its long tapering balancers floating behind it as it swims. Depth: 0-4000 feet.

Of the current two dozen taxa in the genus *Diacavolinia* only five occur in the tropical West Atlantic. However, recent studies based on both genetic and morphometric information suggested that they are conspecific. The above description takes into account the morphological differences between these five taxa.

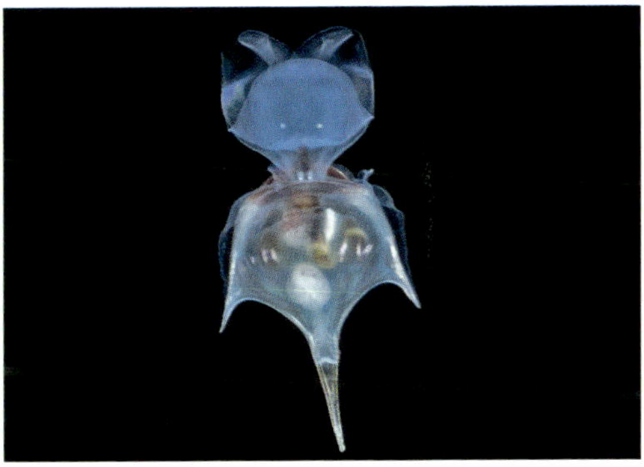

***Diacria major* (Boas, 1886)** **major cavoline**
Size: Shell length to 13 mm. Cavoliniidae
ID: Shell thick, somewhat diamond shaped, front broadly rounded, rear pointed, bilaterally symmetric; ventrally slightly convex with 3 smooth ridges – a very broad median one flanked by 2 narrow ones, dorsally with 5 low ridges; a short, narrow, slightly recurved postero-lateral spine on each side directed outwards and back; a thick-based, very long, sharply-pointed median spine that curves dorsally; protoconch usually broken off; front aperture wide, 1 mm high; narrow slit from front of each lateral spine to front of shell. Animal's parapodia protrude through aperture as broad flat "wings" for drifting/swimming. Shell transparent colorless; **aperture lips brown to reddish-brown to amber**, also sometimes the side slits and concave edges from the side spines to the base of median spine. Orange/brown/white visceral mass clearly visible. Lateral tentacles gold. Depth: 0-1300 feet. Range in West Atlantic is 20°-40° degrees north latitude. East Florida, Gulf of Mexico, Bermuda, Sargasso Sea.

spawning →

Diacria trispinosa **(Blainville, 1821)** **three-spine cavoline**

Size: Shell length to 11 mm. Cavoliniidae

ID: Shell thick; viewed from above somewhat diamond shaped, front rounded, rear pointed, depressed; ventrally convex, a broad median ridge between 2 narrow ones; dorsum with 5 low ridges; a long sharp postero-lateral spine on each side either perpendicular to the shell's longitudinal axis or inclined slightly to the rear; a very long median spine between them arches dorsally, usually with the protoconch broken off; front aperture wide; narrow slit from front of each lateral spine to front of shell. Foot lobes and parapodia can squeeze through each side of the aperture, the latter unfolding as broad flat "wings". Other mantle appendages have cilia and mucous glands, including two long "tentacles" per side that also protrude through the lateral slits. Shell transparent colorless; **aperture lips chestnut brown**, sometimes also the side slits and the concave edges from the side spines to the base of the median spine. Orange/brown/white visceral mass clearly visible. Lateral tentacles gold. Shell often with attached hydroids (e.g. *Laomedea striata*, *Kinetocodium danae, Campaniclava clionis*). Depth:0-726 feet. Found between 40°N and 40°S latitude in all oceans, but most plentiful in the West Atlantic.

mating

with mucous web

Cuvierina atlantica **Bé, MacClintock & Currie, 1972** **cigar pteropod**

Size: Shell length to 10 mm. Cavoliniidae

ID: Shell begins as a slender conical protoconch, sharply pointed at rear. As it grows and the teleoconch develops, the animal relocates to the teleoconch and a septum forms just in front of the protoconch to seal off the posterior. The juvenile's protoconch eventually detaches when the shell reaches its maximum size. The remaining "adult" shell is smooth, almost cylindrical, slightly inflated about one-third from its rear; has faint growth lines, a kidney-shaped aperture with a slight constriction behind it, and a convex posterior septum. Two parapodia (swimming wings) protrude from the aperture; there are no complex external mantle lobes such as in *Cavolinia* species. Shell transparent, viscera clearly visible as an orange, brown and white mass. Parapodia with a median purple blotch. Depth: 0-400 feet, but has been recorded as deep as 825 feet. Found throughout temperate to tropical latitudes of the Atlantic Ocean (44°N to 44°S). Bermuda, Florida, Bahamas, Caribbean. The warm Gulf Stream occasionally transports individuals as far north as Iceland. Mass blooms have been observed. Common.

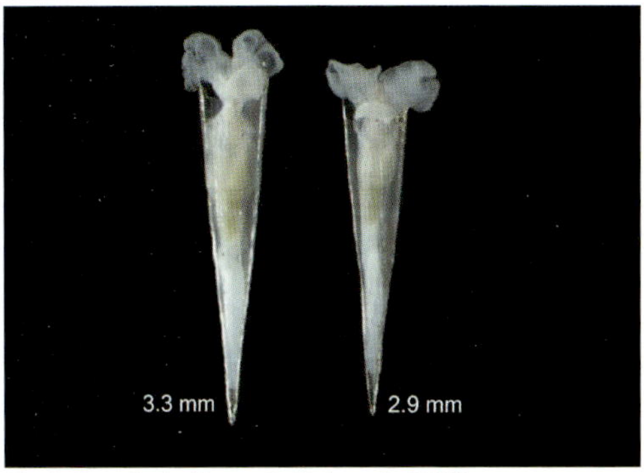

→ preyed on by a pelagic gastropod →

Styliola subula **(Quoy & Gaimard, 1827)** **keeled clio**

Size: Shell length to 13 mm. Creseidae

ID: Shell smooth, thin, conical, slender, faintly striated transversely; aperture circular; protoconch sharply pointed. Two small wavy-edged parapodia project wing-like from aperture with a much smaller, semicircular, more ventral lobe between them. Mouth in depression formed by united bases of the three lobes. No other mantle appendages. Shell transparent. Viscera visible as a yellowish, brownish, and pink mass. Wings pinkish. Depth: 0-5000 feet. Between 44°N and 40°S latitude in warm marine waters. In the West Atlantic it has been reported from North Carolina to Florida to Texas, Bermuda, Bahamas, Antilles, Venezuela, Colombia, Campeche Bank (Mexico),

→ mating →

Clio pyramidata **Linnaeus, 1767** **pyramid clio**

Size: Shell length to 15 mm. Cliidae

ID: Shell thin, an elongate tetrahedron with open base at front and triangular faces meeting at a rear vertex having a slender ovoid tip (protoconch); one long edge nearly straight and two curved; faces slightly concave inwards; ventral face almost flat with a median ridge, the other two with three rounded ridges. Parapodia protrude through aperture and unfold laterally as "wings", foot lobes protrude ventrally. Shell and wings transparent colorless; aperture lips and viscera brown to reddish; intestine crimson. Depth: 0-3000 feet, 40°N to 30°S latitude in warm temperate seas. Florida, Bermuda, Greater Antilles, Texas, Quintana Roo, Venezuela.

Clio recurva **(Children, 1823)** **wavy clio**

Size: Shell length to 30 mm, width to 17 mm. Cliidae

ID: Shell is a fragile partly flattened cone with: a broad front aperture, slender rear vertex that slightly curves dorsally; concentric transverse wavy ridges along its entire length, paired on the ventral face; 3 longitudinal ribs on the dorsal face, the middle one widest. Parapodia protrude through the aperture and unfold laterally as "wings", extended foot lobes are ventral. Musculature of wings and mantle well-developed. Shell and wings are transparent colorless. Reddish to dark brown viscera visible through shell. Depth: 0-1400 feet, 40°N to 60°S latitude in warm temperate marine waters. Reported from Florida, Martinique, Barbados, Brazil.

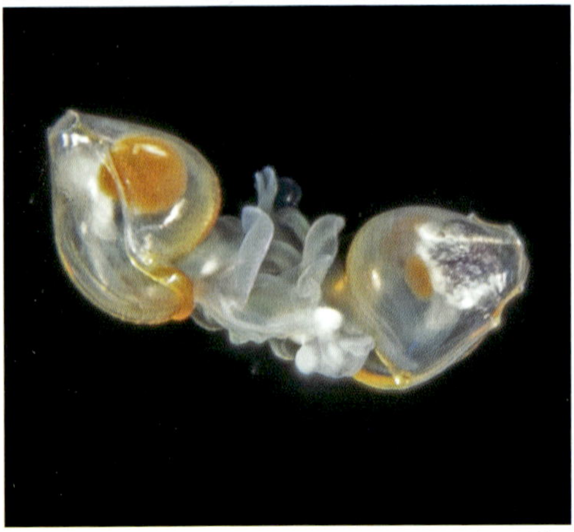

Telodiacria danae (van Leyen & van der Spoel, 1982)
four-spine cavoline

Size: Shell length to 2 mm Cavoliniidae
ID: Viewed from above the shell is oval, longer than wide, bilaterally symmetric, narrowed posteriorly with 4 short triangular lateral spines, rearmost 2 flanking a truncate region. In side view it is moderately convex dorsally, globose ventrally; has a very narrow front aperture and narrow slits along each side. The dorsum has transverse striae and 3 indistinct longitudinal ribs; ventrum with faint concentric striae. Foot lobes and 2 large parapodia protrude from the aperture, the latter as broad flat wings; also 2 long tentacles per side. Shell translucent whitish with reddish-brown pigment on surface, aperture lips, and on the concave edges between spines. Orange, brown, white visceral mass clearly visible. Lateral tentacles gold. Depth: 0-600 feet. 40°N to 40°S. Bermuda, east Florida, Bahamas, Virgin Islands, and Mexico (Yucatan).

Cymbulia peronii Blainville, 1818

Size: Pseudoconch length to 65 mm Cymbuliidae
ID: As the animal matures its small larval shell is replaced by a pseudoconch – stiffly gelatinous, fusiform, rear emarginate to forked, front pointed and with a dorsal pocket, each side with 2 rows of small spines. The parapodia form a single wing plate with a span about half as long as the pseudoconch; rear of wing plate has a small median flat lobe with a projecting tail-like tentacle. Proboscis ventral, extendable, ciliated; mouth at bottom of proboscis. Transparent; viscera visible as a compact white, orange, to black mass; rear tentacle pink; wing plate with fine whitish lines of vascular and branchial networks. Depth: 0-6600 feet. East Atlantic, Mediterranean, Indo-Pacific. East Florida, Gulf of Mexico, Brazil, Argentina.

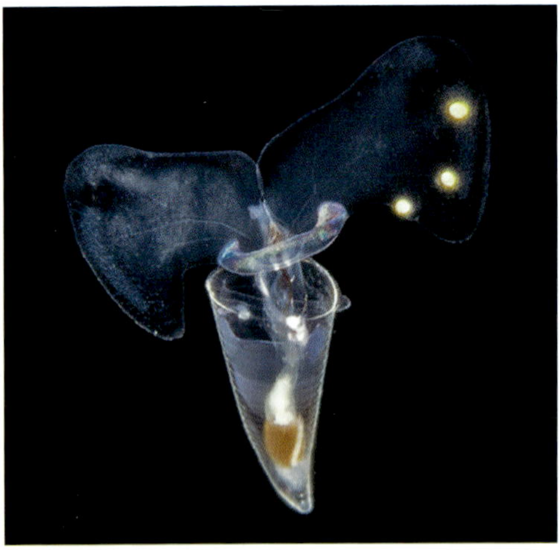

Hyalocylis striata (Rang, 1828) **striate cleo**

Size: Shell length to 8 mm. Hyalocylidae
ID: Shell conical, slightly curved at apex, transparent, fragile; aperture at top and apex at bottom in floating animal; cone angle ~24° off vertical; surface annulate (no whorls); annuli broad and thick, progressively more crowded toward the apex, adults with ~40; microscopic longitudinal striae present between annuli. Viscera clearly visible; 2 broad wing-like swimming appendages protrude from the aperture. The shell's protoconch shrivels and detaches early in the animal's life, leaving the adult shell with a truncate posterior sealed with a small concave septum. Animal transparent; its visceral mass brownish and white. Depth: 30-934 feet. Circumglobal, 40°N to 40°S in warm to temperate seas. Bermuda to Florida, Gulf of Mexico, Antilles, Venezuela, Brazil.

Creseis acicula (Rang, 1828) **straight needle-pteropod**

Size: Shell length to 33 mm. Creseidae
ID: Shell very elongate, narrowly conical (~3°), very thin and fragile, surface smooth, usually straight but sometimes slightly irregular due to growth anomalies; anterior aperture circular, diameter up to 1.4 mm; protoconch ovoid, slightly bulbous, usually not clearly separated from teleoconch. Two small oblong parapodia project wing-like from the aperture, flanking a much smaller, semicircular, more ventral lobe; mouth in the depression formed by the united bases of the three lobes; no other mantle appendages. Transparent glazed shell, yellow internal highlights, viscera a greenish mass. Depth: 0-2500 feet. Circumglobal between latitudes 50°N and 45°S in warm to temperate waters. Throughout the tropical West Atlantic.

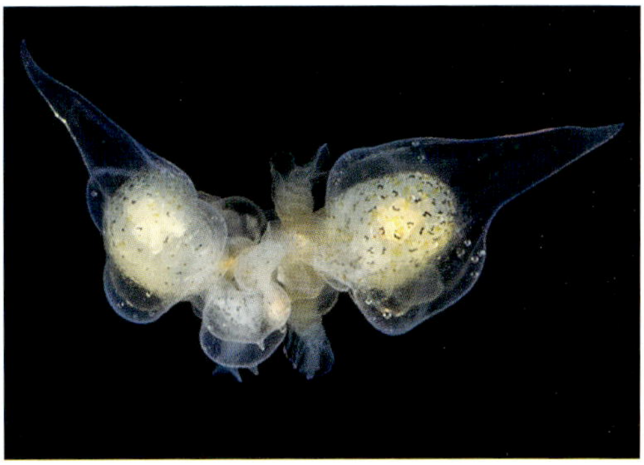

***Fowlerina* cf. *punctata* (Tesch, 1903)**
Size: ~6 mm. Clionidae
ID: Front half of body broadly bulbous, rear half narrow and tapering to a sharp point; large globular visceral mass occupies front half of body; head much less broad, rounded, flattened, with a pair of short conical tentacles; neck very short; wings are short, broad, rounded lobes; no gills. Transparent; skin unpigmented; visceral mass and head yellowish with orange-yellow to blackish spots and short scrawls. Photographed near the surface off Southeast Florida.

all photos are from near the surface off Southeast Florida

mating

spawning

***Notobranchaea* cf. *macdonaldi* Pelseneer, 1886**
Size: ~10 mm. Clionidae
ID: No shell. Head rounded with small conical anterior tentacles; neck long and contracted, parapodial wings moderately large; body broad, inflated at rear, gill posterior; lateral gill inconspicuous (absent?). Head and viscera whitish with reddish-brown spots, viscera usually strikingly orange to reddish. There are very few confirmed records of *N. macdonaldi* in the tropical West Atlantic (specimens collected off Bermuda and Florida), and these photos may be additional records. However, the required features for a positive identification (posterior gill crests, a lateral gill, and foot morphology) are not visible well enough. Specimens have also been described as having jet-black anterior tentacles and parapodial edges, which are also difficult to discern reliably. To remove this uncertainty, analyses of photographed specimens are required.

***Thliptodon* sp.**
Size: ~8 mm
Clionidae
ID: Body fusiform to teardrop-like; posterior truncate to somewhat pointed with a round opening. Head about half the total length, with two very short tentacles; a pair of transverse paddle-like wings at mid-length; viscera compact and well away from the body walls. Transparent, body surface white-dotted, viscera white. Photographed near the surface off Southeast Florida. This is probably *Thliptodon diaphanus* (Meisenheimer, 1902), a circumglobal species that is common around Africa but has rarely been seen in the tropical West Atlantic (Florida, Antilles).

***Hydromyles globulosus* (Rang, 1825)**
Size: to 10 mm
Hydromylidae
ID: Body spherical to ovoid with a thin, flexible mantle; head has 2 long tentacles, each with a short protuberance near the inner base; a paddle-like narrow-stalked wing extends from the foot on each side; wings and tentacles have similar lengths; appendages protrude/expand from antero-ventral transverse slit. Translucent bluish-white with yellow/orange viscera. With appendages retracted it resembles an orange. When disturbed it releases dark brown ink from its anal opening. Epipelagic. Massive Indo-Pacific blooms are common but in the West Atlantic it is rarely seen. Occasionally near the surface, both day and night, off Southeast Florida.

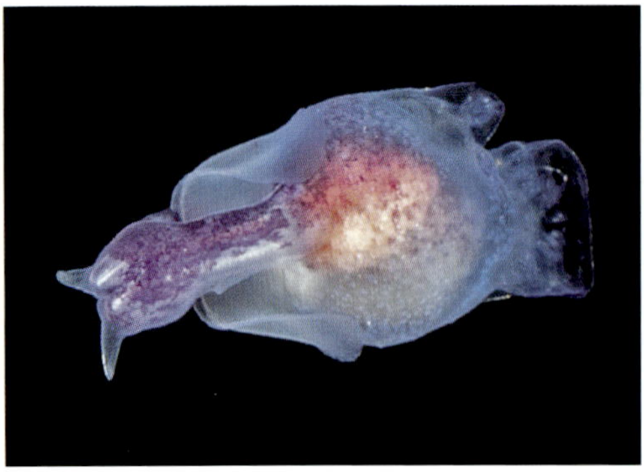
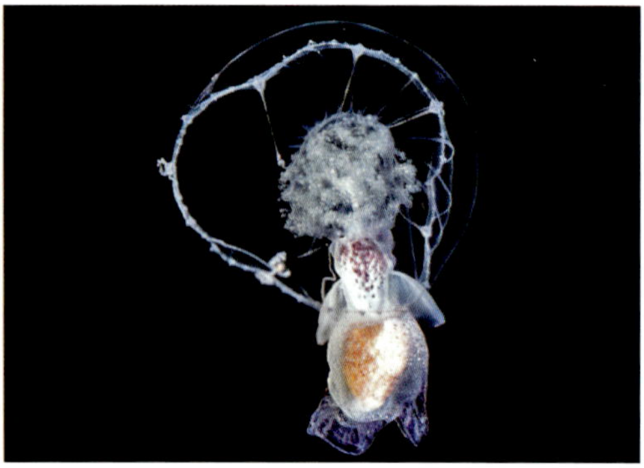

***Pneumoderma* cf. *violaceum* d'Orbigny, 1835**
Size: to 10 mm
Pneumodermatidae
ID: Body oblong, rear half broader; head rounded, slightly depressed, with a pair of conical tentacles; neck long, constricted; small trapezoidal wings; posterior gill an irregular fringed quadrilateral; visceral mass occupies most of body. This is likely *P. violacium*, in which case it also has: a flap-like lateral gill along its right rear side; a long, broad, pointed rear foot lobe; an extendable proboscis flanked by a pair of long sucker-bearing tentacles, visible only when feeding. Ground color translucent whitish; dense violet pigment ventrally on the head, neck, and some body parts; viscera pale orange. Photographed near the surface off Southeast Florida.

dorsal view

ventral view

eating a sea butterfly

***Pneumodermopsis* cf. *canephora* Pruvot-Fol, 1924**

Size: to 12 mm Pneumodermatidae

ID: Elongate; head rounded, with a pair of long narrowly conical tentacles; neck somewhat constricted; parapodial wings of moderate size; body broader than head, vase-shaped, over twice as long as head plus neck; visceral mass spherical, occupies front half of body; two ciliary bands encircle the body – one central, one posterior; rear of body truncate with a circular opening; very long broad lateral gill on right side. If this is indeed *P. canephora* it also has an eversible buccal apparatus with a proboscis and three prey-capture arms – two laterals with a thick cylindrical stalk and one sucker disc, and a larger median one with five suckers. Transparent. Orange flecks on head, white flecks on body, visceral mass orange. Depth: 0-6000 feet.

before eating

eating a sea butterfly

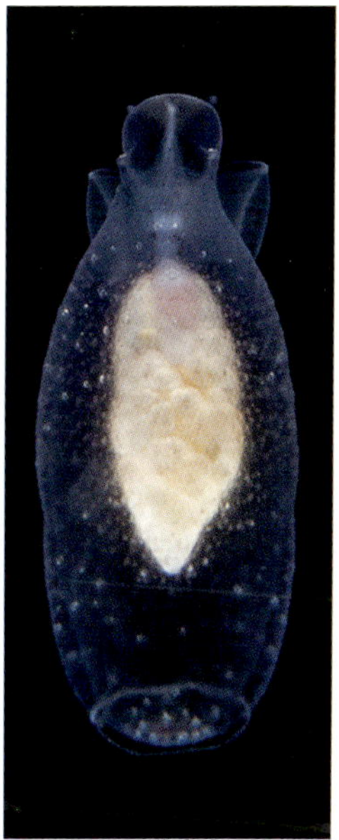

well-fed

***Cliopsis krohnii* Troschel, 1854** **sea cherub**

Size: to 40 mm Cliopsidae

ID: Body elongate, barrel-shaped, thick, gelatinous, rear truncate; head small, two short tentacles; neck short; broad paddle-like wings – one per side behind head; gill a posterior longitudinal tube with a hexagonal vertical crest at its rear tip and two upper and lower membranous linear edges (4 crests) radiating forward; foot reduced to a three-lobed structure below the wings consisting of two small flaps with right-angled rear corners flanking a median plicate tubercle; proboscis can extend to 2-3 times the body length. Adults transparent, sub-adults yellowish to bluish to gray. Viscera clearly visible as a yellowish brown, black, and pinkish mass; head often pinkish internally. The entire body is densely brown-dotted with fatty glands. When it encounters the mucous web of a sea butterfly it immobilizes the animal (up to 3 times its own size) with its long proboscis that is armed with ~60 chitinous "hooks", dismembers it, and then devours it using its cutting teeth. Feeds on *Corolla* species. Depth: surface to at least 5000 feet. Circumglobal in tropical and temperate seas. The above photos were taken near the surface off Southeast Florida.

Family APLYSIIDAE Lamarck, 1809

Body oval to elongate, bulky, usually soft, length up to 750 mm. There are two oral tentacles at the front followed by two rhinophores. Lateral extensions of the foot form large parapodia that fold up over the mantle cavity, but in some species they may be reduced or partially fused dorsally to form inhalant and exhalant siphons for channeling water into and past the gill within the cavity. Mature individuals of most species have a small, somewhat flattish and irregularly-shaped, internal shell remnant hidden between the parapodia above the mantle cavity, but some have only a vestigial fragment or (e.g. *Bursatella*, *Stylocheilus*) no shell at all.

Aplysiids are herbivores that inhabit shallows with a profusion of seagrasses and red, green, or brown algae. Their diet preferences determine their coloration. They feed on algae for nutrients as well as for defensive chemicals they secrete and concentrate within two major glands inside the mantle cavity – the purple gland at the top and the opaline gland at the bottom. The purple gland contains red or purple ink, the opaline gland a thick opaque white fluid (opaline). When attacked, or simply disturbed, it will likely release and mix both secretions within the mantle cavity and then expel a chemical cloud toward the predator. The effects vary: impaired vision; sensory disruption; compromised nervous system; repugnance; "phagomimicry", where the concoction mimics a predator's food source. At worst it provides a distraction that gives the slug time to swim away by vigorously flapping its parapodia.

Being hermaphrodites they have both male and female reproductive organs. The penis is on the right side of the head; the vagina deep inside the mantle cavity. Mating can consist of only two individuals, one acting as a male and the other as a female, but in some species they more commonly exhibit mass insemination where up to 20 of them form a chain with a female at the front, a male at the rear, and those in between acting simultaneously as a sperm donor and sperm receptor. The combination of a large rounded body and long erect rhinophores give them the appearance of a sitting rabbit, hence the common term "seahares" for this group.

adult; Florida

adult; Brazil

juvenile; Guadeloupe

shell; Florida

swimming; Guadeloupe

Aplysia brasiliana **Rang, 1828** **mottled seahare**

Size: to 270 mm Aplysiidae

ID: Body bulbous, soft, strongly contractile; head and neck elongate; oral tentacles short, broad, flat, edges with wavy folds; rhinophores short, slender, conical, rolled, with vertical slit and apical opening, close together; parapodia very broad, edges thin and wavy, widely separated anteriorly and fused low at rear near start of tail; mantle a large thin layer set far back between the parapodia, covers internal shell and mantle cavity; center of mantle with a large foramen in juveniles, reduced to a small papilla in adults; mantle cavity open at each end; mantle section at right rear forms an erect exhalant siphon surrounding the anus, discharging water "inhaled" anteriorly and passed over the gill; tail short. Shell posterior, thin, ovoid, about one-third body length. Translucent yellow, brown, gray, green, reddish, black. Mantle and parapodia usually with green/brown mottling and irregular patches, often with white or yellow spots, sometimes veined and streaked with black or dark brown. Inner parapodial margins often dark with pale vertical streaks. Rhinophores and oral tentacles pale-tipped; foot light to dark brown, shell brownish-yellow. Depth: usually 0-50 feet, but has been dredged from 1200 feet. New Jersey to Brazil, including Bermuda, Florida, Texas, Mexico, Guadeloupe, Aruba, Bonaire, Curacao, Venezuela, Colombia, Costa Rica.

Aplysia cervina **(Dall & C. T. Simpson, 1901)**

Size: to 100 mm

Aplysiidae

ID: Body elongate, soft, smooth, flabby; head and neck relatively small; foot narrow; tail short, triangular to bluntly rounded; parapodia thick, three-quarters body length, front ends widely separated, fused low at rear; rhinophores conical, short, slender, rolled, with vertical slit and apical opening, fairly widely spaced; oral tentacles short, broad, ear-like flaps; mantle cavity open at each end, exhalant siphon at rear, usually erect. Grayish-green to brownish with superficial network of black pigment; surface with many small, round, dark brown to black spots, sometimes with irregular whitish blotches; inner parapodial margins with alternating white and dark irregular bars. Typically found in highly saline warm waters in and near rocky outcroppings or jetties. Preferred foods are green algae such as *Ulva* spp. (sea lettuce). Depth: to 20 feet. South Carolina to Florida, Cuba, Puerto Rico, Anguilla, Curacao, Brazil, Colombia, Panama, and the Bay of Campeche (Mexico).

Aplysia dactylomela **Rang, 1828** — **spotted seahare**

Size: to 250 mm

Aplysiidae

ID: Body bulky, smooth, soft; head and neck large; parapodia broad, rugged, margins usually ruffled, well back from head, usually joined very low at rear, fold up to form inhalant siphon at midbody; exhalant siphon posterior, large, frilled, usually uncovered; mantle covers shell and shallow cavity, central foramen small or (more often) a closed conical papilla; oral tentacles large, fleshy, edges ruffled/flared; rhinophores much smaller, close-set, rolled, with flared top; foot broad, thick, front rounded; tail short and blunt. Shell internal, fragile, rounded, apex oblique, above mantle cavity behind midbody. Ground color is diet-dependent – shades of brown (dark, purplish, greenish), olive-green, yellowish-cream. Body, head, and appendages with thin blackish reticulations. Parapodia with 20-30 variably sized pale-centered blackish rings connected by reticulations on outer face, edges pale, inner faces with blackish blotches and branched vertical bars. Mantle yellowish-brown, black-blotched. Rhinophores and oral tentacles pale-tipped. Shell translucent white. Depth: 0-120 feet. Throughout the tropical West Atlantic.

St. Vincent

shell and siphon; St. Vincent

juvenile; Roatan

spawn; Brazil

Aplysia juliana **Quoy & Gaimard, 1832** **walking seahare**

Size: to 160 mm Aplysiidae

ID: Head broad, neck long and strongly contractile; deep mantle cavity at rear contains vital organs protected by a thin convex shell covered by mantle tissue with an oval contractile dorsal opening (foramen); dorsum normally exposed but can be covered by its two broad fleshy parapodia that are low near the front of the head and rise to a peak near midbody; parapodia broadly fused high at rear, can fold up to form two openings – front inhalant, rear exhalant – to channel water over its gills; oral tentacles broad, flat, edges rolled and wavy; rhinophores small, conical, rolled, open at tip; anal siphon a short wide tube. Globally the ground color is highly variable – yellowish-green, reddish-yellow, dark green, olive, brown, orange-brown, ash gray, blackish. West Atlantic individuals usually brown to orange-brown with many irregular white spots/dots. Parapodial edges white; narrow submargins may have dark brown to blackish blotches. Primarily nocturnal. It can shape the rear part of its foot into a temporary sucker to facilitate a pushing/dragging "walk" like that of a leech. Since it lacks a ink gland, when disturbed it releases only a milky white fluid produced by its opaline gland. Depth: 0-60 feet. Seen occasionally throughout most of the tropical West Atlantic.

Florida

Aplysia morio **(A. E. Verrill, 1901)** **Atlantic black seahare**

Size: to 400 mm Aplysiidae

ID: Body large, not bulky; head and neck short and thick; oral tentacles hemispherical, broad, flat, edges curled; rhinophores close together, short, conical, tip pointed; parapodia very wide, rounded, edges wavy, fused at base of foot; mantle behind midbody, covers internal shell, no foramen in adults; anal siphon large, tubular, exposed; foot extends as a short elliptical tail with wavy edges. Shell ovoid, spire absent, growth ridges concentric. Dark maroon to dark brown to almost black, may have longitudinal purplish-black lines on head and sides. Prefers calm rocky bottoms where it feeds by day or night on algae (green, yellow, red); sometimes swims in search of food. Depth: typically < 10 feet. Occasional in most of the tropical West Atlantic.

Florida

Grand Cayman

Florida

color variants; Grand Cayman

mating; Guadeloupe

spawn; Grand Cayman

***Aplysia parvula* Mörch, 1863** **freckled seahare**

Size: to 40 mm Aplysiidae

ID: Body robust, elongate, smooth, wider at midbody visceral hump, narrows gradually to head and tail; visceral hump large (to one-third body length), oval, indistinctly separate from rest of body; mantle at hump, covers shell and mantle cavity, foramen relatively small, central, circular, elevated; parapodia narrow, fused at rear, touch at front, start to rise well behind rhinophores; anal siphon wide, elongate, protrudes above parapodia; rhinophores and oral tentacles elongate and rolled, tentacles with some projections; tail short, rounded. Shell internal, narrow, elongate, prominent concavity at right rear. Ground color cream to brownish-red; white spots and blotches on all surfaces, each blotch an aggregation of small white dots. Parapodial edge white, submargin with unbroken bluish band. Rhinophores and oral tentacles dark-tipped. Foramen with prominent dark edge and radial white streaks. Inhabits sheltered bays having sandy, grassy, and rocky rubble bottoms with mixed algae (its food). Elongates as it moves. Known to mate in chains with a "male" behind a "female", each pair simultaneously a sperm donor (penis near eye) and sperm recipient (vagina on back). Depth: to 40 feet. Florida, Bahamas (maybe), Virgin Islands to northern Brazil, Roatan, Grand Cayman.

Exuma Islands, Bahamas

Southeast Florida

Roatan

Aplysia ghanimii **Golestanii et al, 2019** **Hessam's seahare**

Size: to 15 mm Aplysiidae

ID: A molecular and morphological study of circumtropical specimens of *A. parvula* revealed a complex of ten distinct species, two of which occur in the tropical West Atlantic – *A. parvula* and the new species *A. ghanimii*. The latter can be distinguished from the former by some subtle differences (compare with *A. parvula* description on previous page): body smaller, more slender and therefore looking much wider in region of visceral hump; the hump shorter, more circular, well differentiated from rest of body; foramen large, oval, not elevated, on left rear of hump; parapodia broad, joined high up, form several shallow folds; oral tentacles with conspicuous flat posterior projection; anal siphon narrow; body yellowish to brownish-red, completely covered with white dots, the dots often densely clustered; parapodial submarginal black band has white spots; foramen's dark inner edge is interrupted with white patches but there are no radial white lines. Behavior and habitat are like that of *A. parvula*. Depth: < 15 feet. Unfortunately, some individuals have distinguishing features of both species, making reliable identifications very difficult. So far, this species has been positively identified from the Bahamas, Southeast Florida, and Roatan.

Guadeloupe

Dolabella* cf. *auricularia **([Lightfoot], 1786)**

Size: to 400 mm Aplysiidae

ID: Starting from a narrow low head the body greatly expands cone-like toward the rear, where it slopes abruptly to form an oblique posterior flattened "shield" with a papillate edge; parapodia fused except for two circular openings – a small midbody inhalant siphon and a large posterior exhalant siphon; cephalic tentacles short but usually longer than rhinophores; body with numerous evenly distributed papillae, few to many branched, lengths varying by up to a factor of eight in the same animal. It is mottled in a range of colors – whitish, dark olive, purplish-green, brown, greenish-brown, dark purplish-brown. By day it is found in crevices, under rocks, or in tide pools in protected bays and lagoons; at night it forages on sand and mud bottoms and in grass beds where it feeds on a mix of brown, green, and red macroalgae. When disturbed it releases purple ink from its exhalant siphon. Depth: 3-70 feet. Found primarily in the tropical eastern and Indo-Pacific, temperate Africa and Australasia, and the tropical Atlantic; rare in the Caribbean. The anterior of the 100 mm pictured animal from Guadeloupe matches that of *Dolabella auricularia* but a posterior view would also be required for a positive identification.

Martinique

Florida

Roatan

Florida

Guadeloupe

spawn; Florida

Bursatella leachii **Blainville, 1817** **ragged seahare**

Size: to 120 mm Aplysiidae

ID: Body soft, oval, slightly longer than wide, back with hump that peaks at mid-length; head short, wide; parapodia fused except for a third-body-length dorsal slit with fleshy flaps at mid-hump overlying the mantle cavity; body and appendages with many papillae of highly variable shapes and sizes – thorny-looking to stringy to stout, large ones often highly branched, sometimes only short and unbranched; back has small tubercles between the papillae; rhinophores short, conical, rolled; oral tentacles short, rounded, thick, laterally projecting. Ground color variable – yellowish, tan, brown, green, brownish-green, gray, black, white, lavender. Usually a dark background overlain with a network of pale dots, spots, and patches; papillae and tubercles usually much lighter, grayish to pinkish-gray, sometimes dark-spotted. Some animals with a few bright blue oval spots. Rhinophores and oral tentacles pale with a dense distribution of tiny dark spots. Inhabits sea grass beds, mangrove communities, and muddy bottoms in shallow coastal areas where it grazes on microalgae (e.g. *Lyngbya majuscula*) that form mats on the surfaces of mud, sand, rocks, shells, pilings, **seagrasses**. Depth: intertidal to at least 33 feet. Throughout the tropical West Atlantic.

Southeast Florida

Guadeloupe

Southeast Florida

15 mm; Guadeloupe

post-larval; Florida

pinball seahare

Notarchus punctatus **R. A. Philippi, 1836**

Aplysiidae

Size: to 32 mm

ID: Body pyriform with narrow neck and globose rear, tuberculate; tubercles conical, contractile, variably sized, largest near peak of back and smallest along sides; parapodia almost completely fused over back, form round opening slightly in front of midbody for mantle cavity water circulation; ridge of small tubercles from head to dorsal cavity, continuing more prominently from behind dorsal siphon; rhinophores and oral tentacles long, rolled, and blunt, tentacles shorter; foot shorter than body, very narrow, may protrude as a short blunt tail, well adapted for climbing along blades of seagrass; eye at outer front of each rhinophore; no shell. Ground color translucent gray; overall greenish-brown to orange-brown to orange cast from a dense network of fine lines; scattered white spots and scrawls, white pigment usually on/around projecting tubercles; some tubercles yellowish. Balloon-like to deflated looking as it fills/empties water from its mantle cavity. When disturbed it may curl up into a ball, or try to escape by blasting a jet of water forward as it tumbles backwards like an out-of-control "pinball". Usually found among rocks and sea grasses in relatively shallow water. Depth: to 50 feet. Throughout the tropical West Atlantic.

Grand Cayman

San Andres, Colombia

Roatan

Guadeloupe

Guadeloupe

Grand Cayman

Petalifera petalifera **(Rang, 1828)** **Rang's seahare**

Size: to 40 mm Aplysiidae

ID: Body elongate to oval, narrowed at front, broadest and highest behind mid-length, rounded at rear, degree of flatness changeable; back smooth or with some conical or branched, retractile, small papillae; parapodia fused except for short slit behind midbody to right of dorsal midline, each side of slit with small rounded central flap, flaps fold to form two round openings – front inhalant siphon and rear exhalant siphon for mantle cavity water circulation; rhinophores widely spaced, solid, tips rolled and blunt; oral tentacles rolled, tips flared, longer than rhinophores; rhinophores and oral tentacles often with white papillae; prominent seminal groove from common genital aperture near front of dorsal slit to male gonopore below right rhinophore. Shell internal, fragile, visible below the parapodial slit. Greenish, grayish, yellowish shades of brown; usually black-edged pink-to-purplish blotches, white papillae, brown/black specks and dashes; sometimes long threads of white dots; seminal groove often with white edged black stripe; sometimes a black or white stripe from rhinophore to tip of adjacent oral tentacle. Ground color matches that of its algal food source: For example: green => *Thalassia*, yellow => *Sargassum*, etc. Depth: 3-30 feet. Throughout most of the tropical West Atlantic.

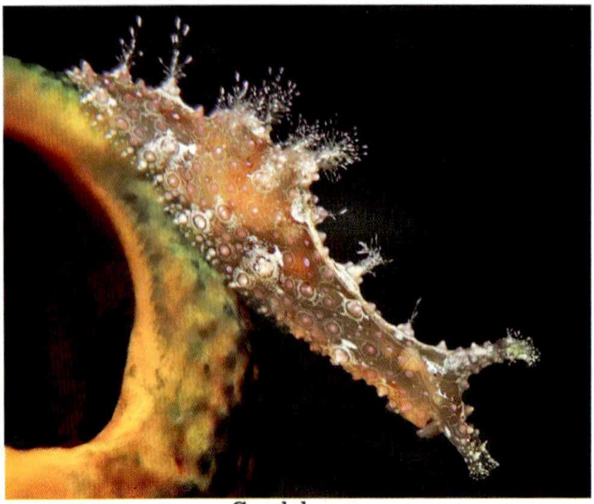

Guadeloupe

Roatan

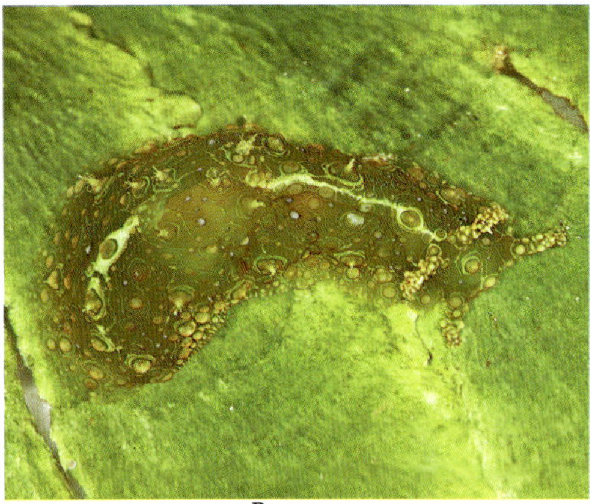

Panama

Dominican Republic

juvenile; Roatan

with spawn; Guadeloupe

***Petalifera ramosa* Baba, 1959** **Baba's seahare**

Size: to 70 mm Aplysiidae

ID: Body elongate, broadest at rear, convex above; many variably sized rounded to conical tubercles; back with larger bushy compound tubercles – small ones clustered around a large one with a highly branched papilla; parapodia fused above but with a short midbody slit offset to the right; flaps of slit edges form two siphons – front inhalant, posterior exhalant – for mantle cavity water circulation; rhinophores and oral tentacles rolled, tips flared; rhinophores longer, sometimes spiky; seminal groove narrow, high left wall, from front of dorsal slit to male gonopore below right rhinophore. Shell thin, broad, teardrop shaped (point at rear), below slit. Color brownish, greenish, pinkish, yellowish. Tubercles often purple, pink, smaller ones blue; white-tipped; bases darker-ringed. Compound tubercles ground color, orange, blackish, whitish; filaments white-tipped, basal ring dark brown. Most tubercles interconnected by fine white lines that sometimes broaden into blotches. Usually found on leaves of seagrass on shallow silty or sandy bottoms with a mix of seaweeds, sponges, algae, ascidians, and bryozoans. Sometimes on rocks. Depth: 6-120 feet. When disturbed it can swim via dorso-ventral flexions of its body. Throughout the tropical West Atlantic.

Guadeloupe

Southeast Florida

Cuba

Guadeloupe

Roatan

St. Vincent

***Dolabrifera* spp.** **warty seahare**

Size: to 100 mm Aplysiidae

ID: *Dolabrifera dolabrifera* (Rang, 1828) has long been considered to be the only member of the genus in the West Atlantic. But recent studies have shown that it is restricted to the Indo-Pacific. The West Atlantic species are *D. ascifera* (Rang, 1828), *D. virens* Verrill, 1901, *D. edmundsi* Valdés et al, 2018. Overlaps in their external variations prevent reliable identifications by appearance alone. They share the following features: body oval to elongate, flattened, tapers from broadly rounded rear to narrow head; highly tuberculate, tubercles rounded, variably sized, retractile papilla at top, large ones usually edged by small ones; parapodial edges tuberculate, fused dorsally except for short rear section with flaps forming two openings for mantle cavity water flow; rhinophores long, rolled; oral tentacles short, wide, rolled; internal shell present. Green, greenish-brown, greenish-gray, light to dark brown, red, sometimes white; with blotches of different colors. In shallow areas with rock/rubble substrates, to which it anchors itself using its broad foot as it moves slowly in search of algae food. Depth:0-30 feet. Confirmed records: *D. ascifera*: Florida, Mexico, U.S. Virgin Islands, Jamaica, Panama, Venezuela, French Guiana; *D. virens*: Bermuda, Florida, Jamaica, Curacao, Panama; *D. edmundsi*: Gulf of Mexico.

Roatan

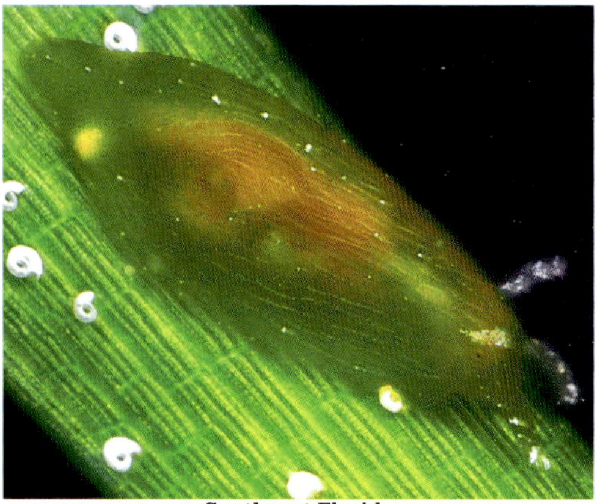
Southeast Florida

Guadeloupe

Roatan

Roatan

Key Largo, Florida

Phyllaplysia engeli **Er. Marcus, 1955** **turtlegrass seahare**

Size: to 20 mm Aplysiidae

ID: Body elongate oval, flattened, smooth or warty due to small papillae; parapodia fused except for right-side aperture extending back one-fourth body length from midbody; two transverse parapodial flaps fold over middle of aperture to create front and rear siphons for mantle cavity water flow; seminal groove from inside front aperture to below right rhinophore; rhinophores elongate, narrow, cylindrical, rolled, flared at tip; oral tentacles somewhat flattened, length twice width, bases separated. Ground color translucent shades of green, sometimes brownish. Back with sparse to dense variably sized spots, dots, patches – white, pink, dark brown, some white spots blue-tinted; evenly spaced white dots usually along mantle edge; back may have crowded thin white lines. White pigment most prominent on left rear of back, varying from a line of white dots to a narrow or broad irregular longitudinal white blotch overlying orange to blackish internal organs. Rhinophores and oral tentacles ground color; sometimes darker; slightly to heavily white-frosted. Among rocks and mixed algae. Depth: to 20 feet. Well camouflaged on seagrasses (e.g. *Thalassia testudinum*, *Halodule* spp.); known to feed on accumulated surface organisms. Throughout the tropical West Atlantic.

***Phyllaplysia* sp. 1**
Size: to 20 mm
Aplysiidae
ID: External morphology similar to that of *P. engeli* (previous page). Translucent pale green. Back with crowded, thin, irregular, often broken and sometimes inconspicuous, longitudinal white lines. Mantle outer edge with line of mostly equally spaced small white spots. Parapodial inner edges with submarginal dark brown line, sometimes with superimposed white papillae. Elongate, narrow, irregular brownish strip along left rear of back contains a few large white papillae ringed with dark brown; a few similar but smaller papillae elsewhere, notably two more prominent ones anteriorly on dorsal midline. Dark brown stripe from base of each rhinophore onto basal two-thirds of adjacent oral tentacle. Rhinophores and tips of oral tentacles white-frosted. Eyes black, white-rimmed. Found among seagrasses off Southeast Florida. Depth: to 15 feet. This is the *Phyllaplysia* sp. described on p. 276 of *Bahamian Seashells* (Redfern, 2013).

***Phyllaplysia* sp. 2**
Size: to 20 mm
Aplysiidae
ID: Body broadly oval, flattened; back and appendages densely covered with conical papillae; oral tentacles short, rounded, fused at base, tops flared, appear tuberculate or papillate; rhinophores similar to oral tentacles but half as long and half as wide. Parapodia fused except for a short slit on right side behind midbody. It was found in 2004 among *Thalassia testudinum* seagrasses in shallow water off the southwest coast of Florida (Tampa). It has several features consistent with *Phyllaplysia smaragda* K. B. Clark, 1977, a slug described from east Florida: surface with small white papillae; rounded oral tentacles; green with gray-pink patches on back, especially at left rear; sometimes with fine white longitudinal lines. Unfortunately, no photographs exist for that species for comparison purposes. As of 1995, it seems to have become extinct.

Bonaire

Brazil

***Stylocheilus longicauda* (Quoy & Gaimard, 1825)** **longtail seacat**
Size: to 60 mm
Aplysiidae
ID: Body elongate, inflated posteriorly; a scattering of short, contractile, branched papillae; narrow foot protrudes as a slender tapering tail, as long as or longer than the body; rhinophores and oral tentacles rolled, long and slender, with or without papillae; small parapodia fused except for a dorsal slit with side flaps forming two openings for mantle cavity water flow. Yellow to green, usually with red-ringed blue spots; sometimes blue-ringed red spots or unringed white/blue spots. Papillae whitish-yellow. Usually in open water on drifting sargassum. Feeds primarily on the filamentous cyanobacterium *Lyngbya majuscula* from which it sequesters noxious chemicals for deterring predators. Discharges a yellow fluid when disturbed. Its long tail and sticky sole are probably adaptations to life on floating weeds. Florida, Bonaire, Trinidad & Tobago, Panama, Jamaica, Brazil. Rare.

black-blotched; St. Vincent

St. Vincent

juvenile; St. Vincent

juvenile; Grand Cayman

mass aggregation; Bonaire

egg string; St. Vincent

Stylocheilus polyomma **(Mörch, 1863)** **striated seahare**

Size: to 15 mm Aplysiidae

ID: Body elongate, slender at front, humped toward rear; narrow foot protrudes as a long conical tail; parapodia fused except for a short dorsal slit at the hump with flaps creating front and rear siphons for mantle cavity water flow; body with some long branched papillae (largest on hump) and shorter simple ones, sizes and shapes variable; rhinophores and oral tentacles elongate, rolled, slender, with short papillae. Ground color greenish-gray to reddish-brown. Body has **thin, longitudinal, crowded black stripes**; some white dots; sometimes blackish blotches; **usually bright-blue black-edged spots inside a broad orange ring**. Papillae cream to white, some with black stripes. Internal organs appear black. Inhabits bottoms with rocky rubble, open sand, and mixed algae, feeding on filamentous cyanobacteria *Lyngbya* spp. from which it sequesters noxious chemicals for defensive purposes. Mass aggregations that appear and disappear rather quickly have often been reported, wherein thousands of individuals crowd together on a sandy bottom and crawl in long lines, sometimes over top of one another. The reasons for this behavior remain a mystery: end-of-life mass mortality? search for food? mating? Depth: 0-100 feet. Throughout the tropical West Atlantic.

Guadeloupe

Martinique

Southeast Florida

Guadeloupe

mantle and shell; Southeast Florida

pair with eggs; Dominica

Syphonota geographica **(A. Adams & Reeve, 1850)** **geographic seahare**

Size: to 170 mm Aplysiidae

ID: Body large, heavy, smooth, ovate; head broad; parapodia narrow at front, broaden to rear, free at both ends, fold up to cover back and create peak behind midbody; mantle at peak, covers shell, has small central foramen; mantle prolonged at rear to form a median rear-pointing siphon that projects between the parapodia; oral tentacles broad, flap-like, rolled, tips flared, point sideways; rhinophores small, thin, rolled, close together, between front ends of parapodia, incline backwards; tail short, rounded; eyes small, in front of rhinophores. Shell one-third body length, quadrate, concave down, flexible. Pale brown, pale green, yellowish-green; parapodia may have irregular white lines and streaks (some branched), random small white spots, and/or an indistinct dense distribution of very fine brown "veins"; usually a white stripe from each rhinophore that passes through the eye and continues onto the rear edge of the oral tentacle. Found in shallows crawling among seagrasses or partly buried in sand, feeding on green and brown algae. Mates in chains of up to 20 individuals, each acting as a male and female. When disturbed it ejects a purple fluid and swims by flapping its parapodia. Depth: to 20 feet. Florida, Dry Tortugas, Guadeloupe, Martinique, St. Lucia, Dominica, Venezuela. Rare.

Family ACTINOCYCLIDAE O'Donoghue, 1929

A small family of dorids in the large superfamily Chromodoridoidea. There are only two genera, *Actinocyclus* and *Hallaxa*, differentiated primarily by their internal anatomies. These sea slugs are characterized by the presence of a postero-dorsal circlet of gill leaves that can be retracted into a protective branchial cavity beneath the mantle's surface.

Body usually ovoid and dorsally convex; more elongate-looking when crawling; mantle smooth to pustulate to tuberculate, broader than foot; rhinophores bulbous, lamellate, retractile; postero-dorsal gill leaves encircle anal papilla; radular teeth denticulate, jaws with rodlets, soft body lacking spicules.

Most are very cryptic nocturnal feeders on their sponge prey. They are found worldwide at all latitudes.

***Hallaxa* cf. *apefae* Er. Marcus, 1957**

Size: 7 mm Actinocyclidae

ID: Body elongate oval, convex; a tuberculate ridge extends from the rhinophores to the gill; a mid-dorsal hump; scattered small tubercles; rhinophores slender, lamellate; gill bushy. Translucent grayish; tubercles brownish, gill leaves gray, rhinophoral lamellae edged with brown, foot edges white, internal organs pinkish. It was found at 6-33 feet off Martinique, crawling on hard substrates and encrusting sponges in areas with sand, rocks, and detritus (*H. apefae* lives on sponges). The egg mass is a white wavy ribbon deposited on a hard surface as a spiral coil of 3 whorls. *H. apefae* is the only member of this genus reported from the Atlantic Ocean and is known from only two specimens: Brazil (1957), West Africa (Ghana, 1997). The pictured animal is probably that species but could also be undescribed.

Family AEGIRIDAE P. Fischer, 1883

A family of about 30 species in two genera – *Aegires* & *Notodoris* – that share the primary diagnostic feature of a non-retractable dorsal gill. They are circumglobal at all latitudes.

Body elongate, somewhat elevated, stiffened by calcareous spicules; mantle skirt reduced, margin indistinct; dorsum with variably shaped bumps, tubercles, and papillae; rhinophores smooth, simple, retractable into usually elevated cylindrical sheaths with a tuberculate rim or into pits with flap-like covers; gill postero-dorsal; simple, bipinnate or tripinnate leaves arranged in an arc in front of the anal papilla; gill leaves not retractable into a protective pocket but are guarded by some protruding tuberculate lobes; radula lacks a rachidian tooth, lateral teeth hook-shaped.

Inhabit marine habitats with calcareous sponges on which they feed exclusively. Lengths in the tropics range from 15-140 mm.

Grand Cayman

***Aegires corrugatus* Ortea, Moro & Espinosa, 2015**

wrinkled aegires

Size: to 12 mm Aegiridae

ID: Body narrows to a bluntly pointed tail; rounded humps on back, flanks, in front of rhinophores; dorsal humps align to form 2 keels that start united at gill cover, separate and run forward to rejoin behind rhinophores, and bifurcate once again to end at a small anterior tubercle; usually 3 broad-based swellings per side: below rhinophores, at mid-length, behind gill; rhinophores narrow, smooth, sheaths low; 3-leaved gill near mid-length, shielded anteriorly by high trilobate cover, triangular middle lobe largest. **Yellow, yellowish-white, brownish-white, with small reddish-brown spots on back, sides, tail, and bases of rhinophoral sheaths. Rhinophores pale brown to greenish-brown.** Gill leaves translucent, tips white. Depth: 16-66 feet. Rare, so far reported only from Bermuda, Cuba, Bonaire, Guadeloupe, Martinique, Roatan, Grand Cayman. Like other Aegiridae, it probably feeds on calcareous sponges such as *Clathrina* species.

Bonaire

6 mm; Roatan

***Aegires* cf. *sublaevis* Odhner, 1932**

Size: to 12 mm Aegiridae

ID: This appears to be identical to *Aegires sublaevis* (see next page) the rhinophores do not have a subapical brown band.

Aegires sublaevis **Odhner, 1932**
Size: to 12 mm Aegiridae
ID: Body elongate, narrows to bluntly pointed tail; head rounded; a low narrow median ridge extends from between the eyes, broadens abruptly near midbody, continues to the gill; ridge ends as a rear-pointing trilobate protective cover that overhangs the gill, central lobe widest; ridge edges are crests, ridge bifurcated in front of eyes; gill posterior, with 3 tripinnate leaves; body with some stalked dorsal tubercles and a scattering of rounded bumps, especially along the sides and behind the gills; rhinophores simple, smooth, sheath with an antero-lateral mantle flap. Ground color usually translucent lemon-yellow, sometimes pale cream with a dense scattering of minute brown dots. Several large brownish spots on back and tail, pale individuals may have large dark brown spots ringed with white. Rhinophores pale brown, with or without a subapical brown ring. Gill leaves whitish or light yellow with brown dots. Depth: 3-80 feet. Bermuda, Florida, Bahamas, Virgin Islands, Guadeloupe, Bonaire, Panama, Costa Rica. Rare.

pale cream variant; Southeast Florida

Southeast Florida

Martinique

Destin, Florida

Aegires gomezi **Ortea, Luque & Templado, 1990** **mushroom aegires**
Size: to 5 mm Aegiridae
ID: Body elongate, slender, smooth-sided, firm; mantle reduced, skirt indistinct; four longitudinal rows of thick-stalked flat-topped mushroom-shaped tubercles – a row along each upper side and two rows on the back between the rhinophores and gill; a single median row of tubercles on the tail; sometimes two side-by-side tubercles in front of the rhinophores and a median one just behind them; rhinophores cylindrical, narrowed at the tip; rhinophoral sheaths low, rim with **only one outer lobe**; gill with three leaves, shielded anteriorly by three rear-pointing large tubercles. Ground color usually translucent whitish, rarely yellowish. Body with a dense distribution of superficial bright white dots. Tubercles translucent with white dusting on stalks, tops with a small brown spot (sometimes indistinct). Rhinophores and sheaths white. Gill leaves transparent with tiny white spots. A shallow water species (3-23 feet) found on hard substrates and rubble bottoms. Unlike its congeners, it has been seen exhibiting rapid body movements. Florida, Cuba, Cayman Islands, Costa Rica, Martinique. Rare.

Southeast Florida

Bermuda

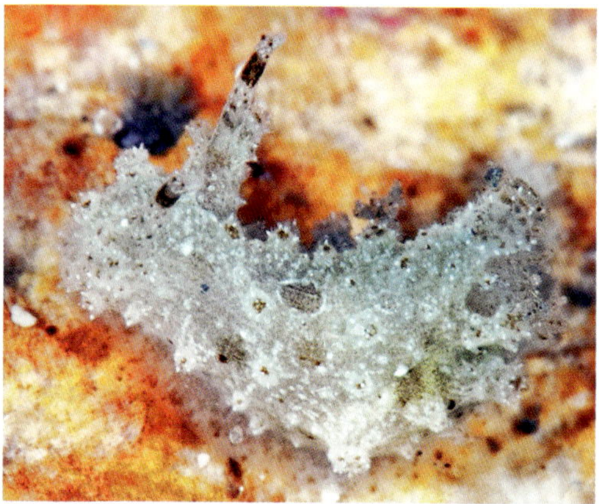

Southeast Florida

Colombia

Aegires ortizi **Templado, Luque & Ortea, 1987**

warty aegires

Aegiridae

Size: to 8 mm

ID: Body firm, elongate, narrows toward rear; many tubercles – conical, flat-topped, rarely mushroom-shaped (possibly changeable), tips may have projecting spicules; dorsal tubercles approximately in 4 longitudinal rows: one along each side and two dorsal ones merging as arcs behind the rhinophores and in front of the gill; 2-3 large tubercles in front of the rhinophores; the rhinophores retractable, smooth, cylindrical, basal sheaths **usually trilobate**; gill three-leaved, non-retractable, shielded anteriorly by three large tubercles. Ground color creamy-yellow. Scattered white spots, white dots along rows of tubercles, brownish between rows. Some pale-edged brown spots, largest beside tubercles in front of gill and behind rhinophores; tubercles with dark brown tips, sometimes may be indistinct. Rhinophores light brown, tips cream, sometimes a brown band and 2 dark brown spots below tips. Gill cream to brown. Usually on *Cladophora* and *Sargassum* algae, occasionally on coral rubble and sandy or grassy patches. Since all other members of the family Aegiridae are known to feed exclusively on calcareous sponges, this species probably has similar prey. Depth: 3-30 feet. Southeast Florida, Bahamas, Cuba, Cayman Islands, Venezuela, Panama. Rare.

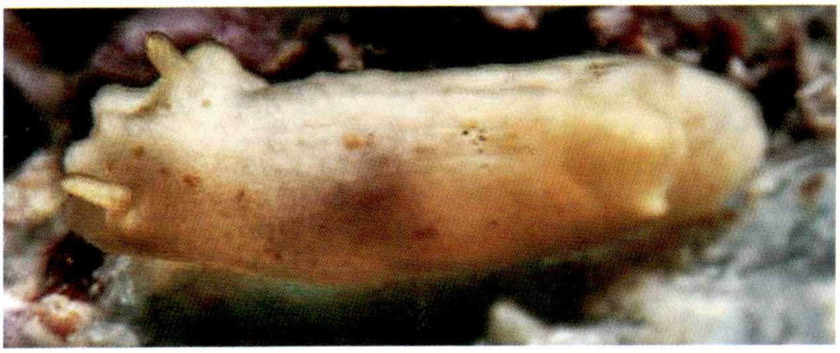

Aegires acauda **Ortea, Moro & Espinosa, 2015**
Size: 6 mm Aegiridae
ID: Body elongate, almost cylindrical, rear inclines abruptly at 45-60 degrees; back smooth, longitudinal spicules embedded in surface; cephalic veil rounded; two tubercles in front of rhinophores flank a median hump; mantle broadens slightly at mid-side; rhinophoral sheaths smooth, rims thick, outer half elevated; rhinophores cylindrical, smooth; gill posterior, behind a protective concave hump between two small crests; gill has 2-4 leaves, branched distally. Pale yellow to white with some small brown spots and rings; rhinophores yellow; gill leaves transparent with white tips. So far, only one specimen has been found on the wall of a shallow cave off northwest Cuba. Its forward movement was described as very slow – "twisting as it advances and with a slight oscillation on its main axis".

Aegires ochum **Ortea, Espinosa & Caballer, 2013**
Size: 7 mm Aegiridae
ID: Body elongate, tapers to rear; some lateral humps, a dense grid of low ridges; head rounded, two thick rounded tubercles; each rhinophore arises from the inner base of a large rounded tubercle, its tip slightly beveled; gill just behind mid-length, with five bipinnate leaves, behind a protective barrier of 2 tubercles flanking a larger one. Yellow with coppery spots and blotches; spots less dense on the head; rhinophores yellow, coppery spots more faded; gill leaves white. One animal found at 60 feet in a shallow cave off Guadeloupe. Animal's movements very slow and very stiff.

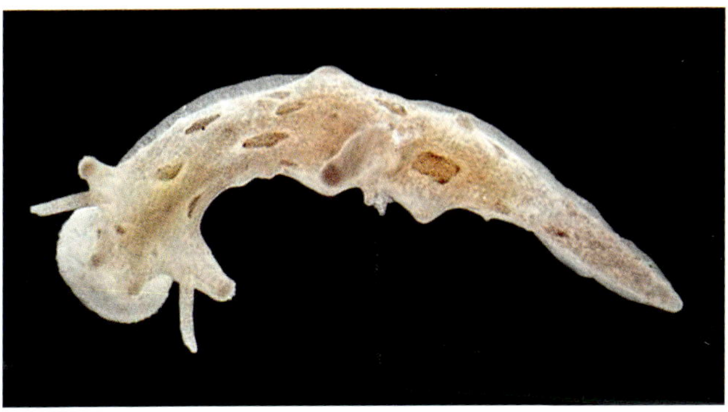

Aegires gracilis **Ortea, Moro & Espinosa, 2015**
Size: to 11 mm Aegiridae
ID: Body elongate; 3-4 broad-based tubercles per side, largest near midbody; broadly rounded oral veil; tail long, tapers to blunt tip; rhinophores slender, tips beveled; sheaths simple, conical, rim's inner edge low but outer edge very high; gill at mid-length with 3 long, thin, unipinnate leaves, behind the protective cover of a large conical median lobe two much smaller lateral lobes; two low ridges extend forward from gill, converge behind the rhinophores, continuing between them and bifurcating with each branch ending at a tubercle near the front of the head. Ground color cinnamon-hazel. Back with dark brown white-haloed rhomboid patches, notably two behind the gill, two behind the rhinophores, a larger one medianly in front of gill, and a longitudinal row of three along each side extending back from midbody. Tail with small brown spots and patches; two tubercles near front of head have chestnut tips; rhinophores gray to hazel; gill cover brown-tipped; gill leaves hazel with white tips. Specimens were found under rocks in shallows and on the walls of a cave in deep water. When moving forward it angles its rhinophores forward and outward with tips arched back. Depth: to 60 feet. Cuba, Costa Rica. Differences between this species and *A. sublaevis* have been noted (gill positions, rhinophore morphologies, radulae) but their similarities warrant genetic comparisons to confirm that they are distinct species.

Family CADLINIDAE Bergh, 1891

In 2011, based on the results of molecular phylogenetic studies, the genera *Aldisa* and *Cadlina* were transferred from the large family Chromodorididae to the newly-created small family Cadlinidae. The approximately 60 species are found worldwide at all latitudes, some in polar and sub-tropical regions but most in temperate waters.

Body ovate, flattened to slightly convex; mantle larger than foot, spiculose, slightly tuberculate, with a scattering of small glands, sometimes with large glands submarginally; rhinophores lamellate, retractable into pockets; gill simple to elaborate, retractable, leaves encircle anus; primary diagnostic features are internal: teeth highly denticulate, penis armed.

They feed on sponges in intertidal to abyssal depths.

Roatan

Guadeloupe

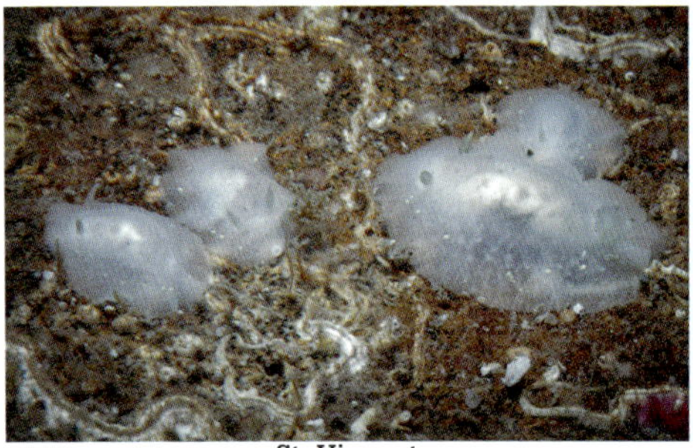

St. Vincent

Cadlina rumia **Er. Marcus, 1955** **pale cadlina**

Size: to 13 mm Cadlinidae

ID: Body oval, flat, covered with tiny pointed tubercles; large visible spicules on back randomly oriented and simply linear but at the broad mantle margin they are radially oriented and have 2-5 projections at one end; rhinophores slender, conical, retractile, with 6-8 lamellae; gill posterior, with 3-8 retractile unipinnate leaves. Ground color translucent white to cream. Prominent lemon-yellow spots (mantle glands) around inner edge of mantle skirt from rhinophores to rear. Rhinophores white to brown, gill leaves white to yellowish-brown with pale yellow tips. Found in shallows (to 15 feet) under rocks and on/near sponges that it feeds on. Preys non-preferentially on diverse spiculate and non-spiculate species, e.g. *Scopalina*, *Haliclona*, *Callyspongia*, *Dysidea*. Secretes toxic chemicals from its food and accumulates them in its mantle glands for defensive purposes. Amphiatlantic. Florida, Bahamas, Antilles, Venezuela, Panama, Brazil.

St. Vincent

Destin, Florida

Aldisa **cf.** ***smaragdina*** **Ortea, J. M. Pérez & Llera, 1982**

Size: ~22 mm Cadlinidae

ID: Body oval; mantle covered with variably sized rounded tubercles; rhinophores with ~12 oblique lamellae; gill has 5 pinnate leaves, sheath rim jagged. Bright red; 2 smooth round dark areas between rhinophores and gill, ringed with paler red; elongate transverse white patch on each side of anterior dark area; scattered white dots, some in patches at rear, others encircling tubercles; gill leaves red with white tips. Northwest Florida, St. Vincent, St. Lucia, Puerto Ric; on hard substrates, algae, and sponges to at least 70 feet. Probably the East Atlantic *Aldisa smaragdina* but identification is tentative.

Destin, Florida

Aldisa **sp.**

Size: 20 mm Cadlinidae

ID: Bright orange, however the small rhinophores, low bushy gill leaves, and many tubercles are white. Several round dark patches behind rhinophores and near gill.

Family CHROMODORIDIDAE Bergh, 1891

A large family of colorful dorids currently consisting of over 400 described species, and many more awaiting description, in about 16 genera. They occur circumglobally in temperate to tropical waters, most diverse in shallow sub-tropical to tropical reef habitats.

Body elongately oval, low to relatively high, surface usually soft and smooth but some species have dorsal tubercles, papillae, ridges, and epidermal spicules; mantle margin can be very reduced but is usually broad and extends down over the sides of the body to overlap the foot, sometimes creating elaborate folds; mantle glands that secrete noxious to toxic chemicals from their prey for the purpose of deterring predators are usually present along the margin, their shapes, sizes, ramifications, and distributions varying by species; foot can extend rearwards considerably beyond the body, especially when the animal crawls; rhinophores club-like, lamellate, retractile into pockets; gills postero-dorsal, leaves usually unbranched, sometimes bi- or tripinnate, retractile into pockets, leaves usually arranged as arcs or circles that wrap around the anus; radular ribbon present, narrow to broad, usually with denticulate lateral teeth and no rachidian; penis unarmed.

Most species feed on sponges containing chemicals they can extract and store in their mantle glands for defensive purposes. Sponge preferences vary according to the dorid species, sometimes very specialized. Body lengths are usually in the range 10-30 mm.

Utila

Belize

Costa Rica

Saba

Cozumel

juveniles; Bahamas

Chromolaichma hemera **Layton & N. G. Wilson, 2024** — **red-tipped sea goddess**

Size: to 65 mm — Chromodorididae

ID: Mantle smooth, edge ruffled; short thick rhinophoral stalk supports a club with ~20 oblique lamellae, apex pointed, tip with small mucron; gill posterior, with retractable pinnate leaves. Translucent white to bluish, sometimes dusted with orange-red on back and sides of foot. Mantle with yellow margin and a narrower red submargin; foot edge similarly banded but the red band can be very thin to absent. Rhinophoral stalk translucent white; lamellae bright red with tapered vertical groove at rear, sides sometimes whitish. Gill leaves bright red with white stalks. Depth: 2-95 feet. Found near gorgonians and algae on reefs, in rocky areas, and sponge-covered mangrove roots. Feeds on spiculated sponges. Wiggles its gill leaves rhythmically when moving forward. South Florida, Bahamas, Turks and Caicos, Antilles, Colombia to Cozumel. The similar *C. sedna* (Ev. Marcus & Er. Marcus, 1967) is found in the Gulf of California and the Eastern Pacific.

St. Vincent

Southeast Florida

Roatan

Southeast Florida

juvenile; St. Vincent

***Felimare acriba* (Ev. Marcus and Er. Marcus, 1967)**

gold-crowned sea goddess

Size: to 60 mm

Chromodorididae

ID: Body elongate, stocky; mantle with highly ruffled edges; foot sole narrow; tail long, tapers to point; rhinophores pointed, with ~30 lamellae, retractable, high-sheathed; gill posterior, retractable, ~11 unbranched leaves. Ground color blue. Back, tail, sides of foot, and rhinophoral sheaths with yellow/gold/orange reticulations (especially dense medianly on back); a row of large irregularly-shaped brilliant blue blotches along each edge of back. Mantle skirt white to pinkish; margin thin, white to pale yellow; submargin thin, red to orange; small irregularly-spaced black spots along inner edge of skirt, next to the blue blotches. Rhinophores purplish-blue. Gill leaves white to gold to pale orange with dark ribs. The pictured orange animal is presumed to be a juvenile based on its close proximity to adults. Depth: 10-60 feet. Inhabits reefs where it feeds on sponges that manufacture toxic defensive chemicals. After ingesting these chemicals, the animal secretes them in its mantle glands for its own defensive purposes. Southeast Florida, most of the Antilles, Venezuela (Islas Los Roques), Costa Rica, Roatan, Quintana Roo.

photos from St.John, USVI

Felimare cf. *acriba* (Ev. Marcus and Er. Marcus, 1967)
black-barred sea goddess

Size: 10 mm Chromodorididae

ID: May be a color morph of *Felimare acriba* (previous page), differing by having the black spots on the mantle margin replaced by thin black transverse bars. A positive identification will require specimen analysis. Depth: 10-50 feet. Also reported from Puerto Rico and south Florida. Rare.

Guadeloupe

gill retracted; Bonaire

Martinique

gill retracted; Anguilla

Felimare alaini **(Ortea, Espinosa & Buske, 2013)**
Alain's sea goddess

Size: to 45 mm Chromodorididae

ID: Body elongate, thick, smooth; tail long, tapering; mantle margin wide; rhinophores pointed, retractable, >30 nearly horizontal lamellae, sheaths elevated; gill retractable, 13-16 unbranched leaves. Back, sides, and tail pale blue with irregular yellow lattice; four orange blotches on back – first on the head, second and third between the rhinophores and gill, fourth at front of the tail. Mantle margin is a pink-edged broad white band with scattered black spots (not just along its inner edge as in the similar *F. acriba*). The band's inner edge is bordered by orange scrawls, and between the middle pair of orange dorsal blotches it is crossed by an irregular broad orange bar that darkens to an elongate black blotch on the margin's outer edge. Rhinophores dark blue to purple; gill leaves white with black ribs; foot edge pale blue with black spots. Depth: 30-85 feet. Has been found on a coral rock slope and on a sandy bottom with rocky and coral debris. Presumably feeds on sponges like other members of this genus. Deposits a rigidly-walled coiled ribbon containing large orange eggs embedded in jelly. Sint Maarten, Anguilla, Guadeloupe, Martinique, Bonaire. Rare.

Key Biscayne, Florida

inset: spawn; Riviera Beach, Florida

Riviera Beach, Florida

Felimare bayeri **Ev. Marcus & Er. Marcus, 1967**

black-spotted sea goddess

Size: to 60 mm

Chromodorididae

ID: Body elongate, narrow, stocky; tail moderately long, pointed; rhinophores conical, pointed, retractable, ~20 lamellae; gill posterior, retractable, ~8 unbranched leaves. Ground color blue. Back with 6-10 thin yellowish-orange, black-edged, longitudinal stripes from gill to front of head, some (or many) being interrupted and/or interconnecting. Sides of back with thin yellowish-orange margin, broad white submarginal band, and numerous variably sized black spots along the band's inner edge. Rhinophores dark blue, sheath bases ringed with yellowish-orange; gill leaves whitish, ribs yellowish-orange with black stripe; sides of foot and tail with yellowish-orange stripes and black spots; tail with blue edges. Depth: 10-60 feet. Inhabits reefs where it feeds on sponges of the genus *Dysidea*. Southeast and Northwest Florida, northern Cuba, Guadeloupe, Curacao, Colombia, Panama, Roatan, Belize, Bay of Campeche and Isla Mujeres (Mexico). Common in Miami's Biscayne Bay, uncommon to rare elsewhere.

photos from Colombia

Felimare **cf.** ***bayeri*** **Ev. Marcus & Er. Marcus, 1967**

Size: to 60 mm

Chromodorididae

ID: Ground color blue to green, gill leaves yellow/orange with white tips, yellow stripes and blotches as for *Felimare bayeri*. It could be a color variant of that species.

Guadeloupe

Martinique

juvenile; Guadeloupe

Martinique

Felimare espinosai **(Ortea & A. Valdés, 1996)** **Espinosa's sea goddess**

Size: to 12 mm Chromodorididae

ID: Body elongate; foot narrow; tail short, triangular, pointed; rhinophores somewhat flattened laterally, taper to a point, ~15 lamellae, retractable; sheaths low, rims circular and smooth; gill posterior, retractable, ~9 narrow unbranched leaves with pointed tips. Ground color dark blue to blackish. Mantle with: yellow/orange submarginal band; yellow/orange median stripe from front of rhinophores to gill and wraps around its base; a similar stripe (often broken) on each side of the median stripe extends forward from the gill's side onto the rim of a rhinophoral sheath, and proceeds to form a continuous arc in front of the rhinophores. The median stripe does not meet this arc. Large iridescent blue/green spots form rows between the stripes, in particular 4-8 near the mantle margin and onto the head. Yellow/orange markings behind the gill and along the foot edges. Rhinophores blue, white vertical stripe on front and rear faces. Gill leaves grayish-blue with white ribs. Inhabits coral reefs to depths of at least 35 feet, where it feeds on sponges of the family Dysideidae (e.g. *Dysidea etheria*). Scattered locations in Lesser Antilles, also Quintana Roo (Mexico). Rare.

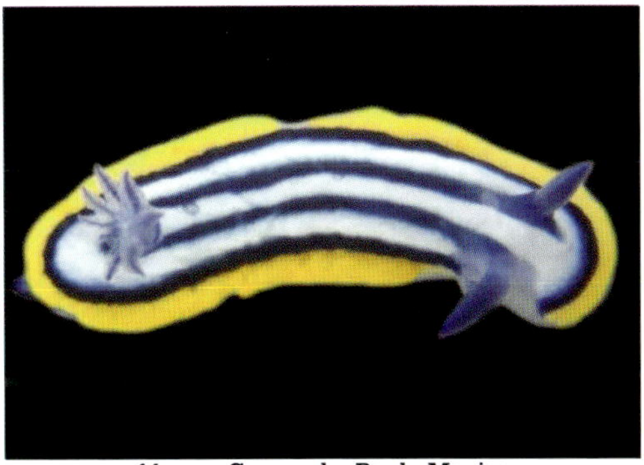

11 mm; Campeche Bank, Mexico

12 mm; Campeche Bank, Mexico

Neptunazurea sisalensis **(Ortigosa & Á. Valdés, 2012)**

Size: to 12 mm Chromodorididae

ID: Body elongately oval; mantle smooth-edged, band of small rounded glands along margin (except at front); rhinophores conical, pointed, lamellate, retractable, sheaths low; gill posterior, retractable, 9 unipinnate leaves; tail short, broadly triangular. Ground color opaque white. Mantle with broad yellow margin, a narrower blue submargin, and **a blue stripe from each rhinophore to behind the gill where they curve and unite**. Gill leaves and rhinophores blue. Foot sole white with blue edge. Known only from 3 specimens collected on the Madagascar Reef, Campeche Bank (Mexico) at a depth of 23 feet. No ecological information is available. In the similarly colored *Neptuneazurea molloi* (Ortea & Valdés, 1996), described from a single Venezuelan specimen and never seen again, the two dorsal stripes also unite to form an arc in front of the rhinophores. Genus changed from *Felimare* in 2025.

Bonaire

Southeast Florida

Southeast Florida

Southeast Florida

St. Kitts

juvenile; Guadeloupe

feeding on *Dysidea* sponge; Bonaire

Felimare fregona **(Ortea & Caballer, 2013)** **sunset sea goddess**

Size: to 40 mm Chromodorididae

ID: Body narrow, high, up to 9 times longer than wide; mantle and foot narrow; rhinophores consist of a thick stalk supporting a club with ~15 oblique lamellae, sheaths relatively low; gill posterior, retractable, with ~10 lamellate leaves; tail moderately long, broad, posteriorly rounded. Ground color white. Back with a few bluish-green, blackish, or grayish blotches between front and gill; three thin, parallel, dorsal, continuous yellow stripes – a median one flanked by the other two – that become orange anteriorly, posteriorly, and at midbody. The two side-stripes form an orange collar on the base of each rhinophoral sheath, continue forward and arc inward to meet the median stripe on top of the head. Between these stripes, there may be a few similarly colored broken stripes which, in large individuals, may all be broken, randomly interconnected, and disorganized looking. The three-line stripe pattern continues onto the bluish/purplish tail with the side stripes sometimes broken; tail usually has some black spots. Mantle edge is white, followed by a thin yellow to pale orange margin with a **deep orange blotch at midbody**, then a **submarginal row of round black dots** and a **large dark blotch** next to the orange blotch. Along the body's flanks there are also three parallel yellow/orange stripes and black spots crossing regions of white, blue, and bluish-green. Rhinophores white with purple side-stripes. Gill leaves white, stalks white below and dark blue above. Depth: 10-115 feet. Southeast Florida, Bahamas, Antilles, Venezuela, Panama, Quintana Roo (Mexico). Uncommon.

all photos from Southeast Florida

feeding

everted buccal mass

Felimare* cf. *fregona **(Ortea & Caballer, 2013)**
Size: to 40 mm Chromodorididae
ID: Like *F. fregona* but the body's dorsal and lateral lines are different – they are more numerous, more continuous, and uniformly red-orange (no yellow). This could be a color variant but comparative specimen analyses would be required in order to confirm the identification.

Roatan

Bahamas

Felimare nyalya **(Ev. Marcus & Er. Marcus, 1967)** **red-line blue sea goddess**
Size: to 35 mm Chromodorididae
ID: Body elongate, narrower at midbody, head wide, front and rear rounded; tail short, bluntly pointed; rhinophores conical, with ~11 lamellae, sheaths low; gill posterior, 6-7 unbranched leaves. Ground color bright blue. Mantle margin red or orange, sometimes red with an inner orange fringe, submargin white; rhinophoral lamellae, gill leaf ribs, and tip of tail dark blue. Inhabits reefs, rocky bottoms, undersides of rocks, and areas with mixed algae. Feeds on sponges (e.g. *Dysidea etheria*), sometimes in small groups. Depth: 6-30 feet. Uncommon in East Florida, Bahamas, northern Cuba, Martinique, Roatan, Bay of Campeche (Mexico). Locally common in Bimini.

all photos from Southeast Florida

feeding aggregation

juvenile

juvenile

egg ribbon

Felimare juliae **(S. DaCosta, Padula & Schrödl, 2010)** **Julia's sea goddess**

Size: to 57 mm Chromodorididae

ID: Body elongate; mantle skirt undulating; rhinophores conical, retractile, ~20 lamellae, sheaths moderately elevated; gill with ~10 multipinnate leaves; tail broadly triangular. Ground color pale blue to pale green with small iridescent blue spots and patches. **Thin yellow/orange mostly continuous crowded stripes on back, flanks, and tail;** stripes irregular, broken, and interconnecting on the head. Mantle skirt with thin orange margin, thin white submargin, followed by a broad grayish band with a line of **roundish pale blue spots**; an **orange-spotted black patch** on each side next to a **central transverse dark patch**; a **black patch at front of head** where stripes terminate; at least one transverse dark patch in front of and behind the central one. **Foot sole pale blue, edged with crowded transverse orange dashes**. Rhinophores blue to dark blue; sheath rims colored by dorsal stripes. Gill leaves whitish, ribs dark blue to black with orange edges. Inhabits bottoms with mixed algae, sand, and grass, feeding on the blue sponge *Dysidea etheria*. It deposits a bright orange spiral egg ribbon on adjacent algae. Depth: to 25 feet. Amphiatlantic. Southeast Florida, Bahamas, Grenada, Brazil. Rare.

Saba

Bonaire

Bonaire

Dominica

trailing; Guadeloupe

with egg coil; Roatan

Brazil

Neptunazurea kempfi **(Ev. Marcus, 1971)** **purple-crowned sea goddess**

Size: to 60 mm Chromodorididae

ID: Somewhat broadly rectangular with rounded ends; mantle broad, slightly ruffled; rhinophores conical, retractable, ~12 lamellae; gill posterior, retractable, ~9 unipinnate leaves; tail triangular, may protrude slightly when the animal is moving forward. Ground color whitish. An **opaque white median stripe** extends from rhinophores to gill. **Mantle margin broad, orange; submargin with irregular black spots, often as transverse bars**; regions flanking the median stripe are darker. **Rhinophores blue**; gill stalks blue to purple, leaves whitish; foot sole blue. Inhabits reefs with algae and sponges, feeding on the sponges. Exhibits "trailing" behavior where two or more individuals connect front-to-back, possibly as a precursor to mating. Since its genital opening is on the right side, the mating configuration would be adjacent right sides. Depth: 15-130 feet. South Florida, Antilles, Venezuela to Quintana Roo (Mexico), Brazil. The original description was based on blue Brazilian specimens, which have a color pattern that is quite different from Caribbean specimens. However, due to the existence of some intermediate-looking Brazilian variants it has been assumed, thus far, that the Caribbean populations are indeed the same species. This was *Felimare kempfi* until 2025.

Utila

Roatan

juvenile; Roatan

Roatan

Felimare marci (Ev. Marcus, 1971) purple-spotted sea goddess

Size: to 26 mm Chromodorididae

ID: Body elongate, mantle moderately broad, slightly constricted behind head; rhinophores conical, retractable, ~20 lamellae, sheaths low; gill posterior, retractable, ~15 unipinnate leaves, tail moderately long, triangular. Ground color blue. **Head and back with yellowish-orange reticulations delineating numerous, irregular, variably sized, blue spots**; those near the outer perimeter of the reticulated area tend to be **iridescent blue, sometimes coalescing into transverse dashes**; reticulated area with **4-5 diffuse dark-orange/brown dorsal blotches**: 1 on head, 1 behind gill, 2-3 on back. **Mantle's broad white orange-edged margin contains a row of large dark brown spots, the one at body mid-length especially large and blackish**. Flanks with similar markings. **Rhinophores white with purplish/blue tips**; sheaths reticulated, rims yellow/orange. Gill leaves white, ribs dark blue. Often seen crawling over algae-covered dead corals. Feeds on *Dysidea* sponges. Depth: 12-130 feet. Puerto Rico, Cayman Islands, Jamaica, Bonaire, Isla Los Roques (Venezuela), Panama, Belize, Quintana Roo (Mexico), northern Brazil. Common in the Bay Islands of Honduras.

Saba

Bonaire

juvenile; St. Vincent

juvenile; St. Vincent

Felimare fortunensis **(Ortea, Espinosa & Buske, 2013)** **fortune sea goddess**

Size: to 11 mm Chromodorididae

ID: Body elongate; tail broadly triangular, short; mantle edge broad; rhinophores conical, ~12 lamellae, sheaths somewhat elevated; gill posterior, ~10 unipinnate unbranched leaves. Back with a bright yellowish-orange irregular network on a bright blue background. Mantle with narrow orange margin and a broad white-to-pale-bluish submargin with many black spots along its inner perimeter. Flanks pale blue with black spots and a yellow longitudinal line that breaks up at the tail. Tail black-spotted with median orange streak and blue tip. Rhinophores bright blue/violet, gill leaves white with purplish-black stalks. Early juveniles white with short orange median stripe, few black spots, vague orange network; blue rhinophores and flanks, fewer gill leaves. Rare sightings were in rocky/sandy/muddy areas at 10-70 feet. Saba, St. Eustatius, Guadeloupe, St. Vincent, Bonaire, Curacao.

Buzios, Brazil

Arraial do Cabo, Brazil

Felimare lajensis **(Troncoso, F. J. Garcia & Urgorri, 1998)**

Size: to 50 mm Chromodorididae

ID: Body elongate, higher than wide, smooth; mantle margin undulating; rhinophores conical, lamellate along entire length (~20 lamellae in 30 mm animal), retractable, sheaths moderately elevated; gill posterior, ~9 unipinnate or bipinnate leaves encircle anus, retractable into a low sheath; tail triangular, one fifth the body length, stout. Ground color blue-violet. Narrow longitudinal yellow stripes: 3-9 on back from front of rhinophores to rear of mantle, inner ones only to gill sheath; 4-5 on sides and tail. Some stripes fragmented. Mantle margin either white with some yellow at mid-length and front/rear, or completely yellow; submarginal band pale blue with large diffuse dark blue spots that may alternate with whitish spots. Rhinophores usually darker than ground color, sheaths yellow-rimmed; gill ribs dark, leaves paler. Foot sole pale blue. Found on shallow reefs crawling on coral heads and rocks with ascidians, hydroids, and sponges, where it feeds on *Dysidea* sponges (e.g. *Dysidea etheria*) that manufacture toxic defensive chemicals. After ingesting these chemicals, it secretes them in its mantle dermal formations (MDFs) for its own defensive purposes. The MDFs are spherical opaque white glands distributed along the mantle edge (sometimes visible), most concentrated near the rhinophores and gill but absent at midbody. Depth: 0-80 feet. southern Brazil. Not yet reported in the Caribbean.

Southeast Florida

Bahamas

Southeast Florida

Southeast Florida

Felimare olgae **(Ortea & Bacallado, 2007)** **Olga's sea goddess**

Size: to 38 mm Chromodorididae

ID: Body elongate oval; mantle moderately broad, constricted behind head; rhinophores conical, retractable, 11-19 lamellae, sheaths low; gill posterior, retractable, 9-11 mostly unipinnate leaves, some branched near the tip; tail broadly triangular. Ground color orange to greenish-orange. **Back with irregular pale blue spots edged with small black dots**, the spots tending to iridescent and larger near the edges of the back; the back often has 2-3 darker diffuse transverse blotches. **Mantle edge pale orange, margin white with band of small black spots along inner edge** (missing in juveniles); band interrupted at rear and midbody by **transverse orange blotch with larger black spots**, sometimes also at front of head. Similar coloration on flanks, foot, and top of tail; foot sole blue. Rhinophores white to yellowish, tip and basal peduncle blackish-purple, sheath rim orange. Gill leaves white, stalks purplish-blue, sheath orange with blue spots basally. Found crawling on live coral, under chunks of dead coral, on drifting algae, and on the sponge *Dysidea etheria*, which is probably its prey. Deposits a coiled ribbon ~5 mm high containing relatively large eggs (>0.3 mm in diameter), a characteristic suggestive of direct development (i.e. within the egg capsule), which is known in only two other Caribbean Chromodorids – *Felimare nyalya* and *Felimare lilyeveae*. Depth:10-50 feet. Southeast Florida, Bahamas, Cuba, Bonaire, Roatan, east coast of Yucatan (Mexico). Rare.

St. Vincent

Grand Cayman

St. Vincent

Southeast Florida

Saba

juvenile; St. Vincent

juvenile; Southeast Florida

adults and juvenile feeding; St. Kitts

*Felimare ruthae (*Ev. Marcus & Hughes, 1974) **gold-lined sea goddess**

Size: to 27 mm Chromodorididae

ID: Body elongate, narrow, thick; rhinophores retractable, robust, torpedo-shaped, up to 14 lamellae, sheaths low; gill posterior, retractable, ~9 unipinnate leaves, some branched near the tip; tail triangular. Ground color dark blue to almost black. From head to gill there are **5-6 mostly continuous gold stripes**; **between them bright blue streaks and spots and often thinner gold line segments**. The 2 most lateral stripes skirt the rhinophores and unite to form a median arc at the front of the head. The number of stripes increases with age: 1, then 3, then 3 plus 2 faint ones, etc. **Head with a dark brown transverse blotch, mantle with several more,** the largest at midbody and spanning the entire mantle. The broad **mantle margin is white with a thin gold stripe along its outer perimeter**. Sides of body and foot are similarly patterned but the gold stripes are more fragmented. **Rhinophores dark blue with white (sometimes yellow) vertical stripe on front and back surfaces**; sheath rims gold. **Gill leaves white with dark blue stalks (sometimes just near tips)**. Found on living reefs crawling on coral heads, among algae, under rocks, on sandy or mucky bottoms. Feeds on blue *Dysidea* sponges. Depth: 3-40 feet. Florida, Bahamas, and most regions of the Caribbean.

adult

color pattern variants

late juvenile

early juvenile

feeding

spawn

Felimare picta **(R. A. Philippi, 1836)** **Florida regal sea goddess**

Size: to 130 mm Chromodorididae

ID: Body elongate, broadly rectangular, thick; mantle moderately broad, margin undulating; rhinophores conical, lamellate along entire length, retractable; sheaths reach nearly halfway up; gill posterior, retractable, up to 14 leaves, sheath elevated; foot as wide as mantle, raised laterally; tail one-fifth body length, triangular. Ground color **dark blue** to almost black. Adult's mantle with **6-7 often fragmented thin yellow stripes usually bracketing partial rows of yellow spots, rings, dashes, and line segments**; pale blue margin, broader yellow submargin. Head with prominent convoluted stripe extensions, some yellow spots. A stripe encircles the rim of rhinophoral sheath and continues onto the head. Foot, top of tail, and sides of body with yellow line segments, spots, and rings. Mantle waviness, stripe number and degree of fragmentation increase with age (see juveniles above). **Rhinophores dark blue; gill leaves dark blue with yellow or yellow-spotted stalks**. Yellow stripes run up gill sheath. Found on reefs under rocks, in crevices, and crawling on coral heads while feeding on *Dysidea* sponges (e.g. *D. fragilis*) that manufacture toxic defensive chemicals it can harness for its own defensive purposes. Depth: 0-80 feet. North Carolina, Georgia, Florida, Puerto Rico, Venezuela, Yucatan (Mexico), Brazil. All of the above photos are from Southeast Florida. *F. picta* is generally regarded as a complex of 6 amphiatlantic subspecies with color patterns and morphologies varying with locality. The above description is for the West Indies subspecies.

Felimare lalique **(Ortea & Caballer, 2013)**

Size: 18 mm Chromodorididae

ID: Body elongate oval, slightly narrowed behind the rhinophores, broadly rounded anteriorly and posteriorly. Foot broad, tail narrowly triangular, projects up to one-fifth body length. Rhinophores retractable, a short thick stalk supporting a pointed cone with ~14 slightly oblique lamellae; sheaths a low ring along the edge of the rhinophoral cavity. Gill posterior, small, robust; 9 unipinnate leaves project out and down; the sheath also a thin ring. The ground color is shades of grayish blue. Three irregular continuous stripes extend from the rhinophores to the gill: one median and one behind each rhinophore; the stripes are yellow with orange segments near mid-length, and flank two similarly-colored fragmented stripes. An orange arc on the head. The broad white mantle margin is crossed by a large dark blotch at mid-length, with 2 thin orange lines along its outer edge. The body flanks have narrowing and widening continuous lines – yellow upper, orange lower – that merge at the tail. Rhinophoral stalk and internal axis dark blue, lamellae translucent orange, sheath ring yellow. Gill leaves white, stalks purplish-black, sheath ring yellow. The foot is pale blue; the tail is dark blue above and violet blue below, the edge orange. Only one animal has ever been found, at ~90 feet on a rocky bottom near **Guadeloupe**. *F. ruthae,* at a similar size, has twice as many stripes, half the number of gill leaves, and dark blue lamellae on the rhinophores

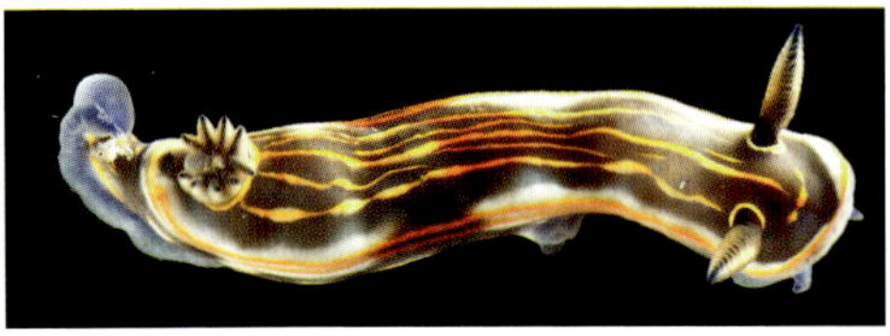

Felimare lilyeveae (Alejandrino & A. Valdés, 2006)
Size: to 26 mm Chromodorididae
ID: Body elongate, slender, height increases rearwards to peak near start of tail; its median dorsal ridge has 3 pairs of rounded lateral swellings, a pair's left and right segments may be irregular and sometimes offset longitudinally. Head broader than body, front bilobed. Foot sides tuberculate, tail triangular. Rhinophores club-like, 10 lamellae, tips pointed. Gill posterior, has 7 unipinnate leaves. Ground color orange-reddish with a grayish median dorsal strip. A black-dotted white patch from front of head to end of tail covers the head, narrows to a stripe behind the rhinophores and follows the ridge, expanding to cover the swellings. Grayish strip also black-dotted. A row of blue dots along each side of the mantle. Foot sides blue with large white spots and small black spots; tubercles white-spotted. Rhinophores white with purple line on rear of stalk; 3 purple vertical stripes on club converge at tip. Gill leaves white with purple-sided stalks. Has been found on reefs with mixed algae. Depth: 25-60 feet. Bahamas, Puerto Rico, Virgin Islands, Guadeloupe. Rare.

Guadeloupe

Arraial do Cabo, Brazil

Cabo Frio, Brazil

Cabo Frio, Brazil

Felimare sp. 1
Size: ~30 mm Chromodorididae
ID: Body elongate, narrow, sides parallel, rounded anteriorly and posteriorly, slight constriction behind head; rhinophores with sturdy stalk supporting a club with ~12 oblique lamellae, tip pointed; gill posterior, 7 lamellate leaves; tail narrowly triangular, pointed, moderately long. Ground color white. Back with a series of very thin orange longitudinal lines extending from the rhinophores to (and beyond) the gill. Three of the lines – one median and two lateral – are mostly unbroken and bracket short line segments of the same color. There is a transverse orange blotch near midbody. The mantle margin contains a series of what appear to be dark blue conical papillae, each set on top of a broad diffuse gray base. The rhinophores are white with a dark blue stripe on the inner and outer sides. The gill leaves are white with dark blue stalks. The tail is white with a median orange stripe and blue sides. Several described _Felimare_ species are partial look-alikes, so this could be a variant of one of those, or an undescribed species. They have been found among algae in relatively shallow water (to 20 feet) off southeast Brazil. Not yet reported from the Caribbean.

Trinidad and Tobago

***Felimare* sp. 2**

Size: 30 mm Chromodorididae

ID: Body elongate and narrow; tail moderately long and pointed; rhinophores club-like, lamellate; the gill posterior and retractable, with ~8 unbranched leaves. Ground color blue-gray. The mantle edge is orange; a whitish submarginal band densely filled with small black spots; rest of mantle covered with large black spots and narrow, longitudinal, orange stripes composed of irregular line segments of various lengths. Foot with similar stripes and spots. Rhinophores dark blue, sheath rims orange. Gill leaves whitish, their black stalks edged with orange. Found on a silty seagrass bottom at a depth of ~6 feet. It could be a color variant of a known species (e.g. *F. bayeri*, *F. samueli*) but the differences in color pattern are quite significant.

all photos from Panama

***Felimare* sp. 3**

Size: to ~20 mm Chromodorididae

ID: Body elongate, tail triangular, moderately long; mantle broad, margins somewhat ruffled; rhinophores conical, lamellate; gill posterior, ~10 unipinnate leaves. Its back and head are orange with a scattering of dark blue irregular spots and black-edged iridescent blue spots mostly along the perimeter of the orange region. Mantle margin white with a yellow outer edge and a row of small black spots centered in dusky patches along its inner perimeter; a large blackish blotch near mid-length. Rhinophores dark blue; gill leaves brownish-white with black ribs; tail dark blue with orange scrawls. The smaller reddish individual with the median pair of broad dark blue dorsal stripes, found in the same general area, is probably a juvenile of this species. Depth: < 30 feet. Bastimentos Island (Panama). This could be a variant of a described species but a specimen analysis is necessary for the confirmation of that possibility.

***Felimare samueli* (Caballer & Ortea, 2012)**

Size: to 85 mm Chromodorididae

ID: Body elongate, wider than high; foot broad, front margin bilobed; tail broadly triangular, pointed, moderately long; mantle margin broad and wavy in adults, 7-8 mantle glands anteriorly and 10 posteriorly (rearmost 3 very bulky); rhinophores conical, with up to 30 lamellae, retractable, sheaths low; gill posterior, 10 unipinnate retractable leaves. Adult ground color greenish-gray. Mantle with ~10 longitudinal yellow lines from in front of rhinophores to gill – a continuous central one, others consisting of line fragments and spots; large bluish-black blotch at mid-length, as wide as body; margin broad, gray to white, with two rows of black spots; mantle edge has a thin reddish-orange stripe, except at rear. Below mantle, along each side, a single continuous yellow line flanked by fragmented ones (1-2 above, 1-6 below), continuous ones converging at tail. Rhinophores uniformly dark blue; each sheath has a neighboring yellow dorsal line bifurcate at its base's outer side, each branch wrapping partially around the base, and the two pairs of branches continuing transversely inwards to merge and form a loop ("bridge") that connects the rhinophores. Gill leaves grayish-blue, stalks black with orange to orange-red edges. Foot and tail with irregular yellow lines. Juvenile color differences: ground color blue-green; foot bright blue; mantle has 3 longitudinal yellow lines but no black blotch; mantle margin whitish-green with 9 large black spots per side; no yellow bridge between rhinophores. Three individuals were found near *Thalassia testudinum* and *Halimeda* species on a sandy/muddy channel at the mouth of a shallow (6 feet) lagoon. When disturbed it retracts its gill and secretes a white fluid along its mantle edge. **Trinidad & Tobago, Venezuela**. Rare.

adult

adult

adult

intermediate

juvenile

all photos from Bermuda

juvenile

spawn

Felimare zebra **(Heilprin, 1889)** **zebra sea goddess**

Size: to 180 mm Chromodorididae

ID: Body broadly elongate; mantle skirt reduced, rounded at the ends; foot narrower than mantle, tail broadly triangular; rhinophores conical, long, retractile, bottom one-third a short broad-based stalk narrowing upwards, pinched under a club with ~28 lamellae; sheaths low, rims smooth; gill dorsal, posterior, up to 14 leaves, retractile into a high sheath. Ground color blue; rhinophores dark blue with yellow/orange (YO) sheaths at all stages. **Juvenile** (5 mm): mantle margin white with YO edge, submarginal blackish spots on pale blue background touching YO stripe from rhinophore to gill sheath, similar median stripe on head, tail blue with median YO stripe, gill leaves white with dark blue stalks. **Intermediate** (20 mm): convoluted YO dorsal pattern with pale blue spots are forming; smaller marginal spots, more numerous; tail with extra stripes. **Adult** (>50 mm): mantle crowded with YO irregular/interconnecting stripes; orange margin, blue submargin, no spots; gill leaves bluish, stalks dark blue with YO edges. Lives among seagrasses, hides under rocks. Grazes on blue sponge *Dysidae etherea*, secreting sesquiterpenes that make it unpalatable to predators. Its bright coloration warns attackers of its taste, negating the need to hide as it forages. Spawns year-round, releasing long coiled ribbons of red eggs adhering to substrates and hatching into free-swimming larvae. Depth: 1-65 feet. Found only near Bermuda.

St. Kitts

Dominica

Southeast Florida

juvenile, Southeast Florida

6 mm juvenile; Southeast Florida

with spawn; Southeast Florida

Felimida binza **(Ev. Marcus & Er. Marcus, 1963)** **harlequin sea goddess**

Size: to 30 mm Chromodorididae

ID: Body broadly elongate, foot narrow; tail short, triangular; rhinophores blunt-tipped, retractable, club with ~10-15 lamellae; gill posterior, retractable, with up to 12 leaves that form a circlet around the anus. Ground color pale blue. Mantle with narrow yellow/orange margin and broad whitish submargin. Back appears orange/red with numerous blue spots of variable size; orange/red region with many triangular outward projections, usually sharply pointed, often quite long, sometimes dark-tipped; two disjoint narrow elongate yellowish patches between rhinophores and gill enclose large blue spots along the dorsal midline – one extending back from between the rhinophores, the other extending forward from the gill. The patches' interspace is a broad transverse orange/red band. Rhinophores blue to purplish with an anterior dark purple vertical stripe; apex clear. Gill leaves whitish-gray with purple stalks. Foot pale blue, tail blue with purplish median streak. Found under rocks or crawling over rubble and empty shells in protected rocky and sandy areas. Feeds on sponges, e.g. *Chelonaplysilla erecta*. Deposits a spirally coiled flat jelly-like ribbon standing on its narrow edge with transverse rows of white/yellow kidney-shaped eggs. Depth: 0-100 feet. Southeast Florida, Bahamas, Cuba, Jamaica, Lesser Antilles, Colombia, Costa Rica, Isla Mujeres (Mexico), Brazil.

Southeast Florida

Grand Cayman

Southeast Florida

juveniles

5 mm juvenile

with spawn

Felimida clenchi **(H. D. Russell, 1935)**

harlequin blue sea goddess

Chromodorididae

Size: to 30 mm

ID: Mantle broadly elongate; foot narrower; tail short, triangular; rhinophoral stalks short and robust, clubs long and **pointed**, ~17 lamellae, retractable; gill posterior, retractable, 9-12 leaves, some tips bifid. Ground color pale blue. Mantle with narrow orange-red margin, broad white submargin. Back with elongate orange-red "H" pattern, the transverse cross member mid-dorsal, side members being irregular stripes along mantle's submargin; stripes usually with short blunt lateral projections bounding a few blue spots and dashes. Two median narrow pale yellow-cream patches – one in front of and one behind the "H"cross-member. The red/orange side stripes skirt the rhinophores to loop around and unite in front of them, where they join the front pale patch with one or two red lines, forming two blue spots. Rhinophores purplish with three dark purple vertical stripes meeting at the apex. Gill leaves whitish-gray with purple stalks. Tail blue/purplish with orange to dark purple median streak. Found under rocks or crawling over rubble and empty shells in protected rocky and sandy areas. Feeds on sponges. Spawn is a spirally coiled flat upright ribbon crowded with capsules each containing one white egg. Eggs turn reddish in 7-8 days, just before hatching. Juveniles are assumed to be so by their proximity to adults. Depth: 0-100 feet. Bermuda, Florida, Bahamas, Antilles, Venezuela to Bay of Campeche. Brazil.

all photos from Southeast Florida

***Felimida* cf. *clenchi* (H. D. Russell, 1935)**

Size: to 30 mm

Chromodorididae

ID: The external morphology of this Florida dorid is the same as that of *Felimida clenchi* (previous page) but the color pattern is very different. Its ground color is pale blue to purplish. There are reddish-edged yellow reticulations on the entire back, extending as irregular short lines toward the edge of the mantle. The mantle has a narrow red/orange margin and a broad white submargin. Rhinophores are white with blue or purple tips and an anterior purplish blotch near the base. The **tail has a median white stripe**. Inhabits patch reefs with sand and rocks in depths to ~10 feet. In *Caribbean Sea Slugs* (Valdés et al, 2006) this animal is identified as *Chromodoris neona* (Er. Marcus, 1955) and not as a color form of *Felimida clenchi*. However, some taxonomists currently consider *C. neona* to be conspecific with *F. clenchi* which, if true, would identify the pictured animals as *F. clenchi*. However, in his original description of *Chromodoris neona*, Marcus described the tail as having a **bright red midline stripe**, not the white stripe seen in one of the above photos. That could be an intraspecific color variability for *Felimida clenchi* but that discrepancy needs to be resolved to permit positive identification.

Guadeloupe

Bonaire

Destin, Florida

Puerto Rico

St. Vincent

juvenile; Martinique

Felimida regalis **(Ortea, Caballer & Moro, 2001)** **royal sea goddess**

Size: to 24 mm Chromodorididae

ID: Body elongate, narrow; mantle with many minute conical papillae and a few larger ones; mantle margin wavy, with many gland-like dermal formations – a band of small rounded ones along its outer edge and an irregular line of large more conical ones roughly between the band and the outermost dorsal papillae. Foot narrow, front corners triangular, tail moderately long and pointed. Rhinophores sturdy, club-like, 12-14 lamellae, tip pointed, retractable into a somewhat elevated sheath. Gill with 7-8 unipinnate retractable leaves that encircle the anal papilla. Ground color salmon pink to orange. Mantle with fine network of reddish-orange lines, white papillae; margin with a thin orange outer edge followed by a narrow yellow stripe with small dermal formations and a broad white band of large dermal formations; margin may be crossed by dark transverse streaks. Rhinophores translucent salmon-pink, lamellae orange-edged. Gill leaves translucent pink to pale orange to whitish with red stalks. Tail bluish/purplish with a yellow margin. Anal papilla white. Juveniles purplish with white rhinophores. Found on patch reefs and rocky bottoms where it feeds on toxin-producing sponges, ingesting and accumulating the toxins for its own defensive purposes. Depth: 6-30 feet. Northwest Florida, Puerto Rico, St. Kitts, Guadeloupe, Martinique, St. Vincent, Bonaire, Costa Rica, Yucatan. Uncommon.

all photos from Southeast Florida

spawn

Rudmania fentoni **(Á. Valdés, Gatdula, Sheridan & Herrera, 2011)** **strawberry sea goddess**

Size: to 16 mm Chromodorididae

ID: Body elongate, wide, low; mantle margin broad, ruffled; rhinophores conical, pointed, lamellate, retractable, sheaths low and smooth; gill posterior, unipinnate leaves arranged in semi-circle, retractable. Ground color whitish. Back with dense but irregular covering of reddish pigment and orange-centered yellow spots, the distribution of spots being more concentrated near the mantle's sides. Mantle with translucent grayish marginal band, white submarginal stripe, followed by a thin yellow/orange line. Rhinophores white with white-tipped purple apex, sheath purple with orange spots. Gill leaves white with reddish or yellow/orange spots, sometimes with purplish tips. Has been observed on the red sponge *Igernella notabilis*, which it probably feeds on. Depth: to at least 30 ft. Known only from Southwest and Southeast Florida. Rare. Name changed from *Felimida fentoni* in a forthcoming paper.

Southeast Florida

Puerto Rico

juvenile and adult; Puerto Rico

juvenile; Puerto Rico

Martinique

Rudmania grahami **(T. E. Thompson, 1980)** **Graham's sea goddess**

Size: to 25 mm Chromodorididae

ID: Body elongate oval, soft; mantle moderately broad; mantle margin thick and fleshy, undulate and flared in places; tail moderately long, broadly triangular; rhinophores club-like, ~12 lamellae, retractable, sheaths low and rims smooth; gill posterior, retractable, ~14 unipinnate leaves, sheath rim slightly raised and flared. Ground color white. **Dorsum diffusely mottled with salmon-pink (adults) to bluish-pink (juveniles)**, with round to elongate **red spots somewhat aligned in three irregular longitudinal rows**. Mantle with a broad white margin, a narrow yellow-orange submarginal stripe (may be absent), sometimes with a yellow border and a thinner red submarginal stripe. Rhinophores with purple clubs, white speckles, and white tips. Gill leaves whitish below, purple above, tips white. Foot translucent grayish-white with a broad white border; tail with a broad median white stripe containing a thin dark or yellow line (sometimes broken). Known from sand, sponges, and under coral debris in calm protected areas. Spawn is an almost 3-whorled spiral of flat jelly-like ribbon standing on edge on a hard substrate, densely filled with small white egg capsules. Depth: 1-36 feet. Southeast Florida, Puerto Rico, St. Maarten, Saint Barthélemy, Martinique, St. Lucia, St. Vincent, Venezuela, Colombia, Panama, Costa Rica, Jamaica. (Note: In a forthcoming paper the name *Felimida grahami* is changed to *Rudmania grahami* and the Brazilian species *Felimida paulomarcioi* is synonymized with *R. grahami*)

continued on next page

↑ Colombia ↓

Cabo Frio, Brazil

Baía das Tartaruga, Brazil

Saco do Engenho, Brazil

Florida

Florida

Felimida punctilucens **(Bergh, 1890)**

Size: to 35 mm Chromodorididae

ID: Body elongate oval, length twice width, sides elevated and back somewhat convex (height equals width); mantle broader than body, margin ruffled with pronounced vertical folds; tail moderately long, broadly triangular, robust; rhinophores club-like, ~30 lamellae, retractable; gill posterior, retractable, several multi-pinnate leaves, gill opening transversely oval, surrounds a cylindrical flat-topped anal papilla. Dorsum dark gray to blackish with numerous white to yellowish spots, some with faint bluish halos. Mantle's folds with yellowish-white top followed by a black line and a blue line. Rhinophoral lamellae black, upper front yellowish; rim of rhinophoral sheath yellowish-white. Gill leaves yellowish-white with pale blue highlights. No ecological information available. Depth: to 224 feet. Gulf of Mexico west of Dry Tortugas. Very rare.

Roatan

***Felimida* sp. 1**

Size: 8 mm Chromodorididae

ID: Body elongate oval; mantle moderately broad; tail moderately long, broadly triangular; rhinophores club-like, ~7 lamellae; gill posterior, ~3 contractile leaves. White everywhere, mantle with an orange submarginal band and a few small orange spots. Tail translucent white with opaque white median stripe. No other information is available. Depth: ~30 feet. Possibly a juvenile of *Rudmania grahami* (q.v.).

Southeast Florida

***Felimida* sp. 2**

Size: 15 mm Chromodorididae

ID: Body elongate, narrow; foot narrow, tail short and pointed; rhinophores conical, pointed, lamellate, with low sheaths. Ground color translucent to pale blue. A dark purple blotch crosses the back at midbody. A few irregular longitudinal, reddish-edged orange and yellow stripes and scrawls (some branched, some linked) extend from the behind gill to the middle of the head, the most lateral of which have downward-directed short blunt protrusions bracketing a series of blue spots. There are two large purple blotches directly in front of the rhinophores, each with a bright blue spot at its anterior margin. The mantle has a narrow yellow margin and a broad white submargin. **The rhinophores are purplish-white, gill leaves white, both with dark purple tips**; tail dark purple with a short median yellow stripe. Found on a patch reef with sand and rocks at a depth of ~15 feet. This could be a variant of *Felimida clenchi*, *Felimida* cf. *clenchi*, or *Felimida binza*, or an undescribed species. Specimens need to be analyzed for clarification.

Roatan

***Felimida* sp. 3**

Chromodorididae

Size: 6 mm

ID: Body elongate, moderately broad; rhinophores conical, pointed, with about 8 oblique lamellae; the gill is posterior, with 3 leaves. Ground color is purplish. The head and back have mostly linear, irregular, yellow markings that partially wrap around iridescent blue spots. The mantle's edge is yellow and its margin is broad and white. The rhinophores are purplish, with a purple band at their bases; The white gill leaves have dark purple stalks. It was found close to shore on rubble at a depth of ~17 feet. It could also be the juvenile phase of a previously described species.

Southeast Florida

Felimida sp. 4

Size: 8 mm Chromodorididae

ID: Body elongate oval; mantle broad; the rhinophores are club-like with about 14 lamellae; gill posterior, with a circlet of ~10 long bushy leaves. The mantle is mottled white, its back has a few irregular purplish blotches; there is a narrow yellow edge and a broad diffuse pale orange wash submarginally. The rhinophores are brownish orange with white tips. The gill leaves are yellowish-orange. It was found crawling on a hard substrate at a depth of about 10 feet.

Exuma Islands, Bahamas

Felimida sp. 5

Size: 10-15 mm Chromodorididae

ID: Body elongate, moderately broad; dorsum tuberculate – many small, a few larger; rhinophores conical, pointed, ~8 oblique lamellae; gill posterior, ~9 conical leaves; mantle margin broad, thick, ruffled. Dorsum dark brown to blackish, tubercles white-tipped. Rhinophores and gill leaves white with orange tips. Mantle with narrow translucent edge, followed by a broader pale yellow band, then a very broad white band containing what may be transverse white glands, and then a series of large black spots along the junction of the mantle margin and the dark dorsum. Pale blue pigment appears between some of the black spots. Has been found on or near sponges. Depth: ~25 feet. Known only from the Exuma Islands (Bahamas).

Felimida sp. 6

Size: 15 mm Chromodorididae

ID: The body is elongate, moderately broad; mantle margin broad, indented anteriorly; the rhinophores are conical, pointed; gill posterior with short leaves. Dorsum grayish-white with brownish-orange mottling. The mantle margin has a narrow orange stripe and a broader submarginal white band with diffuse purple patches along its inner edge. Rhinophores and gill leaves are whitish below, purple above. Found on a hard substrate. Depth: ~25 feet. Destin, Florida.

Destin, Florida

Isla Mujeres, Mexico

Felimida sp. 7

Size: ~8 mm Chromodorididae

ID: Body elongate oval; mantle moderately broad, edges slightly ruffled; rhinophores club-like, shaft pedunculate and broader below, pit openings wide and snug, 11 lamellae, apical mucron; gill posterior with several short leaves. Ground color pale blue with pinkish cast. Dorsally with numerous round to oval pink-edged white spots; two large irregular white spots behind the rhinophores, two others connected by a narrow isthmus at front edge of gill opening. Rhinophores with white shaft and pinkish club, pit openings pink-edged; gill leaves white proximally, pinkish distally, tips whitish. Mantle with yellow edge and submarginal row of opaque white branched mantle glands. Found on a hard substrate at a depth of ~33 feet.

Puerto Rico

Mexichromis multituberculata **(Baba, 1953)**

tuberculate sea goddess

Size: 30 mm Chromodorididae

ID: Body elongate, rectangular, strongly tuberculate; tubercles dorsal, elevated, bluntly conical, variably sized (some very large); mantle broad, edge ruffled; oral tentacles very short, obscured by mantle; rhinophores long, narrow, lamellate, pointed, well separated; gill posterior, with several leaves; foot narrower than body, front corners short and recurved, tail moderately long, broadly triangular. Ground color white to yellowish. Tubercles purple-tipped, mantle edge with purple spots or transverse blotches; gill stalks, foot margin, and tips of oral tentacles purple; rhinophores purple on upper half. Usually found on sand/silt bottoms near scattered coral heads in calm waters. Known prey include sponges in the genera *Dysidea* and *Euryspongia*. Depth: 15-70 feet. This is a well-known Western Pacific species. This individual may have been introduced to Puerto Rican waters via ballast water of international shipping.

Venezuela

Panama

Brazil

Tyrinna evelinae **(Er. Marcus, 1958)** **Eveline's sea goddess**

Size: 30 mm Chromodorididae

ID: Body elongate, oval, broad at front, narrower at rear; back smooth; tail long and pointed; mantle wide, with two irregular submarginal rows of spherical defensive glands, small and densely packed in outer row, large and sparsely packed in inner row; oral tentacles rolled (groove ventral), stout to slender, concealed by mantle; rhinophores with12-21 lamellae, sheaths smooth and slightly elevated; gill posterior, with 5-9 bipinnate to tripinnate leaves arranged in a semicircle that is open at the rear, the posterior ones smaller. Ground color translucent white to pale bluish-gray. Mantle and tail with a narrow white margin, mantle surface with many small orange/yellow spots dorsally and large orange/yellow-centered white glands submarginally; rhinophores and gills translucent whitish with opaque white tips. Has been found in rocky tide pools, on *Sargassum*, and under rocks in the shallows. Feeds extensively on the sponge *Dysidea etheria*, ingesting its toxic chemicals and accumulating them for its own defensive purposes. Depth: to 16 feet. Known from both sides of the Atlantic and the Eastern Pacific. Jamaica, Puerto Rico, Dominican Republic, Venezuela, Colombia, Panama, Costa Rica, and Brazil.

Key Largo, Florida

Key Largo, Florida

Tarpon Springs, Florida

***Tyrinna* sp.** **Dimitris sea goddess**
Size: 30 mm Chromodorididae
ID: Body elongate, oval; mantle very broad, margin undulate, truncate to slightly concave anteriorly; head and back with a scattering of small tubercles; foot entirely hidden except for moderately long tail that tapers to rounded point; rhinophores conical, bottom third smooth, top two-thirds with ~9 lamellae; gills posterior, stalks conical, tall, tuberculate, some branched; oral tentacles short, smooth above, concealed by mantle. Ground color translucent gray. Except at margin, mantle densely covered with small white spots/dots between large round unpigmented areas containing a central orange spot. Mantle margin lined with **transversely elongate narrow orange stripes** centered in elongate white patches (glands), each long patch separated by two or more short ones. Rhinophores are based in large unpigmented areas, lamellae white. Gill leaves with white speckles. Tail ground color to whitish. Found crawling on a hard substrate in shallow water. Depth: to 10 feet. It has been found only at South and Southeast Florida.

Key Largo, Florida

Family CORAMBIDAE Bergh, 1871

The family consists of about a dozen species in the genera *Corambe* and *Loy* that inhabit littoral to sublittoral marine waters mainly in temperate regions. The genus *Loy* has not been reported from the tropical West Atlantic.

For *Corambe* species: body rounded, flat to slightly convex; mantle fleshy, smooth or slightly bumpy, non-spiculose, covered by a shedding cuticle, has median posterior notch; oral tentacles usually elongate triangular; front of foot bilobed or with tentacular front corners; rhinophores smooth and retractable; gills simple, serial (in a row), on each side of the body below the mantle margin, not completely retractable into gill pockets; anus postero-ventral between the gills at the location of the notch; the well developed buccal pump opens off the buccal bulb; radula lacks rachidian teeth, first laterals triangular-based with a recurved denticulate cusp, other laterals much smaller and non-denticulate.

Corambe species are very well camouflaged on their prey – colonies of bryozoans that encrust substrates like rocks, mangrove roots, and macroalgae. Their size is typically 5-10 mm.

souheast Brazil

Corambe carambola **(Er. Marcus, 1955)**
Size: to 6 mm Corambidae
ID: Body broadly oval, convex; mantle completely covers head and foot; rhinophores well-separated, short, retractile; sheaths low, rims smooth; each rhinophore is complex – a smooth narrow cylindrical shaft partly wrapped by a taco-like flap open at the rear to reveal a narrow vertical keel, this structure then wrapped by a wider and taller loose membrane more broadly open at the rear. Translucent white with black to brown to yellowish-brown spots. Foot white; rhinophores transparent, the visible internal organs are brownish-yellow. Found in the upper littoral zone among algae growing on rocks and their associated bryozoans. Feeds on the bryozoans *Alcyonidium hauffi* and *Biflustra tenuis*, possibly on hydroids. Camouflages well on *A. hauffi*. Depth: upper littoral. Very rare.

Brazil

Corambe testudinaria **H. Fischer, 1889**
Size: to 7 mm Corambidae
ID: Body broadly oval, convex; mantle's **rear margin with a deep median notch**; rhinophores short, retractile; sheaths low, rims smooth; each rhinophore a smooth cylindrical axis with a narrow posterior keel extending two-thirds of the way up the shaft, followed by two vertical postero-lateral expansions that wrap loosely around the front. Translucent whitish, the central part of the mantle having small brown spots on a yellowish reticulate background, the animal's overall darkness varying with the density of brown spots. Rhinophores unpigmented, tips dusted with brown. Under stones and on bryozoans to 7 feet. Amphiatlantic. Rare off south-central Brazil.

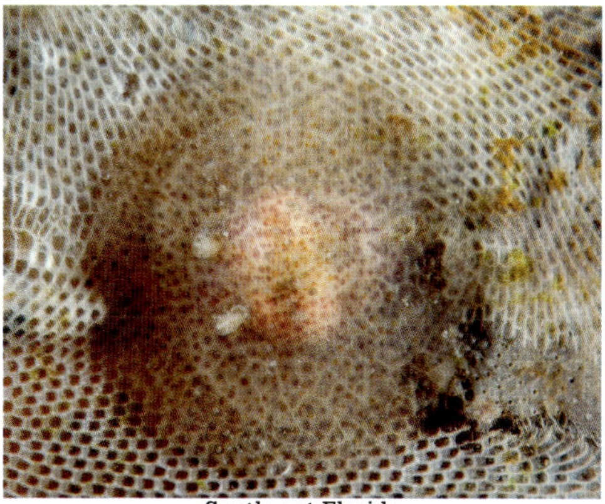

Southeast Florida

Corambe obscura **(A. E. Verrill, 1870) limpet corambe**
Size: to 8 mm Corambidae
ID: Broadly oval, slightly convex; smooth; rhinophores short, tips flared, thin central axis, retractile; foot broad. Translucent whitish; mantle has yellowish to whitish reticulations on a background of black to brown to reddish-brown spots; central internal organs yellowish to orange-brown; rhinophores and foot white. Marine with some tolerance for brackish conditions. Depth: to 25 feet. Under rocks and on floating algae with encrusting bryozoans that it feeds on, e.g. *Alcyonidium verrilli*, and especially the mesh-like looking species of *Membranipora* and *Acanthodesia* that provide it good camouflage. Probably circumglobal. Florida, Bermuda, Gulf of Mexico. Uncommon.

Family DENDRODORIDIDAE O'Donoghue, 1924 (1864)

A family currently composed of two genera, *Dendrodoris* and *Doriopsilla*, with a total of nearly 80 described species and others likely to be soon discovered via molecular analyses as cryptic species within species complexes. They occur circumglobally in temperate to tropical seas, with several species found in the tropical West Atlantic.

The body is elongate and soft for *Dendrodoris* but spiculose and stiff for *Doriopsilla*; mantle smooth or tuberculate, margin usually broad, wavy, delicate; oral tentacles usually absent or very reduced, sometimes fused, usually inconspicuous; front of foot may have a deep transverse furrow leading to a recessed mouth; rhinophores with a cylindrical stalk and a distal lamellate club, contractile into pockets, sheath present or absent; gill postero-dorsal, tripinnate leaves retractile into a cavity & form a circlet often completed by the anus; anus medianly posterior (*Dendrodoris*) or eccentric to the left (*Doriopsilla*); radula and jaws are absent.

They feed by using glandular secretions to dissolve the tissues of sponges in order to ingest the remaining fluids using a long protrusible suctorial tube. Many can also synthesize chemicals to store in their mantles for defensive purposes. The close morphological and anatomical similarities of many species, combined with considerable variability in intra-species colors and patterns, often make it very difficult to distinguish between species. It is in taxa like this that the probability of discovering cryptic species through molecular analyses is quite high. Lengths of Caribbean species are typically 20-120 mm.

Guadeloupe

Dendrodoris karukeraensis **Ortea, Espinosa, Caballer & Buske, 2012** **Karukera dendrodoris**

Size: 27 mm Dendrodorididae

ID: Body elongate, width about one-third length; mantle margin broad, thin, undulate; rhinophores stout, club-like, upper half with ~11 oblique lamellae, tip with short thin knob; rhinophoral sheath low, tightly wraps the stalk; gill very posterior, with 6 bi- to tripinnate leaves grouped in two symmetric sets of three leaves each; the foot protrudes slightly at the rear as a broadly rounded tail. White everywhere except for variably sized blackish-brown circular blotches on the back, sides, and head; the largest blotches occur medianly on the back; some are fragmented and/or overlapping. The rhinophoral stalk is transparent. Viscera visible as a diffuse orange mass. Has been encountered on rocky bottoms. Easily alarmed by lights during night dives. Depth: to at least 50 feet. So far, it is known only from Guadeloupe. The species name derives from the ancestral Arawak term for the island of Guadeloupe – "Karukera", which translates as "island of beautiful waters".

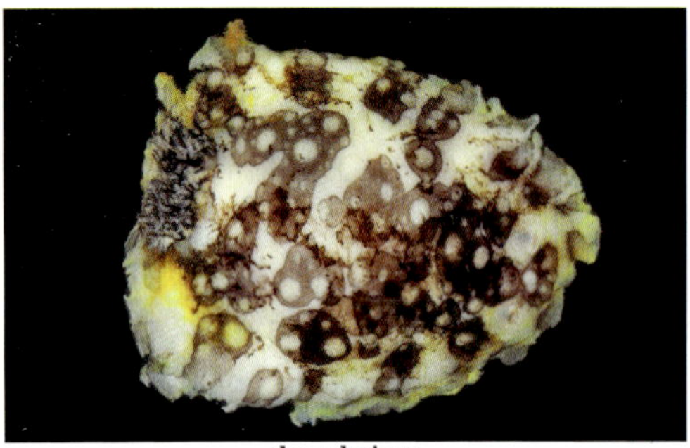

dorsal view

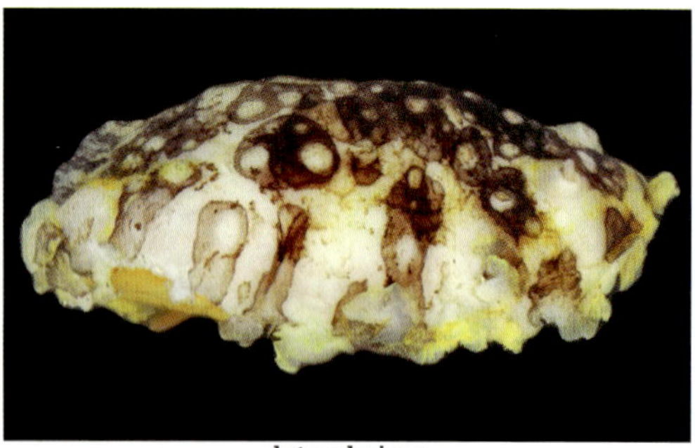

lateral view

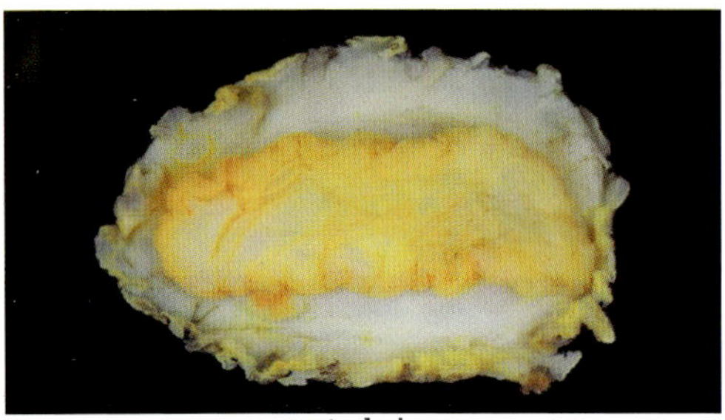

ventral view

Dendrodoris sinusensis **Galià-Camps, A. Enguídanos, Cobb-Fletcher, E. F. García & A. Valdés, 2025**

Gulf dendrodoris

Size: to at least 100 mm Dendrodorididae

ID: Body oval to elongate; mantle broad, with a scattering of low, rounded, variously sized tubercles; rhinophores elongate, tentacular, 14 lamellae; gill posterior with 5 large tripinnate leaves; the sheaths of both are smooth and slightly elevated. Cream centrally, grading to yellow at the mantle margins, with numerous scattered brownish-gray blotches surrounding single or groups of cream-yellow tubercles. Rhinophores cream; gill leaves mainly white, stalks and tips with some black patches. Mantle margin opaque white ventrally; foot yellow. So far, known only from a depth of 330 feet. **Endemic to the northeast Gulf of Mexic**o, hence the name sinusensis (Latin for "from the gulf").

Bonaire

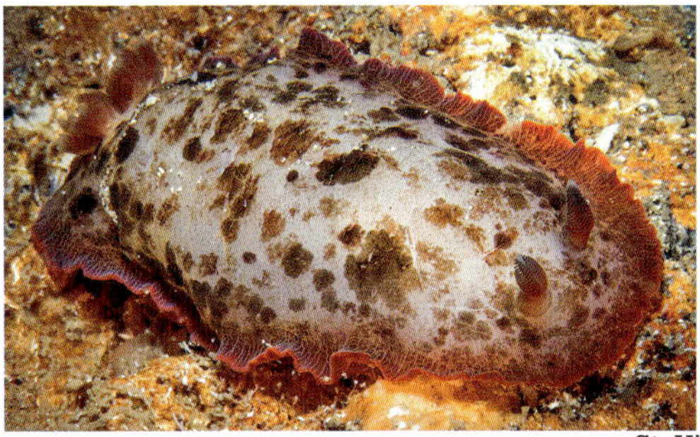

St. Vincent

Cuba

Bonaire

Roatan

with spawn; Belize

Dendrodoris krebsii **(Mörch, 1863)** **slimy dendrodoris**

Size: to 120 mm Dendrodorididae

ID: Body oval to elongate, thick, soft, slimy; mantle smooth to wrinkled, sometimes (when disturbed) with soft warts and blisters of irregular shapes, sizes, and separations; mantle skirt wide, thin, undulating, margin finely frilled with vein-like radial striations; foot shorter and narrower than mantle; rhinophores stout, broad-based, distally club-like, ~20 lamellae, tip conical; gill posterior, 6-9 tripinnate leaves, retractable. Color highly variable – white, gray, yellow, orange, red, green, brown, black, blue. May be spotted with red, black, gray, yellow, or white. Pale individuals usually with dark blotches, dark ones with pale spots/blotches. Mantle margin often with radial brown streaks; sometimes reddish. Rhinophores dark brown to black (darker than back) with white tips. Gill leaves usually body color, ribs darker, tips white. **Juveniles uniformly pink to red, sometimes gray with diffuse dark spots.** Hides under coral debris and rocks near sponges to suck nutrients from (e.g. *Haliclona* species). In one hour it can deposit a white to yellow spirally coiled (4-5 whorls) flat egg ribbon standing on edge (5 mm high, 0.4 mm thick) with a wavy top, crowded with vertical rows of ~20 spherical capsules containing one egg each. Depth: 1-80 feet. Georgia, Florida, Bahamas, most of the Caribbean.

Dendrodoris magagnai Ortea & Espinosa, 2001
magagna dendrodoris

Size: to 26 mm Dendrodorididae

ID: Body flat, elongate oval (width one-fourth length); tail protrudes one-fourth body length, broadly rounded; mantle margin thick, undulating; rhinophores short, stout, 8-9 highly inclined (almost vertical) lamellae, base narrowed, no sheath, highly retractable; gill posterior, 7 bipinnate and tripinnate leaves. Ground color whitish. Dorsum covered with small/large irregular clusters of black pigment cells and roundish gold/yellow/orange spots. Head, side of foot, and tail white with black spots. Mantle edge with thin yellowish-gold stripe, somewhat interrupted by undulations. Rhinophores white with dense black spotting, tips white. Gill leaves white with large black spots on ribs and some tufts; gill appears uniformly black when contracted. Little is known about its ecology. Assumed to feed by sucking fluids from sponges, like other members of this genus. Depth: to 65 feet. So far, it is known only from the Caribbean coast of Costa Rica.

Costa Rica

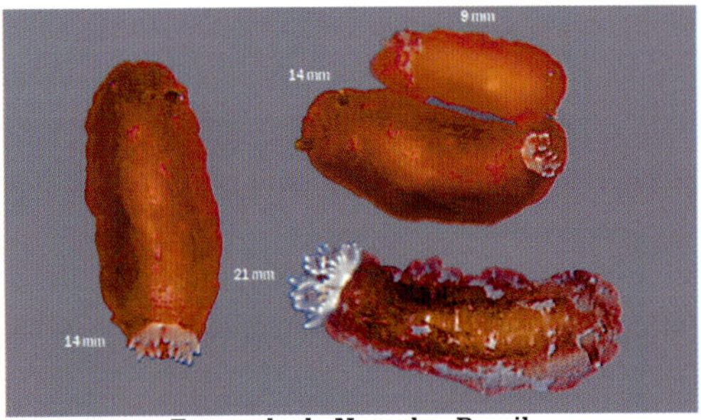

Fernando de Noronha, Brazil

Dendrodoris senegalensis Bouchet, 1975

Size: to 21 mm Dendrodorididae

ID: Body elongate, width about half length, soft, smooth; mantle margin wide, delicate, slightly striated; rhinophores with cylindrical stalk and lamellate club; gill posterior, five moderately long tripinnate leaves encircle the anal papilla. The dorsal surface and mantle margin are either red-brown with irregular white areas (21 mm specimen in photo) or uniformly red in smaller animals. Rhinophores red with white tips; gill leaves and anal papilla white. Found under rocks in association with sponges and ascidians. Spawn is a spiral coil with many longitudinal rows of spherical capsules within larger polyhedral capsules, each inner capsule containing a single yellow egg. Depth: 0-46 feet. Known from the west coast of Africa (Senegal, Cape Verde Islands), but its range was recently extended (2002) to Fernando de Noronha, Brazil. More study is needed to establish the extent of its range in the tropical West Atlantic.

Destin, Florida

100 mm; Destin, Florida

Gray's Reef, Georgia

Dendrodoris warta Ev. Marcus & Gallagher, 1976
warty dendrodoris

Size: to 120 mm Dendrodorididae

ID: Body oval, elevated; back covered with irregularly-distributed and variably sized tubercles/warts, many almost touching; mantle margin narrow, ruffled, striated, no warts; rhinophores stout, club-like, 24-30 lamellae, retractile; gill posterior, numerous multi-pinnate retractile leaves. Color variable: usually brown, sometimes gray to green to blue, even white with dark spotting. Warts generally paler than background. In dark brown to gray individuals warts are pinkish-beige and gill leaves brown with white edges. In blue individuals warts are bluish-white, gill leaves blue with whitish stalks. Mantle margin pale yellow to whitish above, paler beneath. Rhinophoral stalk brown to blackish, lamellae whitish, tip white. Originally found intertidally off west Florida, attached to the undersides of *Crassostrea virginica* oyster clusters that were covered with bright orange boring sponges (its prey). Feeds by sucking fluids from the sponges. Its egg ribbon is a ruffled spiral coil with round capsules each containing one yellow/orange egg. Hatching occurs in ~6 days. Depth: intertidal to 55 feet. South Carolina to Northeast and Northwest Florida. Also Venezuela. Rare.

Doriopsilla nigrolineata K. B. Meyer, 1977

blackline doriopsilla

Size: to 25 mm Dendrodorididae

ID: Body oval, low; mantle stiff; back covered with simply rounded tubercles, largest/tallest near midline and progressively smaller/shorter toward mantle edge; mantle margin broad, slightly wavy; rhinophores cylindrical, tips pointed, up to 12 lamellae; gill dorsal, posterior, circlet of four tripinnate leaves; oral tentacles small, fused, only tips are free. Ground color translucent yellowish-gray to shades of orange. Back with numerous, thin, irregular black line segments of varying lengths between tubercles, giving the appearance of a "disjointed network" of black lines. Tubercles ground color or slightly paler, sometimes with fine reddish scrawls (see photo) and bases ringed with white dots. Tiny white speckles everywhere except in areas with large tubercles and along mantle margin. Rhinophores pale yellow, tips white. Gill pale yellow. Little ecological information is available, but as a dendrodorid it likely feeds suctorially on sponges. Depth: 6-33 feet. Panama, Guanaja (Honduras). Rare.

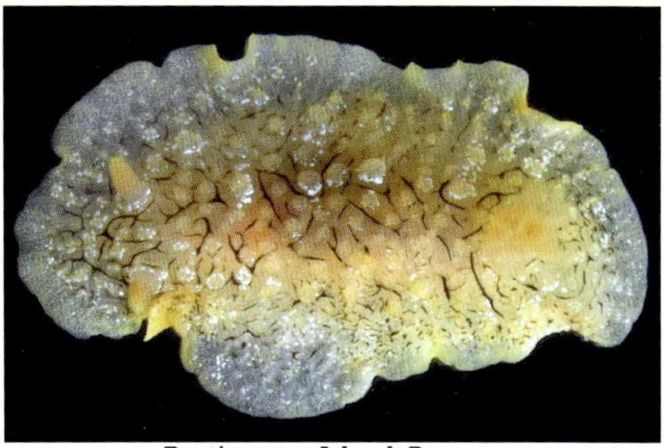

Bastimentos Island, Panama

Roatan

Roatan

Doriopsilla espinosai Á. Valdés & Ortea, 1998

Size: to 15 mm Dendrodorididae

ID: Body elongate, oval, slightly elevated, mantle stiffened by subepidermal spicules; back covered by low conical stiffened tubercles, decreasing in size toward mantle edges; mantle margin broad, slightly wavy; rhinophores with ~9 lamellae, taper to a rounded tip; gill posterior, 2-3 tripinnate leaves; oral tentacles minute, fused, hidden below front of mantle; foot shorter than mantle. Ground color translucent white to yellowish, yellowness increasing with size. Back with a scattering of small but prominent red spots and a few large opaque white patches; internal organs show through skin as a pink mass; rhinophores pale yellowish with a white stalk; gill leaves translucent white. Found under rocks and feeds on sponges. The animal's rhinophores point backwards when it is stationary but forward when it advances. During forward motion the body elongates to approximately twice its width. Spawn is a short spiral ribbon (to 1.5 whorls) of large yellow eggs. Depth: to at least 6 feet. North and south coasts of Cuba, Southeast Florida, Bahamas, Roatan. Rare.

Doriopsilla cf. _espinosai_

Á. Valdés & Ortea, 1998

Dendrodorididae

Size: ~ 10 mm

ID: The body is oval; the mantle is spiculose, covered with long narrow conical tubercles; rhinophores are lamellate; with the gill posterior. The ground color is translucent gray. Mantle has a scattering of small white dots and many yellow or orange spots. Tubercles, rhinophores, and gill are whitish. Viscera are white to cream. Have been found in coral rubble. Depth: ~10 feet. It may be *D. espinosai*, but the color of the dorsal spots and the shape of the tubercles do not match the original description. It could also be an undescribed cryptic member of a *Doriopsilla espinosai* species complex.

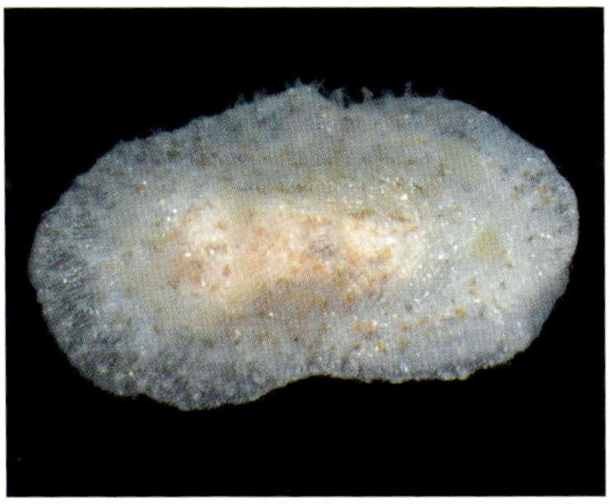

Riviera Beach, Florida

Puerto Rico

***Doriopsilla* cf. *areolata* Bergh, 1880**

Size: ~15 mm

Dendrodorididae

ID: Body elongate, oval, elevated; back covered with low, simply rounded, scattered tubercles; mantle margin broad, slightly wavy; rhinophores conical, tips rounded, ~13 lamellae. Orange everywhere except for rings of white speckles surrounding the bases of some tubercles. There is little ecological information available, but as a dendrodorid it likely feeds on sponges. Depth: to 20 feet. This seems to be a color morph of the east Atlantic/Mediterranean species *Doriopsilla areolata*, which is known to have considerable color variability. It could, however, be an undescribed species.

St. Vincent

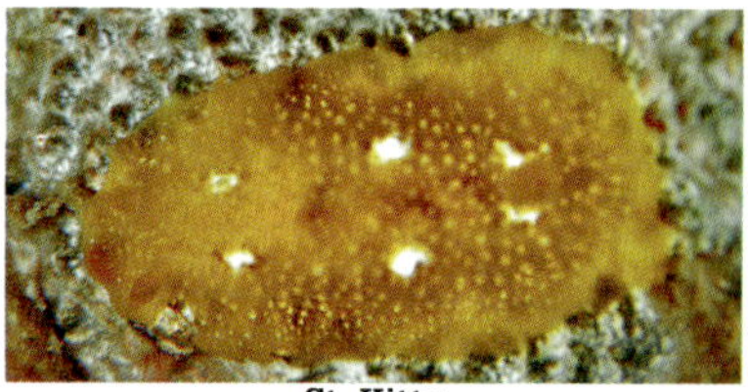

St. Kitts

Puerto Rico

***Doriopsilla elitae* Á. Valdés & Hamann, 2008**

whitespot doriopsilla

Size: to 25 mm

Dendrodorididae

ID: Body oval, flat, mantle hard due to a subepidermal web of stiff calcareous spicules, back covered by short, stiff, conical, simple tubercles; mantle margin wide, slightly undulating; rhinophores with up to 13 lamellae; gill posterior, with 4 tripinnate leaves in a circle closed by the anus, which is left of center; oral tentacles very short, bases fused around central mouth; front margin of foot slightly indented. Ground color yellow to dark orange. Back with variously sized opaque white round patches centered on tubercles, either densely/broadly distributed across the dorsum or just a few large ones on the more sizable tubercles bordering the visceral hump. The rhinophores and gill are ground color. Like other members of this genus, it is probably a suction feeder on sponges. Depth: to 100 feet. So far, reported from Puerto Rico, St. Kitts, Nevis, Martinique, St. Lucia, St. Vincent and the Grenadines, Aruba. Rare.

Cordelia Banks, Honduras

Doriopsilla tishae **Á. Valdés & Hamann, 2008** **Tisha's doriopsilla**

Size: to 49 mm Dendrodorididae

ID: Body oval, low; mantle stiff; back covered with stiff hemispherical tubercles, largest near midline and progressively smaller toward mantle edge; mantle margin broad, slightly wavy; rhinophores cylindrical, tips pointed, up to 15 lamellae; gill dorsal, posterior, four tripinnate leaves, anus at left edge of cluster; oral tentacles small, fused, inconspicuous. Ground color translucent yellowish-white. Back with numerous, thick, irregular, black line segments of varying lengths between tubercles, giving the appearance of a "disjointed network" of black lines. A dense distribution of tiny white speckles near the mantle margin; especially concentrated on tubercles, which appear uniformly whitish. Tubercles sometimes with fine reddish scrawls. Rhinophores and gill pale yellow. The pictured slug was found on a reef with staghorn and lettuce corals. As a dendrodorid, it likely feeds suctorially on sponges. Depth: 17-100 feet. So far, known only from Roatan and Guanaja (Honduras). The similar *D. nigrolineata* differs by having: thinner, shorter, less conspicuous black lines; much smaller tubercles; white dots surrounding only tubercle bases instead of covering their entire surface.

Southeast Florida

Doriopsilla **sp. 1**

Size: ~15 mm Dendrodorididae

ID: Mantle with dense network of spicules, small conical papillae at some points of intersection; rhinophores lamellate; gill leaves with long filaments. Orange; spicules white, papillae white-tipped. Rhinophores orange, their tips white; gill leaves orange, their filaments white. On hard coral at ~10 feet.

Roatan

Doriopsilla **sp. 2**

Size: ~10 mm Dendrodorididae

ID: Flat; mantle with linear spicules and small conical tubercles; rhinophores with ~9 oblique lamellae and a broad stalk; gill posterior. Ground color translucent yellow to pale orange to gray. Spicules white. Rhinophores and gill ground color. Viscera orange. Found on a hard coral substrate at ~10 feet.

Roatan

Doriopsilla **sp. 3**

Size: ~15 mm Dendrodorididae

ID: Mantle covered with slender conical tubercles, smaller toward the edges; rhinophores with 8 oblique lamellae; gill posterior, several leaves. Mantle and appendages uniformly orange-yellow. Many tubercles tipped with clusters of white projections. Found in coral rubble. When disturbed, the animal retracts the tubercles' white projections. Depth: ~10 feet.

Riviera Beach, Florida

Doriopsilla **sp. 4**

Size: ~15 mm Dendrodorididae

ID: Oval, convex. Mantle spiculate; crowded with small rounded tubercles. Rhinophores with ~8 oblique lamellae; gill posterior, with several small leaves. Ground color whitish, many brownish-orange spots; tubercles white; rhinophoral stalk white, lamellae pale orange; gill leaves pale orange. Found on a hard substrate at ~10 feet. Perhaps a variant of *D. espinosai*.

Family DISCODORIDIDAE Bergh, 1891

A huge family consisting of hundreds of species in dozens of genera distributed worldwide at most latitudes. They are cryptobranch dorids (cryptobranch = "hidden gill") that can retract their dorsal gill and hide it in a pocket. All species have a radula.

Body oval to elongate, usually small and depressed but sometimes slightly convex; mantle thick, broad, smooth to tuberculate and/or ridged; tubercles variously shaped (conical, rounded, elongate) and variously textured – simple/smooth to complex raised "caryophyllidia" that have a crown of projecting spicules encircling a sensory ciliated apex; rhinophores lamellate, retractable into pits, often with elevated sheaths; gill dorsal, posterior, retractable, leaves usually multipinnate; foot large, front edge transversely grooved to produce an upper and lower lip, upper lip with median vertical notch; small mouth flanked by conical or tentacular oral tentacles.

Most inhabit rocky bottoms in moderate depths (often intertidal), feeding on their preferred sponges by night and hiding in shadows or under rocks by day. Some are known to discard parts of their mantle margin when disturbed.

Roatan

Southeast Florida

rhinophore close-up; Roatan

Atagema browni **T. E. Thompson, 1980** **Brown's dorid**

Size: to 24 mm Discodorididae

ID: Body elongate, oval, flat; mantle margin broad, edge shiny with calcareous spicules; back with a prominent mid-dorsal longitudinal tuberculate ridge; tubercles mostly simple, some compound, some quite large; rest of back densely covered with tiny tubercles, producing a "granular" look; rhinophores stout, ~12 lamellae, retractile, sheaths elevated, tuberculate, including the rims; gill dorsal, posterior, backwards pointing, tripinnate, retractile, with 3 flap-like valves that guard the gill pocket; oral tentacles small, conical, recurved. Ground color pale grayish-brown. Back with many small dark brown irregular spots. Rhinophores, gill, and tubercles on mid-dorsal ridge variously pigmented with white to yellowish. Reported from shallow areas under coral rubble, near sponges that it probably feeds on. Has been observed crawling along a hard substrate with a side-to-side rocking motion. Depth: to 15 feet. South Carolina to Southeast Florida, Costa Rica, Roatan, Jamaica, Grand Cayman. Rare.

Diaulula farmersi
Á. Valdés, 2004

Riviera Beach, Florida

Guadeloupe

Farmer's dorid
Size: 16 mm Discodorididae
ID: The mantle is covered by small caryophyllidia; rhinophores are robust, ~15 lamellae, sheaths elevated; gill with ~9 bipinnate leaves. Yellow with brown speckling; rhinophores dark to light brown, tips white; gill leaves brownish below, white above. Mantle margin with opaque white glands. Usually found in shallow water on a hard substrate. Probably feeds on nearby sponges. Depth: to 6 feet. Key Largo (Florida), Bahamas, Puerto Rico, Virgin Islands, Guadeloupe. Rare. Its original description, based on a single specimen from Florida, describes the gill leaves as being uniformly white. Since the accompanying photos show leaves that are darker below and white above, the possibility remains that it represents some other species, possibly undescribed.

St. Vincent

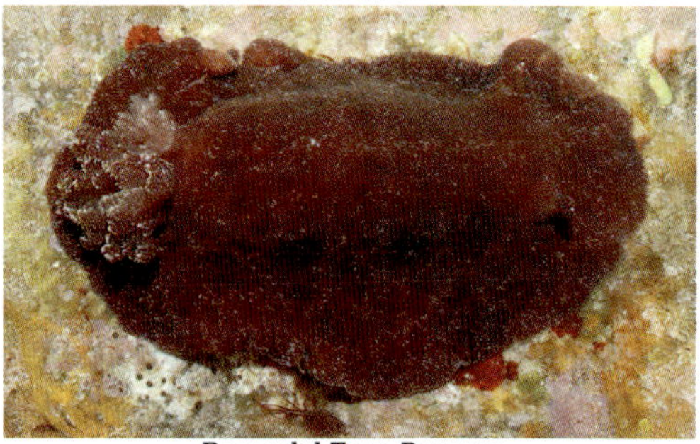

Bocas del Toro, Panama

Bocas del Toro, Panama

Diaulula phoca (Ev. Marcus & Er. Marcus, 1967)
Size: to 55 mm Discodorididae
ID: The body is oval and convex (height about one-third length); mantle rigid, covered with small caryophyllidia; rhinophores retractable, club-like, ~20 lamellae; oral tentacles are cylindrical with pointed tips; gill dorsal and posterior, with six retractable tripinnate leaves. The mantle, rhinophores, and gill are dark purplish-brown to purple with many small opaque white spots and dots, sometimes prominent along the broad mantle margin. Rhinophores are white-tipped. Little is known about their ecology. It was originally found in a shallow area among mangroves, but it has also been found around rocks and detritus. Feeds on sponges. Depth: to 20 feet. Florida, Cuba, Martinique, St. Vincent, Panama, Costa Rica, Honduras. Rare.

Martinique

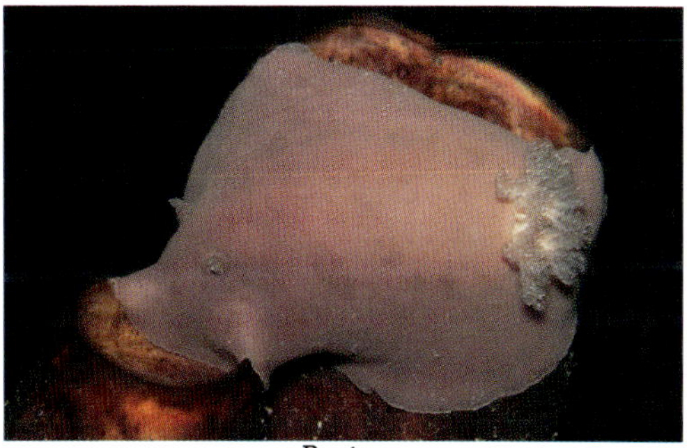

Roatan

Diaulula cf. *phoca* (Ev. Marcus & Er. Marcus, 1967)
Size: to 100 mm Discodorididae
ID: Body oval, convex; mantle densely covered with very small tubercles; rhinophores club-like, ~16 lamellae, retractable; gill dorsal, posterior, ~6 tripinnate leaves. Body uniformly light purplish except for a few very indistinct darker spots and blotches. Mantle edge, rhinophores, and gill leaves with white dots. Rhinophores brownish, gill leaves pale brownish. At Roatan it was found feeding on *Neopetrosia rosariensis* sponges. Depth: to 20 feet. Possibly a variant of *Diaulula phoca* (q.v.), which is typically much darker and densely white-spotted, sometimes with a white net-like pattern.

Guadeloupe

St. Vincent

Florida Keys

ventral view; Florida Keys

rhinophores; Florida Keys

Diaulula greeleyi **(MacFarland, 1909)** **blotched diaulula**

Discodorididae

Size: to 47 mm

ID: Body oval to elongate, flat; mantle crowded with minute caryophyllidia, each with 6-7 spicules protruding up from the base of the tubercle and a long ciliary tuft extending up from its tip; retractable rhinophores robust, club-like, with up to 15 oblique lamellae, a vertical groove at front and rear, apical mucron; rhinophoral sheaths elevated and highly spiculose; gill dorsal, posterior, ~12 unipinnate retractable leaves encircling the anal cone, gill sheath prominent; foot narrower than mantle, its posterior usually concealed by mantle but sometimes projects as a short rounded tail. The ground color is yellow to cream to orange. Mantle with a dense random distribution of variously sized roundish brown blotches, large ones usually darker than small ones, mantle margin sometimes with scattered small spots. Rhinophores dark brown, tips white; gill leaves yellow to dark brown, tips white-speckled; mantle margin with a row of opaque white glands. Found on sand/grass bottoms, on hard substrates, or on sponges it feeds on, (*Haliclona* sp., *Lissondendoryx isodictialis).* Spawn is a pink spiral ribbon of 2 counterclockwise whorls, with a wavy top edge; many rows of capsules with 1 egg each. Depth: to 6 feet. South Carolina, Florida, Bahamas, Guadeloupe, Costa Rica, Brazil.

all images from the Florida Keys

***Diaulula* cf. *alba* (K. White, 1952)**

Size: ~32 mm Discodorididae

ID: The body is broadly oval and flat, its width being half its length. The mantle is densely covered with small papillae that appear to be caryophyllidia; its edge is thin. The rhinophores have ~16 lamellae and an apical mucron; the rhinophoral and gill sheaths are low and their rims are crenulate. Ground color translucent white. The mantle has numerous small irregular clusters of opaque white papillae, especially concentrated along its margins. The rhinophores are whitish to yellowish-white with a white apical mucron. The animal is usually found crawling along a hard substrate in shallow water. Few records exist for *Diaulula alba*. It was first described by White based on a single living animal from the Dry Tortugas (Florida) and then briefly redescribed by Valdés et al in the 2006 book *Caribbean Sea Slugs* (Valdés et al, 2006) based on another specimen from the British Virgin Islands. Externally, the animal shown here appears to be a good match for the one shown on page 190 of the book, so it may be the same species. However, a positive identification would require an examination of its internal anatomy.

Roatan

Destin, Florida

***Diaulula* sp. 1**

Size: ~15 mm Discodorididae

ID: Body elongate oval, flat, covered with minute rounded papillae; rhinophores robust, ~12 lamellae, sheaths somewhat elevated; gill dorsal, posterior, ~9 leaves. Ground color yellow. Purplish blotches on head and back, prominently concentrated on sides of rhinophoral sheaths and along dorsal midline from rhinophores to gill. Rhinophores yellow, edges of lamellae purplish. Gill leaves brownish-purplish, tips white. Found on a rubble bottom. Depth: 5-10 feet. Roatan (Honduras) and Northwest Florida.

Roatan

Diaulula sp. 2

Size: ~100 mm Discodorididae

ID: Body oval, quite convex; mantle densely covered with very small tubercles; rhinophores conical, lamellate, sheath very low; gill dorsal, posterior, bushy-looking with several multipinnate leaves. Whitish-cream with numerous small irregular brownish spots and blotches. Mantle margin white-dotted. Rhinophores brownish, tips white, a few other white dots. Gill leaves grayish-white, tips with white dots. Found feeding on a sponge. Depth: ~20 feet. Could be a variant of *Diaulula phoca*.

all photos from Bonaire

Diaulula sp. 3

Size: ~100 mm Discodorididae

ID: Body oval, convex; mantle broad, densely covered with small rounded tubercles, margins thin and undulating; rhinophores with a thick smooth stalk supporting a conical club having ~20 lamellae and an apical mucron; sheath very low, rim tuberculate; gill posterior, low, with 6 bushy leaves. Mantle pinkish-brown, paler at the margins. Rhinophoral stalks colored like the mantle, lamellae brown with some white dots, apical mucron white. Gill leaves have the mantle's color below, cream above. Found crawling on dead coral. Depth: ~20 feet.

Guadeloupe

Riviera Beach, Florida

Carminodoris hansrosaorum (Dominguez, F. J. Garcia & Troncoso, 2006) **Hans and Rosa's dorid**

Size: to 27 mm Discodorididae

ID: Body oval, slightly depressed; mantle stiff, densely covered with randomly arranged, variably sized, simple, rounded tubercles; **mantle's tubercles shortest medianly and near edges**, 2-3 times taller in between; rhinophores long, club-like, 15-22 lamellae, apical mucron; sheath slightly elevated, rim tuberculate; gill dorsal, posterior, 6-8 tripinnate retractile leaves, sheath slightly raised, rim with two rows of tubercles; foot elongate, one-fourth mantle width, front bilobed and notched. **Broad median strip of mantle orange-brown to orange to red, sides pale orange to yellowish, edges whitish**; perimeter may have dark blotches. **Tubercles orange-brown with white basal ring and pale orange/whitish tip**. Rhinophores orange to reddish, lamellae white, mucron pale cream. Proximally the gill's front leaves are yellowish, the others translucent orange to reddish; distally white. Found on sandy and rubble bottoms under debris. Probably feeds on sponges. Depth: to 15 feet. Riviera Beach (Florida), Guadeloupe, Martinique, Bonaire, Panama, Brazil. Rare. Described only twice, based on specimens from the same region of Brazil – first in 2006 from a single specimen, then in 2013 from three additional specimens. Its intraspecific variability in the West Atlantic is not well known.

Riviera Beach, Florida

Martinique

Riviera Beach, Florida

Jupiter, Florida

Carminodoris madibenthos (Ortea, 2016) **Madibenthos dorid**

Size: to 50 mm Discodorididae

ID: Mantle crowded with **thimble-shaped tubercles having bluntly rounded tops**; tubercles of various sizes – **largest near dorsal midline**, smallest along mantle edge; rhinophores with long robust stalk, shorter club, ~30 lamellae; **apical mucron very small to absent**; gill dorsal, posterior, 6 leaves (largest at front); rhinophore and gill sheaths both slightly elevated with ragged rims. Ground color **violet-gray**. Mantle with a scattering of large dark maroon spots mostly obscured by the tubercles; mid-dorsal region above the viscera is brownish-orange. **Tubercles violet to pale brown to brownish-orange with opaque white basal ring**. Rhinophoral stalks violet-gray to pale orange to whitish, lamellae brownish with pale edges. Frontmost gill leaves violet-gray, middle ones whitish on upper half, rearmost ones opaque white. Large specimens were found around mangrove roots, small ones among rocks and debris on a sandy bottom. Spawn is a lobed white ribbon deposited on a hard substrate as a spiral coil of ~3 turns. Depth: to 7 feet. Martinique, Southeast Florida. Rare. Named for the 2016 MADIBENTHOS expedition, a 6-week coastal survey of marine benthic invertebrates around the island of Martinique led by the Muséum National d'Histoire Naturelle (France).

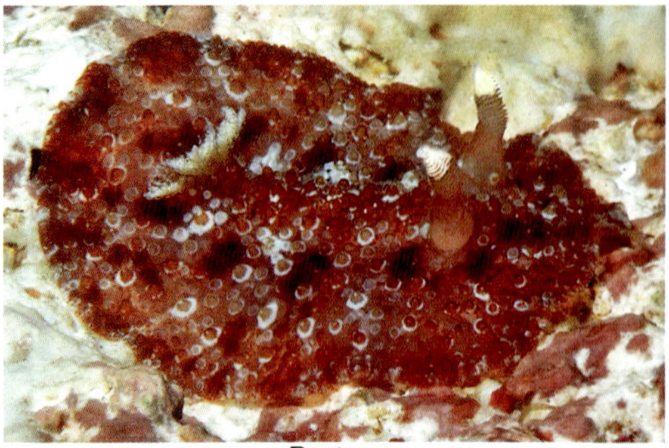

Roatan

Grand Cayman

Grand Cayman

***Carminodoris* sp. 1**

Size: ~7 mm Discodorididae

ID: The external morphology is very similar to that of *Carminodoris hansroseaorum* (q.v.), but the rhinophores are somewhat different — relatively wider, have less horizontal lamellae (14), and the upper half of the lamellate club is strikingly opaque white. Also, the mantle's edge does not have the continuous row of very small crowded white tubercles that is present in *C. hansrosaorum*. Nevertheless, it could be a variant of that species. Depth: to ~15 feet. Grand Cayman, Jamaica, Roatan, Bonaire. It also seems to be the *Hoplodoris* sp. in the book *Caribbean Sea Slugs* (Valdés et al, 2006).

all photos from Florida Keys

gill

***Carminodoris* sp. 2**

Size: ~9 mm Discodorididae

ID: Mantle broadly oval, covered with rounded tubercles; those in a median strip from the gill to beyond the rhinophores are crowded and mostly of similar size, the rest are variably sized and relatively sparsely distributed. Rhinophores with thick base, short robust stalks, club with ~15 lamellae, short mucron at tip; sheath low, rim lined with small tubercles. Gill dorsal, posterior, leaves tripinnate. Ground color pale purplish-gray, almost white. All tubercles brownish with white basal ring but the crowded ones in the median strip are slightly darker, giving the strip a prominently brown appearance. Large brownish-purple blotches encircle the median region and there are some smaller ones irregularly distributed near the mantle edges. Rhinophoral stalk brownish, lamellae and mucron whitish. Gill leaves pale orange/brown below, white above. Found crawling on coral rubble. Probably feeds on sponges. Depth: to ~15 feet. Florida Keys. The whitish color is unusual, but this could be another variant of *Carminodoris hansrosaorum*.

St. Vincent

Riviera Beach, Florida

Roatan

Tayuva lilacina **(A. A. Gould, 1852)** **spotted foot nudibranch**

Size: to 85 mm Discodorididae

ID: Body elongate oval, longitudinally convex; mantle broad, densely covered with small conical tubercles that are usually tallest between mid-dorsum and the mantle edge but occasionally in the middle of the back; rhinophores long, club-like, up to 30 oblique lamellae, retractile, apical mucron, sheath rims with small tubercles; gill dorsal, posterior, 6 tripinnate or tetrapinnate retractile leaves encircle the anus, sheath with small tubercles. Ground color tan to grayish white. Mantle with many variably sized brownish ovate blotches, each a cluster of small spots. Blotches especially crowded dorsally but less so on the sides to produce a paler appearance; some of the largest tubercles at mid-back are white. Usually there is a longitudinal row of large dark brown blotches along each side of the median hump. Mantle's underside and foot are cream-to-white with brown blotches, smaller and denser on the foot. Rhinophores ground color with brownish-yellow white-spotted lamellae and a white tip; largest sheath tubercles white. Gill leaves like the rhinophores plus many brown dots. Found under rocks and in seagrass beds. Feeds on sponges such as *Haliclona* spp. Depth: to 20 feet. Throughout much of the Caribbean and Brazil. This species and *Discodoris branneri* (q.v.) are very similar.

Riviera Beach, Florida

Colombia

St. Vincent

Roatan

Boca del Drago, Panama

egg mass; Jupiter, Florida

Tayuva branneri (MacFarland, 1909) **brown doris**

Size: to 110 mm Discodorididae

ID: Body oval, slightly depressed; mantle broad, crowded with variably sized conical tubercles; their tips mammillate, 5-12 slightly protruding spicules, tuft of cilia at apex; rhinophores long, stout, club-like, retractile, shaft tuberculate proximally, distal club with 19-30 oblique lamellae, apical mucron; sheath low, broad-based, irregular, rim tuberculate; gill dorsal, posterior, 6 retractile tri- or tetrapinnate leaves, sheath tuberculate; anal papilla high, between rearmost 2 gill leaves. Ground color cream, beige, pale olive green, orange, shades of brown, purplish-brown, orange. Mantle with pale to dark brown round blotches, smallest near margins; sometimes a row of blackish ones along each midside; often some whitish median blotches and many white striations along the margins. Rhinophores with white tips, white tubercles on shaft, yellowish-brown white-spotted lamellae. Gill leaves ground color, brown-dotted, tips white. Hides under coral debris and rocks in areas with sponges to suck nutrients from. When disturbed it can discard pieces of its mantle. Depth: 0-30 feet. Common in most of the Caribbean. The similar *T. lilacina* has blunt caryophyllidia (some with white pigment) vs. *T. branneri*'s conical tubercles (many white-tipped), as well as a narrower range of ground colors.

Roatan

Tayuva* cf. *branneri **(MacFarland, 1909)**

Size: ~60 mm

Discodorididae

ID: See *Tayuva branneri* (previous page). Ground color pale brown. Tubercles white-tipped, aggregating to form four white blotches along the dorsal midline and a scattering of much smaller ones, especially near the mantle edges. Rhinophoral lamellae dark brown with white edges, apical mucron white, shaft brown with white dots, sheath rim white. Gill leaves whitish. Foot grayish, densely spotted with small but variably sized brown blotches. Found under coral rubble. Depth: ~4 feet. This seems to be a color variant of *Tayuva branneri* that lacks dark brown to blackish dorsal blotches. Compare this with the dark specimen of *T. branneri* from the same location, which also has some pale blotching along the dorsal midline.

Riviera Beach, Florida

Guadeloupe

Bocas del Toro, Panama

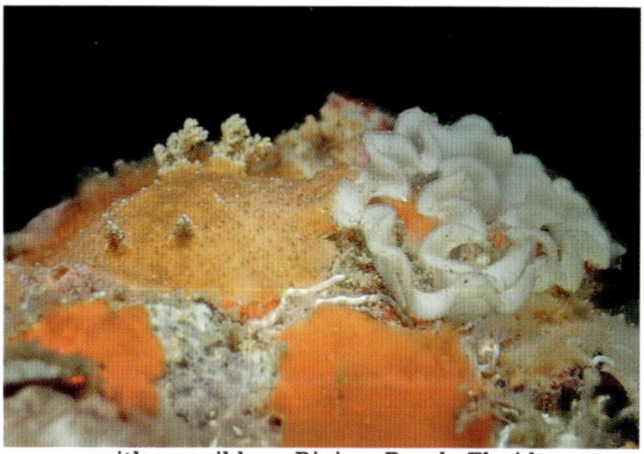

with egg ribbon; Riviera Beach, Florida

Geitodoris immunda **Bergh, 1894**

network geitodoris

Size: to 43 mm

Discodorididae

ID: Body oval; foot narrow, shorter than mantle. Mantle moderately stiff, densely covered with minute nodules; a complex network of low ridges and a scattering of stiff **conical tubercles** at the ridge intersections; tubercles sometimes with protruding spicules. Rhinophores club-like, **about 30 oblique lamellae**; sheaths low, rims jagged. Gill dorsal, posterior, 8 tripinnate retractile leaves, **edge of sheath wavy**. Anal papilla elevated; front border of foot grooved and notched. Ground color grayish-brown to yellowish-orange. Mantle with diffuse dark brown blotches and spots, sometimes with many white dots; tubercles white. **Rhinophores and gill leaves brown with white tips**. Found on rocky substrates and under coral rubble. Probably feeds on sponges. Spawn is a wavy white ribbon deposited on a hard substrate as a counterclockwise spiral coil of ~2 whorls. Depth: to 18 feet. East and west Florida, Cuba, St. Vincent, Venezuela, Panama, Costa Rica, Brazil. Rare.

all photos from Southeast Florida

Geitodoris pusae (Er. Marcus, 1955)

Size: to 60 mm Discodorididae

ID: In the scientific literature the descriptions of *G. pusae* and *G. immunda* are so similar that differentiating between them based only on external appearance is problematic. Several distinguishing features have been identified: (i) *pusae*'s rhinophores have ~12–20 lamellae; *immunda*'s have ~30 (ii) *pusae*'s gill has beige anterior leaves and translucent orange posterior leaves; *immunda*'s leaves are brown with white tips (iii) rounded tubercles in *pusae*; conical in *immunda* (iv) a wavy gill sheath perimeter in *immunda* but not wavy in *pusae*. Unfortunately, these features can be indistinct and feature overlaps or partial inconsistencies can occur, leading to tentative identifications. More detailed comparative studies are needed for these species to more reliably differentiate them based on external appearance. It is usually found on rubble or on the undersides of rocks near sponges that it feeds on. It is capable of swimming by dorso-ventral contortions of the whole body, without flapping its mantle. When alarmed it can rapidly autotomize all or part of its mantle rim. Spawn is deposited on a hard substrate as a counterclockwise spiral ribbon of 2.5 to 3.5 whorls with a crenulated border, containing crowded longitudinal rows of tiny white egg capsules. Depth: to 20 feet. Recorded so far from Florida, Puerto Rico, Guadeloupe, Martinique, Curacao, Colombia, Costa Rica, Jamaica, Grand Cayman, Brazil, Argentina.

Riviera Beach, Florida

Puerto Rico

Geitodoris sp. 1

Size: ~35 mm Discodorididae

ID: Body oval, flat; mantle spiculose, covered with small conical tubercles; rhinophores club-like, ~12 lamellae, cylindrical apical mucron; gill dorsal, posterior, several bushy leaves. Mantle reddish-brown, crossed by a dense network of yellow lines; rhinophores yellowish to white-spotted brownish, mucron white; gill leaves yellowish with white tips. Found crawling on a hard substrate. Depth: ~15 feet. This appears to be the *G. pusae* pictured in *Caribbean Sea Slugs* (Valdés et al, 2006), and might be a variant of that species.

Puerto Rico

Martinique

***Geitodoris* sp. 2**
Size: ~35 mm Discodorididae
ID: Body oval, flat; mantle spiculose, covered with small conical tubercles; rhinophores club-like, ~16 lamellae, cylindrical apical mucron; gill dorsal, posterior, several bushy leaves. Mantle yellow, a few diffuse faintly reddish-brown patches peppered with darker dots. Tubercle tips reddish-brown, bases with white radial spokes. Rhinophoral lamellae edged with reddish-brown, mucron white. Gill leaves dark reddish-brown with whitish tips. Found on a hard substrate. Depth: ~15 feet.

***Geitodoris* sp. 3**
Size: ~65 mm Discodorididae
ID: Body oval, flat; mantle spiculose, covered with small rounded tubercles and a few larger ones, margin ruffled; rhinophores retractile; gill dorsal, considerably inboard from rear. Ground color pale orange; many diffuse irregular brownish spots; larger tubercles white with white radial spokes at base; **broad white-speckled purplish mantle margin**. Rhinophores brownish-orange, tips white; gill leaves white with purplish blotches. Found on a hard substrate at a rocky drop-off. Depth: ~15 feet.

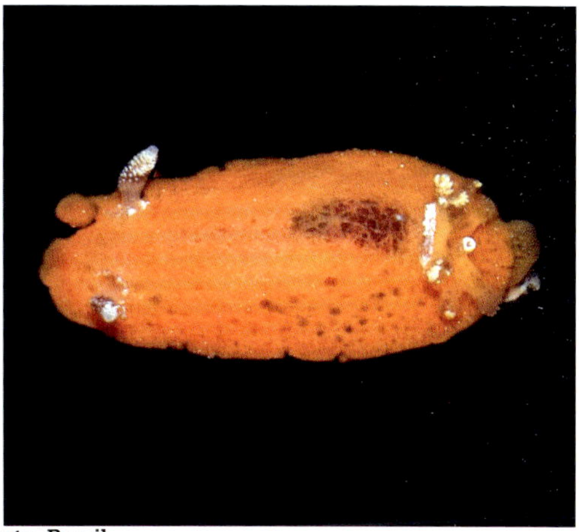

Espirito Santo, Brazil

***Geitodoris* sp. 4**
Size: ~15-20 mm Discodorididae
ID: Body oval, flat; mantle covered with variably sized low rounded tubercles, least dense mid-dorsally, largest at edges of mid-dorsal region with size decreasing toward margin; rhinophores stout, clubs with ~15 lamellae, low sheath rim with several small tubercles; gill dorsal, posterior, ~6 long multi-pinnate leaves. Ground color orange; tubercles dark brown to blackish; some larger ones with white tips and radial white spokes basally. Rhinophoral stalks pale, clubs brown with white-edged lamellae, mucron white. Gill leaves white distally, brown proximally; **front edge of sheath rim prominently white**. On a hard substrate with algae and zoanthids at ~15 feet.

Southeast Florida

***Geitodoris* sp. 5**
Size: ~15 mm Discodorididae
ID: Mantle covered with small rounded tubercles, very spiculose, edge thin and somewhat ruffled; rhinophores club-like, lamellate, retractile; gill dorsal, posterior, several leaves, sheath rim elevated and tuberculate. Ground color red. Mantle tubercles dark with pale tips, many centered on a base of white irregularly-shaped radial spokes (spicules). Rhinophores dark brown with white-edged lamellae. Gill leaves white to brownish. Found on a hard substrate crawling over *Symplegma brakenhielmi* tunicates at ~15 feet. Southeast Florida. It could be a variant of *G. pusae*.

***Geitodoris* cf. *planata* (Alder & Hancock, 1846)**

Size: to 65 mm Discodorididae

ID: Body oval, width about half the length; mantle stiff, densely covered with rounded stalked tubercles. Ground color grayish-brown, fading and becoming more translucent toward the mantle margin. A few dark brown irregular patches. Opaque white pigment surrounds the larger tubercles. Rhinophores and gill leaves are usually the same color as the mantle, with white tips. The pictured specimen was found in coral rubble within a seagrass habitat. Feeds on sponges. Depth not specified. *G. planata* is well known from the coasts of Norway, Sweden, United Kingdom, Ireland, and south to Spain and Portugal, including Spain's north Mediterranean coast. It has also been reported from the continental shelf off the northeast United States (New Jersey, Massachusetts). What may be *G. planata* has been reported from only two Caribbean locations – St. Lucia and Panama.

Bocas del Toro, Panama

all images from Roatan

***Jorunna coloradilla* Ortea & Moro, 2016** **speckled jorunna**

Size: to 18 mm Discodorididae

ID: Body oval; mantle broad, rigid, "fuzzy" – uniformly covered with small similar-sized spiculose tubercles (caryophyllidia); rhinophores club-like, 14-16 oblique lamellae, cylindrical apical mucron, retractile, sheath somewhat elevated; gill dorsal, posterior, 6 tripinnate retractile leaves that (when fully open) fold down onto the back with their tips bent vertically, leaves form a closed circle around the raised anal papilla. Ground color whitish to pearl gray. Mantle, rhinophores, and gill leaves densely and uniformly covered with dark red to chestnut dots. Back with a scattering of variously sized diffuse brownish round spots; visceral region may have pinkish to violet tones. Caryophyllidia translucent with some dark red dots, especially basally; some are opaque white. Mantle edge with small white glands. Rhinophores and gill leaves ground color or slightly darker, sometimes greenish-brown; both are white-tipped. Found on seaweed beds and coral reefs near/on hard substrates; often under rocks. Feeds on sponges such as *Callyspongia pallida*, *Haliclona* spp., and *Chalinula*. spp. A rapid color-changer in response to changing environmental conditions. Species of *Jorunna* can autotomize parts of their mantle when threatened. Depth: 1-43 feet. Cuba, Guadeloupe, St. Vincent, Costa Rica, Roatan. Probably as widespread and as common as *J. spazzola*, with which it has been confused. The species name "coloradilla" references the common Costa Rican tick that produces red spots on the skin, akin to the speckles on this animal.

all photos from Roatan

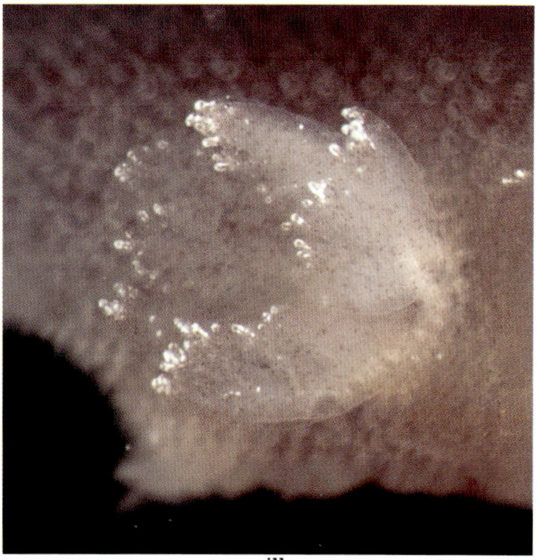

gill

Jorunna davidbowieii **Ortea & Moro, 2016**

Size: to 25 mm Discodorididae

ID: Body oval; the mantle is broad, edge irregular and somewhat wavy, the surface covered with small variably-sized spiculose caryophyllidia, each a "mushroom" with a short peduncle and a flat cap with 6-8 radial spicules that barely protrude beyond the cap's edge, but the caryophyllidia near the mantle edge are biconical; the caryophyllidia distribution pattern has been described as resembling that of mermaid cup algae. The rhinophores have ~20 lamellae; sheath elevated, lobed, covered with smooth globose tubercles; gill posterior, 6 small bipinnate to tripinnate leaves, rearmost two longest, frontmost two shortest but broadest; very long anal papilla at the center of the gill. Ground color uniformly **translucent grayish white**, but when the animal is disturbed **it blushes violet-pink or violet -gray**. The mantle has a row of small white-tipped caryophyllidia along its margin; sometimes a few scattered white spots; the edge may have small yellowish nodules. Viscera appear dark gray. Rhinophores and gill leaves with a brownish dusting and white tips; rhinophores may be yellowish-green, gill leaves may be greenish-brown. Anal papilla white. It has been found under rocks. Species of *Jorunna* can autotomize parts of their mantle when threatened. Depth: usually < 7 feet. So far, reported from Cuba, Martinique, Roatan. Named for the singer David Bowie ("the chameleon") because of its ability to change colors the way he changed costumes.

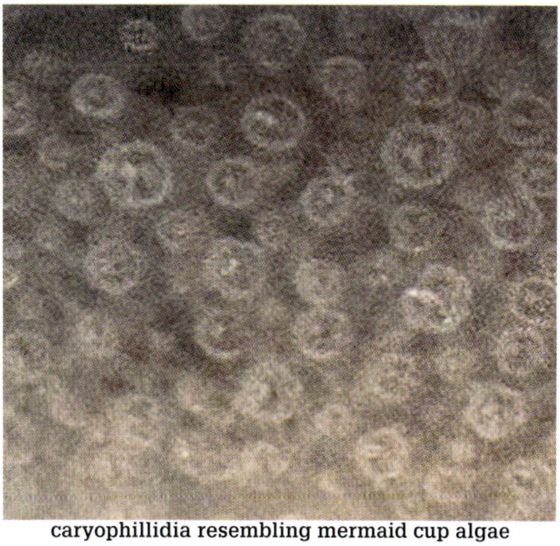

caryophillidia resembling mermaid cup algae

Riviera Beach, Florida

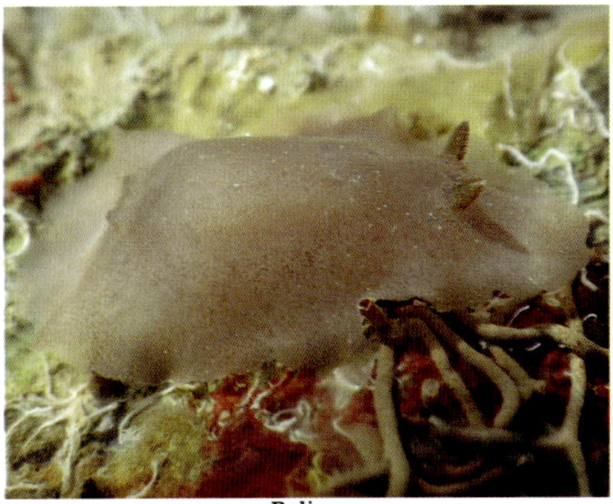

Belize

Cayman Brac

Cayman Brac

Destin, Florida

Jorunna spazzola (Er. Marcus, 1955) **brush jorunna**

Size: to 18 mm Discodorididae

ID: Body elongate oval, moderately convex; mantle broad, rigid, with a few conical papillae but densely covered with uniformly spaced and almost uniformly-sized caryophyllidia; each caryophillidium is a central tubercle with a core of vertical spicules, externally braced by 5-8 vertical curved spicules that fan out rosette-like above the tubercle's oval ciliated apex; the caryophillidia are mostly short (~0.2 mm) and very densely packed, some at the mantle midsides are slightly larger and more widely spaced; rhinophores are long clubs on short stalks, retractile, 8-10 oblique lamellae, cylindrical apical mucron; sheaths wide and low, bases with larger caryophyllidia; gill dorsal, posterior, 5-12 uni- to tripinnate **very short retractile leaves** encircling a raised anal papilla; gill sheath slightly raised, rim ragged. Ground color varies: **ivory, translucent gray, purplish gray, pale yellow, orange**. Mantle may have **scattered, variably sized, diffuse, darker blotches, largest at midside**; **mantle's periphery with tiny white glands**. Rhinophores and gill leaves ground color with dark speckles and white tips, rhinophoral speckles concentrated on edges of lamellae. **White caryophyllidia usually encircle rhinophore and gill sheaths**. Pinkish to reddish-brown viscera may be visible. Found on seaweed beds and coral reefs near/on hard substrates; often under rocks. Feeds on sponges such as *Callyspongia pallida* and species of *Haliclona* and *Chalinula*. A rapid color-changer in response to changing environmental conditions. Depth: usually to 25 feet, but as deep as 130 feet. Throughout most of the tropical West Atlantic.

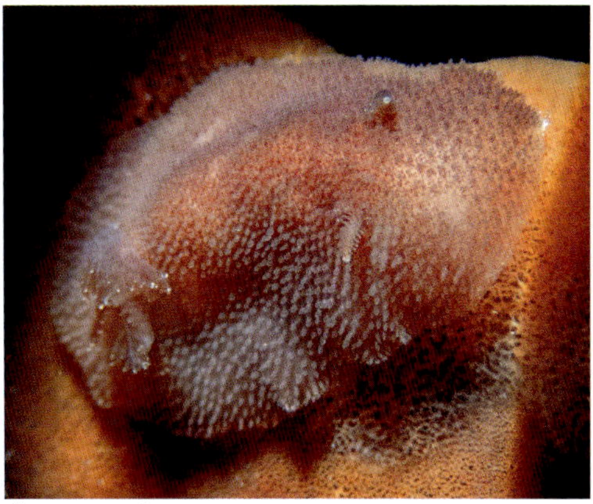

Grand Cayman

***Jorunna* sp. 1**

Size: ~12 mm Discodorididae

ID: This animal's appearance is consistent with some of *J. spazzola's* described features but not some others. **Consistencies**: ground color, densely covered with caryophillidia, long rhinophores with ~10 oblique brown-spotted lamellae, white-tipped rhinophores and gill leaves. **Inconsistencies**: no line of white glands along the mantle's edge, no white spotting along the rhinophoral and gill pit borders, no darker spotting and blotching, caryophillidia are exceptionally long, and the gill leaves are not "very short" as noted by Valdés et al in *Caribbean Sea Slugs* (Valdés et al, 2006).

Grand Cayman

***Jorunna* sp. 2**

Size: 8-10 mm Discodorididae

ID: Body oval; mantle broad, covered with rounded tubercles that are small mid-dorsally and at the margins and largest in between; rhinophores club-like, 8 oblique lamellae, apical mucron small and cylindrical; sheaths low, rims spiculose; gill relatively small. Ground color grayish-white. Large tubercles on each side mostly opaque white with a few sparse clusters of maroon ones aligning approximately in a longitudinal row. Rhinophores and gill ground color; rhinophores with a few distal maroon spots and a white cylindrical apical mucron. Found on a shallow coral reef at ~10 feet.

***Jorunna* sp. 3**

Size: ~40 mm Discodorididae

ID: Body oval; the mantle is broad, its edge irregular and undulate, uniformly covered with small caryophyllidia having radial spicules; the mantle edge seems to be lined with tiny glands; the rhinophores have a thick peduncle, a club with ~18 lamellae, a cylindrical apical mucron, and slightly raised sheaths; the gill has small leaves. The mantle is whitish with tiny brown dots, a pinkish mid-dorsal region, and a white spotted edge. Rhinophores pale brown with dark brown dots, lamellae with dark brown edges, and a white mucron. Gill leaves are white-tipped. This is a reasonably good match for *Jorunna spazzola* except for being **more than twice as long as any examined specimen of that species** and for having ~18 instead of 8-10 rhinophoral lamellae. Specimens of *J. davidbowieii* have been recorded only up to 25 mm long, violet-pink, with up to 20 rhinophoral lamellae, but the rhinophores are yellowish-green. Another possibility is *Discodoris mortenseni*, about which Alvim & Pimenta (2013) state: "belongs to the genus *Jorunna* according to the morphology of caryophyllidia, radula and the reproductive system" and is "distinct from *J. spazzola*". It has been described as up to 28 mm long, pinkish with light-brown spots and white specks, rhinophores dark with 20 lamellae, rhinophores and gill leaves white-tipped. An examination of its internal anatomy is needed for a positive identification.

Roatan

Grand Cayman

***Jorunna* sp. 4**
Size: < 15 mm Discodorididae
ID: Another *J. spazzola* look-alike but its rhinophores and gill leaves are prominently yellow.

Martinique

Paradoris mulciber **(Ev. Marcus, 1971)**
Size: to 60 mm Discodorididae
ID: Body elongately oval, widest near middle, convex (height about one-third length); mantle highly spiculose, margin broad and undulate; dorsum covered with scattered well-spaced rounded tubercles, largest centrally and decreasing in size toward margins; rhinophores thick basally, apical club with 15–21 very thin lamellae, tip with small cylindrical knob; rhinophoral sheaths slightly elevated, sometimes with 1-2 tubercles on each side, rims smooth but undulate; gill posterior with 6 tripinnate leaves, sheath elevated and rim undulate; oral tentacles short, flat, narrowly triangular, grooved near outer edge; foot narrow, front centrally notched and transversely grooved, rear blunt and not projecting. Ground color variable – pale cream, gray, light brown, brownish-violet. Entire dorsum densely covered with minute dark brown dots. Tubercles with small black spots and white specks, apices of larger ones somewhat yellowish. Rhinophores translucent ground color basally, lamellae pale brown with some small dark brown spots and white edging, apical knob white. Gill leaves ground color with random white spots, tips may be cream to yellowish. Ventrum white. Found under rocks and on hard substrates. A sponge feeder. Depth: 6-165 feet. Gulf of Mexico, Jamaica, St. Martin, Guadeloupe, Martinique, St. Lucia, Panama, Costa Rica, Brazil.

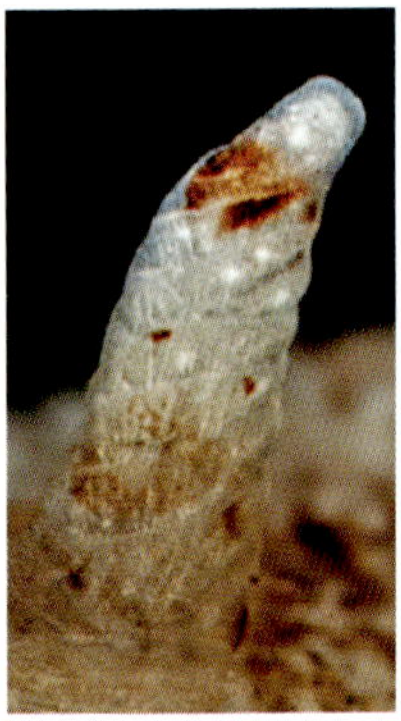

Cuba

Paradoris annularis
Ortea, Espinosa & Moro, 2017
Size: to 10 mm Discodorididae
ID: The body is oval. Mantle's surface and edges covered with variably sized, small, conical, non-spiculate tubercles, no isolated large tubercles. The rhinophores are robust with 8 very low and well-spaced lamellae and a wide apical mucron; there are subdermal elongate vertical spicules; sheaths are low with a smooth rim. The gill is well back on the dorsum; its 6 bipinnate leaves can fold back to reach the rear edge of the mantle, the rearmost two leaves being by far the shortest. Ground color grayish cream. Mantle brown-stippled medianly from rhinophores to gill, small chestnut rings usually encircling tubercles elsewhere. Rhinophoral peduncles cream colored, lamellae spotted with brown and white (especially the upper ones), mucron white. Gill leaves cream colored, tips and outer faces of stalks brown-spotted. Two specimens were found on the wall of a cave at 60 feet off southwest Cuba.

Bocas del Toro, Panama

Paradoris adamsae **Padula & Á. Valdés, 2012**
Size: ~50 mm Discodorididae
ID: The body is oval and elevated; its dorsum is covered with variously sized small holes (having diameters up to 0.1 mm) and many large conical irregular tubercles, the largest and tallest ones often being on the dorsal hump and the small ones along the mantle's margin. The anterior of the foot is grooved and bilabiate. The rhinophores are short, with ~16 lamellae. The gill is posterior, with 6 tripinnate leaves. The ground color is brown. The mantle has a few dark dots and many white to cream small spots; the rhinophores are pale brown and dark-spotted; the gill leaves are grayish-brown, dark-spotted, with orange-cream apices. The animal's underside is pale cream with some brown dots. It was found on a hard substrate. Depth: ~20 feet.

Riviera Beach, Florida

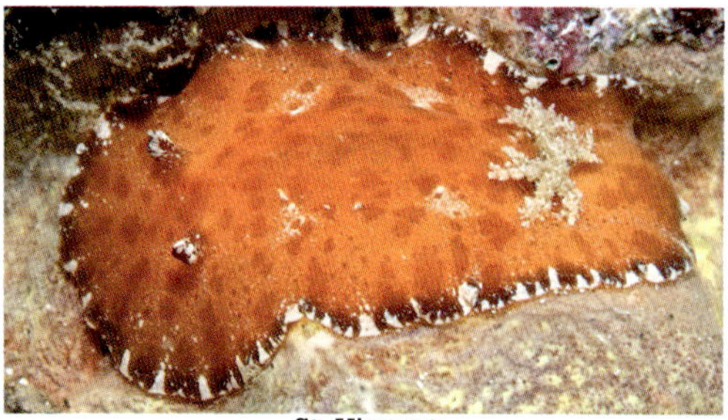
St. Vincent

Martinique

St. Vincent

Dominica

mating; St. Vincent

spawn; St. Vincent

Platydoris angustipes **(Mörch, 1863)** **leather-backed platydoris**

Size: to 150 mm Discodorididae

ID: Body elliptical, **slightly flattened**; mantle stiff, leathery, densely tuberculate; **tubercles small**, rounded, variably wide but uniformly high, with apical tuft of sensory cilia and 5-7 protruding spicules surrounding the base. Rhinophores long, stout, club-like, retractable; up to 25 oblique lamellae; cylindrical tip; sheaths tuberculate, rims jagged. Gill posterior, 6 tripinnate retractable leaves, anal papilla between the two rearmost leaves. Ground color **orange to red to reddish-brown**. Mantle with white specks, often clustered into spots and blotches, sometimes 3-4 prominent white patches on each side; usually mottled with many variably sized indistinct blotches of **darkened ground color**. Some individuals have two white blotches on each side of the midback, **forming the corners of a square**. Mantle margin often frosted with irregular white spots, sometimes a thin white edge bordered by a broad dark brown to purplish submarginal band with **inward-pointing white triangles**. Rhinophoral lamellae dark brown, dusted with white. Gill leaves white to light brown. Inhabits rocky areas of living reefs (including tide pools), hiding under rocks and feeding on sponges. Pairs sometimes engage in head-to-tail trailing behavior where one individual maintains contact with another while following it. Mating occurs with right sides overlapped – the genital opening is under the mantle near the right front. Spawn is a red to orange slightly wavy ribbon coiled into a spiral of ~7-8 whorls, deposited on a hard substrate. Depth: 0-240 feet. Throughout most of the tropical West Atlantic. The Brazilian population of what had long been considered *P. angustipes* was recently determined to be a unique species – *Platydoris guarani* P. Lima & Simone, 2018 – differing primarily through internal anatomical features.

Riviera Beach, Florida

Rostanga byga Er. Marcus, 1958 **fuzzy rostanga**
Size: to 20 mm Discodorididae
ID: Body elongately oval, slightly elevated; foot narrow, tapers to a short triangular tail; mantle densely covered with rounded sensory tubercles (caryophyllidia) with 3-7 protruding spicules basally and a tuft of cilia on top; rhinophores low, wide, 9-16 almost vertical lamellae, narrow apical mucron; gill posterior, up to 10 short unipinnate leaves around a high anal papilla; rhinophoral and gill sheaths covered by tubercles. Animals **orange to orange-red**, either uniformly or with scattered white and red dots, the white dots in larger individuals coalescing into a **line of white patches along the dorsal midline and mantle edges**. Rhinophores with pale orange lamellae and **white "rosette" at tip**. Gill leaves ground color, sometimes white-tipped. In rocky or rubble areas, often under stones in the intertidal zone. Feeds on sponges, its color tending to become that of its prey. e.g. orange animals on the orange sponge ***Mycale microsigmatosa***. Egg mass is a string with 2-3 rows of large red eggs. Depth: 0-213 feet. Bermuda, Northwest Florida, Bahamas, Mexico, Guadeloupe, St. Lucia, Brazil, Argentina. Rare.

Bermuda

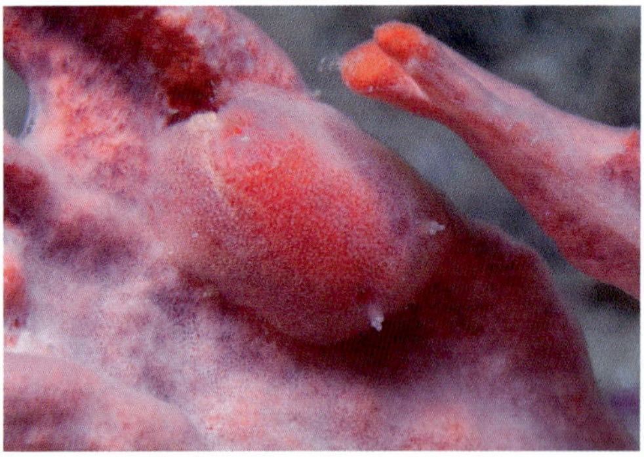

all photos from Destin, Florida

Rostanga cf. *byga* Er. Marcus, 1958
Size: to 20 mm Discodorididae
ID: This is probably a *Rostanga byga* that acquired a pink background color by feeding on the pink sponge apparent in the upper left photo. It may also be a juvenile. However, these are assumptions that need to be confirmed by specimen examinations. Found at a depth of 70 feet off Northwest Florida.

Sanibel Island, Florida

Roatan

Riviera Beach, Florida

Colombia

Roatan

egg ribbon; St. Vincent

Sclerodoris prea **(Ev. Marcus & Er. Marcus, 1967)**

Size: to 40 mm

Discodorididae

ID: Body oval, flat, width about two-thirds length; foot broad. Mantle broad, rigid, edge crenulate, densely covered with variably sized spiculated rounded tubercles, size decreasing toward mantle edge, large tubercles surrounded by irregular clusters of smaller ones; **8-11 of the largest tubercles** and a few smaller ones form two irregular longitudinal rows from the rhinophores to the gill, separated by a strip of small tubercles to create what appears to be a **median dorsal trough between two ragged-looking ridges**; each ridge often with a pair of large tubercles laterally flanking the rhinophoral pit. Rhinophores very long, retractable, up to 30 lamellae; gill dorsal, posterior, 6-7 multipinnate retractable leaves, with a low sheath edged having a few large tubercles. Ground color cream-brown to sandy-brown. Mantle with many dark brown to black spots and specks, sometimes with black patches along the outer flanks of the ridges and in the dorsal trough. Many tubercles with brownish-orange tips, large ridge tubercles sometimes pale with darker tips, adjacent ones usually connected by thread-like white lines. Rhinophores usually cream, lamellae dark-edged. Gill leaves gray to brown with white and black speckles. Usually found on sponge-encrusted hard substrates, often under rocks. Probably feeds on sponges. Very slow-moving, camouflages well by coloration and ability to change shape to match the contours of a rock's surface; adheres very tightly. Spawn is a ruffled white ribbon of eggs deposited as a spiral coil of ~2.5 whorls on a hard substrate (e.g., under a rock). Depth: 10-50 feet. Widespread but rarely seen.

Roatan

Grand Cayman

Guadeloupe

Colombia

Boca del Drago, Panama

juvenile; Roatan

juvenile; Grand Cayman

Sclerodoris worki **(Ev. Marcus & Er. Marcus, 1967)** **Work's dorid**

Size: to 50 mm Discodorididae

ID: Body oval, flattened, width about half length; mantle broad, rough, densely covered with variously sized spiculated tubercles; the tubercles are largest mid-dorsally, smallest near the margin; larger tubercles usually with tightly packed small spheres on sides and a single relatively large one apically, sometimes a large sphere atop a peduncle with 4-8 small round spheres encircling the base; large tubercles in juveniles are of the latter type, sparsely distributed, with their small basal spheres attached to the peduncle by short radial "spokes". Rhinophores long, retractable, up to 25 lamellae, sheath somewhat elevated and covered with medium-sized tubercles. Gill dorsal, posterior, 6 multipinnate retractable leaves; sheath and outer edges of stalks with medium-sized tubercles. Mantle dark brown to cream to yellowish-gray, usually with scattered blackish blotches and a noticeable dusting of brown specks on paler individuals; may be lighter/darker medianly. Large tubercles gray to cream to yellowish, or with a pale orange apex and white basal spheres; smallest tubercles yellowish/whitish. Rhinophores white-tipped; lamellae dark brown to cream, usually white-dotted and with white edges. Gill leaves gray/cream/brown with white dots. Usually on seagrass beds, sand flats, rocky bottoms, under stones. Probably feeds on sponges. Depth: to 40 feet. Georgia to most of the tropical West Atlantic. Rarely seen. Externally, differs from the similar *Sclerodoris prea* primarily by lacking its distinctive mid-dorsal "trough" between two rows of tall tubercles. However, variability in tubercle distributions in both of these species can make it very difficult to distinguish between them.

St. Vincent

***Sclerodoris* sp.**

Size: ~20 mm Discodorididae

ID: Body flat, oval; mantle broad with clusters of variably sized bubble-like tubercles – a narrow row of small clusters along the dorsal midline flanked by two large clusters at midbody, scattered smaller clusters toward the mantle margin; rhinophores club-like, lamellate, with a small cylindrical apex, the base encircled by high tubercles; gill posterior with a large cluster of tubercles at the base. Ground color pale orange. Mantle densely covered with tiny reddish-brown dots that concentrate to form diffuse dark spots, mostly near the mantle margin. Tubercles whitish to yellowish; rhinophores brown with a white apex; gill whitish. Found under a rock. Depth: ~10 feet.

eastern Costa Rica

***Taringa robledales* Ortea, Moro & Espinosa, 2015**

Size: to 22 mm Discodorididae

ID: Body elongate oval; mantle crowded with uniform small tubercles, each with a bundle of slightly-protruding spicules; the head has some large conical tentacles; the foot is narrower than the mantle but longer, its anterior transversely furrowed, its rear protruding when the animal is crawling. The rhinophores are quadrangular in cross-section, have ~16 lamellae, a vertical anterior groove, an apical mucron and a sheath with 3 rings of tubercles around the rim: 2 outer with small tubercles and one inner (at the rim) with ~8-10 large tubercles; the latter have an anvil-like top with a flat inward expansion and a conically pointed outward projection. Gill posterior, 6 outwardly inclined bipinnate leaves, gill sheath configured with tubercles just like the rhinophores. Ground color varies from almost uniformly reddish-purple to orange with reddish spots. Some minute white dots on the tubercles, white granules along the mantle's edge. Head tentacles white. Rhinophores white with an oblique reddish band that spots the sides of the lamellae; sheath's inner tubercles white-spotted. Gill leaves with bright white ribs, many white spots on a clear background and a few purple spots. Specimens were found on purple sponges. Spawn is a ribbon with purplish-pink eggs in spherical capsules, arranged in four layers and coiled as a spiral of up to three whorls. Depth: 60-100 feet. Rare.

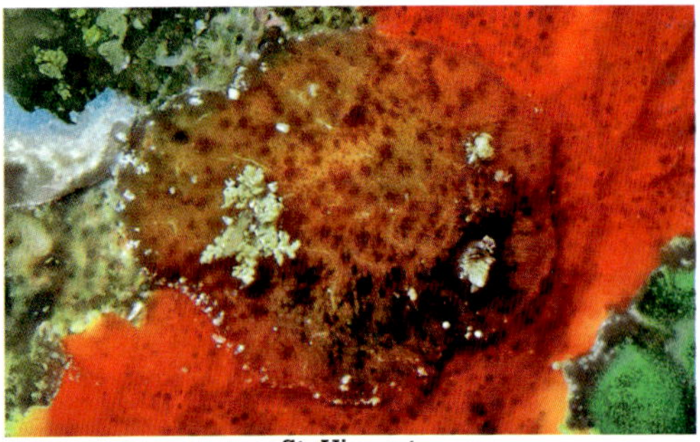

St. Vincent

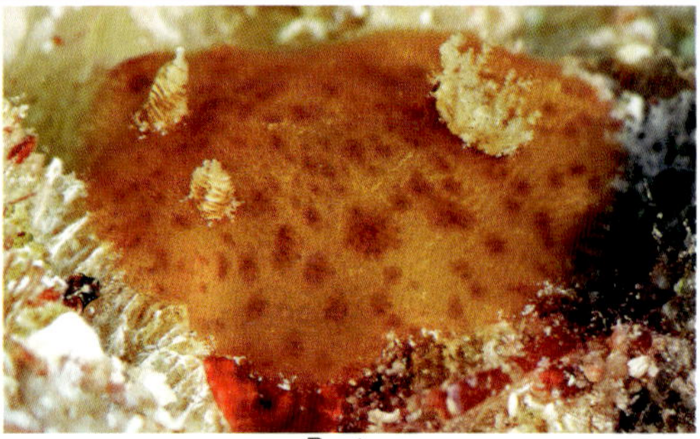

Roatan

Dominica

St. Vincent

Boca del Drago, Panama

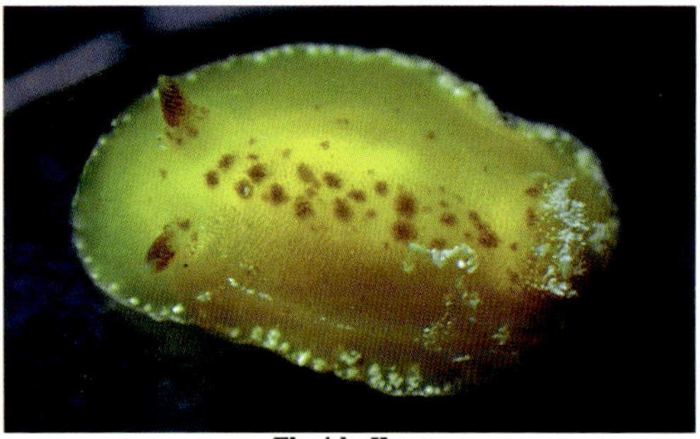

Florida Keys

Roatan

Taringa telopia **Er. Marcus, 1955** **caramel dorid, orangeball taringa**

Size: to 50 mm Discodorididae

ID: Body oval, width about half the length, moderately convex; mantle stiff, densely covered with knob-like spiky tubercles (caryophyllidia), largest at midside, smallest most crowded; rhinophores elongate, vertically grooved in front, spiculate, tip cylindrical, 10-15 oblique lamellae; sheath low, ~8 lobules around its rim; gill dorsal, posterior, 6 tripinnate retractile leaves. Ground color variable: dusky yellow, orange to dark brown, gray, blackish. Mantle usually with variably sized brown/blackish spots/blotches, white/beige flecks, and granules that sometimes coalesce into white spots medianly, along margin, and on rhinophoral/gill sheaths. Lobules on rim of rhinophoral sheath white. Yellowish individuals may be dusted with black specks. Rhinophores and gill leaves yellowish, grayish to brownish, white-tipped; rhinophoral lamellae may be brown with white blotches. Found in shallows under rocks, on rubble bottoms, and coral reefs, usually on hard substrates. Sometimes found in pairs of different colors and sizes. Probably feeds on sponges. Spawn is a spiral ribbon with a crenulated border, 2-3 whorls, ~3 mm high, containing many rows of yellow/orange, white, or rosy egg capsules. Depth: 5-40 feet. Throughout most of the tropical West Atlantic.

Riviera Beach, Florida

West Florida

Bonaire

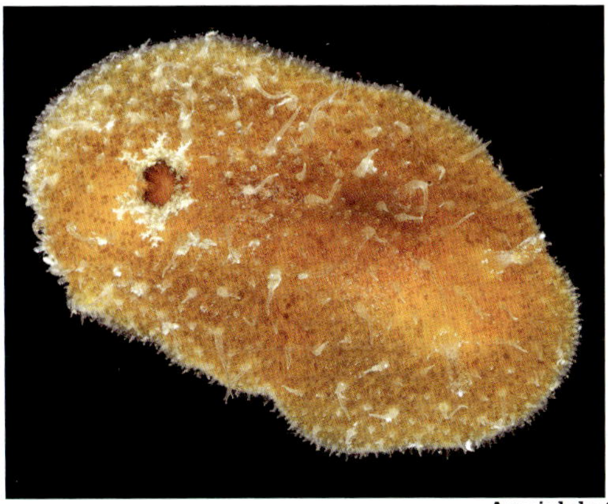

Arraial do Cabo, Brazil

Thordisa diuda **Er. Marcus, 1955**

Size: to 36 mm Discodorididae

ID: Body oval, width one-third to two-thirds length, slightly depressed. Mantle spiculose; covered with equidistant conical to rounded retractile tubercles that are taller than broad; spicules protrude from their sides, largest ones median, otherwise a mix of sizes, most conspicuous ones thin and soft. Rhinophores with 12-13 oblique lamellae and a cylindrical apical mucron; sheath low, tuberculate, sometimes flanked by 2 long tubercles and many small white knobs. Gill posterior; 6 retractile unipinnate to tripinnate leaves encircle the anal cone; sheath low with small tubercles. Foot narrower than mantle, short triangular tail projects when animal glides forward. Ground color yellow to orange to dark orange-brown. Back may have diffuse dark spots and blotches. Tubercles usually paler than mantle, often seemingly encrusted with white debris. Rhinophores and gill leaves dark brown below, shading to white at tips (color gradations less pronounced in juveniles). Found under rocks on sandy bottoms, under coral debris, and more generally on hard substrates. Probably feeds on sponges. Egg ribbon is a slightly wavy counterclockwise coil of 2-4 whorls standing on edge, containing many rows of orange eggs. Depth: 10-65 feet. Florida, Cuba, Bahamas, Guadeloupe, Martinique, St. Vincent, Bonaire, Jamaica, Arraial do Cabo (Brazil). Rare. The above images show variation in external appearance: tubercles very long and thin (Brazil), tubercles low and rounded (Caribbean, retracted?), very much or very little white "frosting" on the mantle.

all photos from Roatan

foot

rhinophore

egg ribbon

its food, *Plakortis halichondrioides*

Thordisa cf. _harrisi_ Chan & Gosliner, 2006

Size: ~20-30 mm

Discodorididae

ID: Body oval, profile low; mantle with long somewhat fuzzy papillae and low rounded tubercles, longest papillae mid-dorsal, some bifurcated basally; rhinophores with ~18 lamellae and small cylindrical apical mucron; gill posterior, sheath low and even, 6 tripinnate leaves also low; foot short, narrow, bilobed at front. Ground color bright orange. Mantle with some white frosting, especially concentrated between rhinophores and gill, and a scattering of small **dark reddish spots**; papillae and tubercles translucent. Rhinophores brownish with some white spotting, mucron white. Gill sheath ground color, leaves whitish with some brown on the stalks. Foot orange, surrounded by a **ring of red dots on the ventral side of the mantle**. Found under coral rubble. Its egg ribbon is a wavy counterclockwise coil (~3 whorls), standing on edge on a hard substrate, containing many rows of orange eggs. Depth: 30-70 feet. Three animals found off Roatan feeding on the sponge *Plakortis halicondrioides*. The pictured individuals may be *Thordisa harrisi*, a species known only from a single specimen collected on the Caribbean side of the Isthmus of Panama in 2006. The only significant points of difference are: (i) the specimen had two elongate median brown blotches between the rhinophores and gill (ii) the specimen's gill was described as opaque brown. It could also be a variant of *Thordisa diuda*.

Thordisa lurca
(Ev. Marcus & Er. Marcus, 1967)
Size: to 36 mm Discodorididae
ID: The body is broadly oval and convex. The foot is narrower than the body, its front transversely grooved. The mantle is spiculose and covered with large and small conical tubercles (intermixed); the tubercles are usually as high as wide and packed with upright spicules, some having considerably projecting mucous filaments. Rhinophores have about 20 oblique lamellae, and sheaths with jagged rims. Gill posterior, 4 tripinnate leaves, sheath rim jagged. The anal papilla is high & blunt. **Uniformly lemon yellow**. Tubercle filaments paler yellow; rhinophores pale brown with blackish tips; gill leaves yellow, tips paler, may have dark edging. On rocky substrates; feeds on sponges. Depth: 10-130 feet. Florida, Cuba, Guadeloupe, Martinique, Colombia, Costa Rica, Mexico, Brazil. Rare.

Cape Salomon, Martinique

Destin, Florida

Thordisa **sp.1**
Size: 10 mm Discodorididae
ID: Body oval; tuberculate, large ones encrusted with white debris. Rhinophores with ~10 oblique lamellae and cylindrical apical mucron; gill with 6 bushy leaves. Mantle orange; rhinophores brown, mucron white; gill leaves white. Found among encrusting orange sponges at a depth of 85 feet.

Baia das Tartarugas, Brazil

Thordisa **sp. 2**
Size: ~20 mm Discodorididae
ID: Elongate oval; many conical papillae, some quite large, smallest along mantle margin. Rhinophores with ~10 oblique lamellae; gill with several small leaves. Mantle orange; papillae transparent, white-tipped; rhinophoral lamellae reddish-brown with white spots and white tip; gill leaves mottled grayish-brown with white tips. Under a rock at 3 feet. Brazil.

Southeast Florida

Thordisa **sp. 3**
Size: ~10 mm Discodorididae
ID: External morphology like that of *Thordisa diuda*, but the coloration is unusual – pale orange mantle, tubercles very encrusted with white, large median black blotch, black rhinophores with white tips.

The following several pages contain numerous examples of discodorids with indeterminate identifications. Some may be color variants of previously described species, some may be new species, some may be variants of one another. Because of significant overlaps in the appearances of known species the uncertainty in the reliable identification of any of them is quite large. Further study is needed for this group.

Pompano Beach, Florida

***Discodorididae* sp. 1**

Size: ~10-15 mm Discodorididae

ID: Translucent white. Viscera visible as a long broad yellowish white dorsal patch; mantle with purplish spots, edges white-dotted; rhinophores pale brown, tips white; gill leaves purplish. On sand at ~10 feet.

St. Vincent

***Discodorididae* sp. 2**

Size: ~10-15 mm Discodorididae

ID: Mantle covered with small caryophyllidia; rhinophores stout, lamellate; gill posterior. Translucent pinkish-white, indistinct dusky spots flank median hump. Rhinophores yellowish. On sand at ~20 feet.

Roatan

***Discodorididae* sp. 3**

Size: ~30 mm Discodorididae

ID: Ground color pale purple, mantle with many variably sized diffuse purplish-brown spots and blotches. Rhinophores brown with white tips; gill brown. Crawling on rubble at ~10 feet.

St. Vincent

***Discodorididae* sp. 4**

Size: to 50 mm Discodorididae

ID: Ground color translucent gray; caryophyllidia white, scattered diffuse brown patches; rhinophores brown, sheaths ground color; gill leaves brownish, white-tipped. On hard substrates/rubble to ~10 feet. St. Vincent, Panama, Roatan. Could be a pale variant of *Tayuva lilacina* but anatomical and/or molecular analyses are required to confirm this.

St. Vincent

Discodorididae sp. 5

Size: ~15 mm Discodorididae

ID: Ground color translucent gray;. covered with irregular white spots, dots, and streaks, more coalesced on mantle sides; a few large dusky spots near edges of median hump; **rhinophores brown** with white spots; gill leaves grayish, white-tipped. Under rocks, to ~20 feet

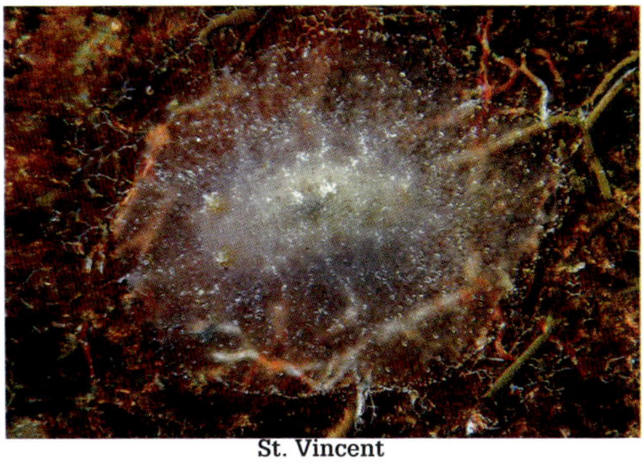

St. Vincent

Discodorididae sp. 6

Size: 10-15 mm Discodorididae

ID: Elongately oval, sides flat, middle humped, mantle covered with small caryophyllidia; the rhinophoral tips pointed; gill posterior. Transparent. Mantle covered with tiny white scrawls, more coalesced in hump region; **rhinophores yellowish**, tips white. Under a rock at 10 feet. Similar to Discodorididae sp. 5.

St. Vincent

Discodorididae sp. 7

Size: 10-15 mm Discodorididae

ID: Broadly oval, sides flat, middle humped, crowded with small caryophyllidia. Pale brown with many variably sized dark brown spots and blotches and opaque white dots and scrawls all over the dorsum. Found under a rock. Depth: ~15 feet.

St. Vincent

St. Vincent

Discodorididae sp. 8

Size: to 30 mm Discodorididae

ID: Broadly oval, sides flat, edges ruffled, medianly humped, mantle covered with small caryophyllidia. Rhinophores with lamellate clubs & cylindrical apical mucron. Ground color translucent grayish white. Mantle with many variably sized pale brown to reddish brown to gray diffuse blotches, smallest near margins. Many blotches darker centrally and/or sprinkled with bright white dots, suggestive of small tubercles. Rhinophores brown, mucrons and sheath rims white. Crawling on sand near sponges to ~15 feet.

St. Vincent

Discodorididae sp. 9

Size: ~30 mm Discodorididae

ID: Oval, sides flat, middle humped, mantle crowded with small caryophyllidia; rhinophores with lamellate clubs and a cylindrical apical mucron, sheaths slightly raised. Pale brown near hump, grayish elsewhere, many variably sized darker brown blotches, largest flanking hump. Rhinophores and sheaths ground color, mucrons white. Under a rock at ~15 feet.

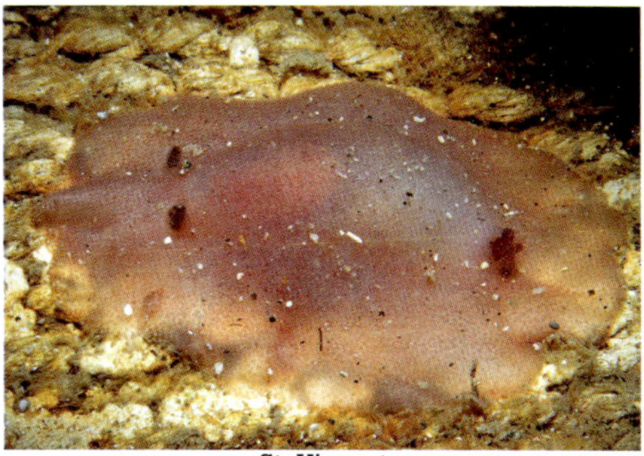

St. Vincent

Discodorididae sp. 10

Size: ~15 mm Discodorididae

ID: Oval, a humped middle, mantle densely covered with small tubercles and papillae; lamellate rhinophores with a cylindrical apical mucron; gill posterior, ~5 leaves. Pinkish, rhinophores & gill leaves dark brown, white tipped. On sand or under stones in grassy areas to ~15 feet. St. Vincent, Roatan. Might be *Discodoris mortenseni* Ev. Marcus & Er. Marcus, 1963.

Riviera Beach, Florida

Discodorididae sp. 11

Size: ~20 mm Discodorididae

ID: Body elongate oval with median hump; dorsum densely covered with low rounded tubercles; rhinophores well separated, stout, sheaths low and tuberculate, ~10 thin oblique lamellae between sheath rim and tip, elongate apical mucron; gill very bushy. Mantle's ground color is yellow, the tubercles orange, rhinophores dark brown, lamellae with white edging, mucron white, sheath rim tubercles white; mantle margin with a row of diffuse reddish-brown transverse streaks. Found under a rock at ~15 feet.

Roatan

rhinophores

foot

Discodorididae sp. 12

Size: ~25 mm Discodorididae

ID: Body elongately oval, flat; mantle broad, margin thick; dorsum covered with rounded tubercles; very small tubercles crowded, large ones sparsely distributed and many with radiating spicules; rhinophores retractable, stalk thick, club with ~12 lamellae; gill posterior, several bushy leaves in a transverse row, sheath elevated. Mantle orange, large tubercles brown with white spicules, many scattered white dots; rhinophoral stalk whitish with brown specks, lamellae brown-edged, tip white; gill sheath and leaves white. Found in coral rubble. Depth: ~15 feet.

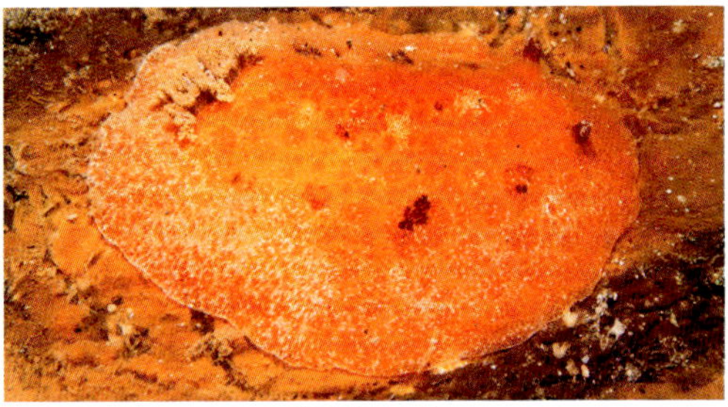

St. Vincent

Discodorididae sp. 13

Size: ~20 mm Discodorididae

ID: Body oval, flat; mantle broad, margin thin; covered with small rounded tubercles; rhinophores retractable, club-like, stalk thick, club with >20 lamellae; gill posterior, wide, 6-7 bushy leaves. Ground color yellowish. Many variably sized diffuse clusters of pale reddish tubercles, a few red and maroon blotches encircle the mid-dorsal region, the outer regions of the mantle are densely covered with white striations. Rhinophores maroon with white tips. Gill stalks maroon, tops yellowish with white spots. Found under a rock. Depth: ~15 feet.

Grand Cayman

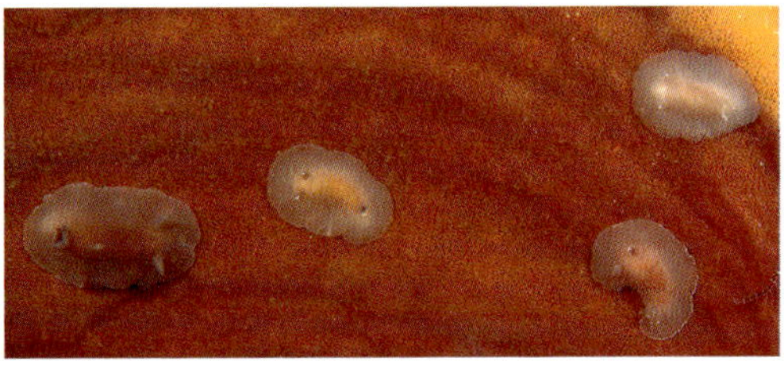

Discodorididae sp. 14

Size: 10-15 mm Discodorididae

ID: Body elongately oval; dorsum slightly elevated medianly; entirely covered with a uniform dense distribution of very small rounded tubercles; rhinophores club-like, stalk short, club with ~12 oblique lamellae, the apex with a short cylindrical knob; the gill is posterior, a circlet of several tall multipinnate leaves. Ground color translucent gray, tubercles whitish-gray, the color of the substrate tends to show through the mantle margins. Viscera, when visible, appear yellow to orange. Rhinophores and gill leaves gray with white tips. Found crawling on hard sponges. Depth: ~15 feet.

Roatan

Guadeloupe↑ Grand Cayman ↓

Discodorididae sp. 15

Size: ~30 mm Discodorididae

ID: Body oval; dorsum slightly elevated medianly; covered with a uniform dense distribution of tiny rounded tubercles; rhinophores club-like, stalk short & thick, club with ~20 oblique lamellae, apical mucron short and cylindrical, sheath low and smooth; gill a circlet of 6-7 highly-branched retractile leaves. Ground color grayish-white. Dorsum with many irregular scattered pale brown spots and patches, many of them slightly darkened centrally but a few with very dark brown centers. Rhinophores and gill leaves pale brown with whitish tips. Found under coral rubble. Depth: to ~20 feet. Guadeloupe, Bonaire, Roatan, Grand Cayman.

St. Vincent

Discodorididae sp. 16

Size: ~20 mm Discodorididae

ID: The external morphology and color pattern are very similar to that of Discodorididae sp. 15, but the gill leaves (which are mostly withdrawn in the photo) are apparently yellow instead of pale brown. It was found under a rock at a depth of ~15 feet.

Riviera Beach, Florida

Discodorididae sp. 17

Size: ~15 mm Discodorididae

ID: Body oval, sides flat, middle humped, mantle covered with small conical tubercles; rhinophores are lamellate with a small cylindrical mucron. Grayish; tubercles brown and white-tipped, their bases ringed with white. Rhinophores yellowish-brown, tips white. Gill leaves white. On and under rocks at ~15 feet.

Roatan

St. Vincent

Discodorididae sp. 18

Size: ~10 mm Discodorididae

ID: Body oval, flat; mantle very broad, appears to be densely covered by small caryophyllidia interconnected by spicules; rhinophores stout, conical, lamellate; gill posterior. Visceral mass white and brownish-orange. Row of 3-5 similar brown spots between each rhinophore and gill. Mantle translucent, caryophyllidia and spicules white. Rhinophores and gill leaves brownish, tips white. On rubble, under rocks at a depth of ~10 feet.

Jupiter, Florida

Discodorididae sp. 19

Size: 10-15 mm Discodorididae

ID: Mantle broad, margins undulate, dorsum is crowded with small rounded tubercles; rhinophoral clubs with ~13 lamellae; gill posterior, very bushy. Pale orange with dense distribution of brown micro-dots. Except for a median strip the dorsal tubercles are prominently white-tipped, giving the mantle a frosted appearance. Rhinophores are pale orange, their stalks and lamellae brown-spotted, tips white-spotted. Gill leaves whitish above. Found crawling on a hard substrate at ~10 feet.

St. Vincent

Discodorididae sp. 20

Size: 10-15 mm Discodorididae

ID: Body oval; mantle broad, margins undulate, densely covered with small rounded tubercles; rhinophores with ~13 lamellae. Ground color white. Mantle with many small brown blotches, darkest mid-dorsally and at front margin, edge white. Rhinophores dark brown, white-tipped. Under a rock at ~10 feet.

St. Vincent

Discodorididae sp. 21

Size: 10-15 mm Discodorididae

ID: Body oval; mantle broad, margins undulate, dorsum densely covered with small rounded tubercles. Translucent gray with net-like white pattern consisting of interconnected blotches. Found under a rock at ~10 feet.

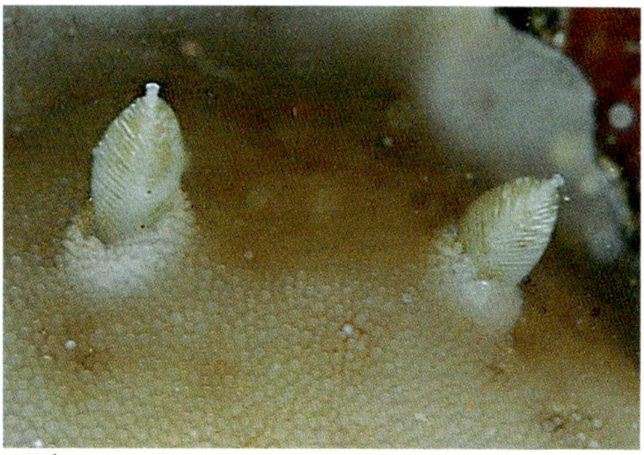

St. Vincent

Discodorididae sp. 22

Size: ~20 mm Discodorididae

ID: Body elongately oval; mantle grainy, margin thin with wavy edge; rhinophores stout, club-shaped, lamellate, tipped with a small cylindrical mucron; sheath slightly elevated, rim ragged; gill posterior. Ground color white, Mantle covered with many small irregular gray patches, some sprinkled with white dots and many with mostly central dark brown spots. Rhinophores and sheaths ground color. On hard substrates or in rubble on shallow coral reefs. Depth: to ~10 feet.

Grand Cayman

Discodorididae sp. 23

Size: ~10 mm Discodorididae

ID: Body oval; slightly elevated medianly; covered uniformly with tiny rounded tubercles; rhinophoral clubs with many oblique lamellae; gill a posterior circlet of 6-7 branched leaves. Reddish brown. Mantle with many white dots, some coalescing into spots medianly and along margins. Rhinophores brownish, tips white. Gill leaves yellow. Found on a sponge at ~15 feet.

St. Vincent

Discodorididae sp. 24

Size: ~15 mm Discodorididae

ID: Clusters of large conical tubercles along dorsal midline from in front of rhinophores to gill, smaller ones along mantle margins; rhinophores are well-spaced, lamellate, with a small cylindrical apical mucron. The mantle and the appendages are grayish-brown; the large tubercles around the rhinophores are slightly darker, Found under a rock at ~15 feet.

St. Vincent

Discodorididae sp. 25

Size: 40-60 mm Discodorididae

ID: Medianly elevated; covered with tiny rounded tubercles; rhinophores with ~15 lamellae; the gill is a circlet of ~5 branched leaves. Whitish medianly, a few clusters of orange tubercles; elsewhere it is white, gray, brown. Gill orange, rhinophores brownish. Under rocks to ~15 feet. Roatan, St. Vincent, Panama.

Guadeloupe

Discodorididae sp. 26

Size: ~20 mm Discodorididae

ID: It is similar to Discodorididae sp. 25 but it is smaller and its rhinophores are blackish with white tips.

Roatan

Discodorididae sp. 27

Size: ~40 mm Discodorididae

ID: Body oval, humped in middle; mantle covered with small tubercles; rhinophores lamellate, retractable, the gill has ~5 retractable leaves. Bluish-gray; mottled with indistinct, diffuse dark gray blotches. Rhinophores and gill leaves blackish with white dots. Found under rubble at ~10 feet.

St. Vincent

Discodorididae sp. 28

Size: ~10 mm Discodorididae

ID: Densely covered with tiny rounded tubercles. Rhinophores with ~10 oblique lamellae. Grayish-white; the internal organs are blackish; rhinophores pale brown, tips white; some random pale brown spots mid-dorsally. Under a rock at 15 feet. Probably a juvenile of one of the other St. Vincent Discodoridae sp.

St. Vincent

spawn

Discodorididae sp. 29

Size: ~20 mm Discodorididae

ID: Body oval; mantle densely covered with small tubercles, margin thin and wavy; rhinophores lamellate, retractable; sheaths snug, somewhat elevated, highly tuberculate; gill posterior, retractable. Mantle tan with close-set brownish tubercles that aggregate to form vague darker patches; mantle edges densely lined with white dots (glands?). Rhinophores have the mantle color, tips white, lamellae with pale edges. Found on rubble at ~10 feet. The egg mass is a ruffled ribbon deposited as a spiral coil of ~3 whorls, densely packed with small white eggs.

St. Vincent

Discodorididae sp. 30

Size: ~40 mm Discodorididae

ID: Oval, covered with small tubercles; rhinophores retractable, lamellate; gill with 6 bushy leaves, rachises thick. Dark orange; scattered roundish brown/blackish blotches. Rhinophores dark brown, tips white-spotted. Gill leaves pale orange with white and black rachises. Found under a rock at ~10 feet.

St. Vincent

Discodorididae sp. 31

Size: ~30 mm Discodorididae

ID: Oval, covered with small tubercles and scattered clusters of larger tubercles and papillae; rhinophores and gill leaves retractable. Mantle black, peppered with tiny white dots, margins with vague transverse white streaks, dorsal patches cream. Found under a rock at ~10 feet.

St. Vincent

Discodorididae sp. 32

Size: ~40 mm Discodorididae

ID: Oval, covered with small tubercles. Brown with an irregular network of brown-spotted white streaks and blotches, oriented somewhat radially on sides of mantle. Rhinophores blackish, white-dotted; gill leaves brown, white-dotted. Found under a rock at ~10 feet.

Colombia

Discodorididae sp. 33

Size: ~20 mm Discodorididae

ID: Oval, finely tuberculate; rhinophores lamellate, small apical mucron; gill with ~6 leaves. Mantle whitish, many irregular pinkish-brown rough white-spotted tubercles, smallest along margins; rhinophores dark brown, mucron white; gill leaves pinkish-brown; internal organs visible as a purplish-black mass.

Baía das Tartarugas, Brazil

Discodorididae sp. 34

Size: ~20 mm Discodorididae

ID: Oval; finely tuberculate; rhinophores lamellate; sheaths elevated, tuberculate; gill with ~10 unipinnate leaves, rachises robust. Mantle orange, reddish-brown stripe from gill to each rhinophore, pale brown blotches along sides, margin white-spotted. Rhinophores dark brown, gill leaves grayish-brown.

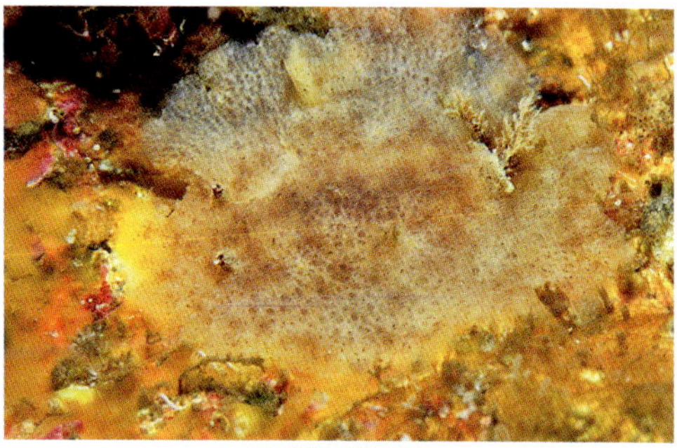

Destin, Florida

Discodorididae sp. 35

Size: ~30 mm Discodorididae

ID: Oval; finely tuberculate; rhinophores lamellate, sheaths elevated and tuberculate; gill leaves branched; sheath elevated, rough. Mantle yellowish; tubercles brownish (darkest medianly), dark brown dot at tip. Rhinophores dark brown, white tip. Gill yellowish-white. Maybe *Geitodoris pusae*. At 75 feet.

Family DORIDIDAE Rafinesque, 1815

A circumglobal family of cryptobranch dorids that is found at all latitudes, most diverse in cold to temperate regions but also well represented in the tropics. Over 90% of the family's species are in the genera *Doris*, *Doriopsis*, and *Aphelodoris*.

Body oval to elongate; mantle thick, stiffened by a network of spicules, margin broad; dorsum covered with tubercles that are conical or rounded but sometimes with protruding spicules; front edge of foot with a transverse groove but no median vertical notch; no true oral tentacles but the mouth is flanked by conical lateral extensions; rhinophores are lamellate, retractable into circular cavities with usually smooth (occasionally papillose) edges, some species with elevated sheaths; gill dorsal, posterior, leaves usually unipinnate, retractable into a cavity that also contains the anal papilla; a radula is present.

Usually found on hard substrates in various habitats, including rocky bottoms and reefs, intertidally and deeper. They feed on sponges on which they are often well camouflaged. Typically 20-70 mm long, but can reach 200 mm.

St. Vincent

Roatan

St. Vincent

Grand Cayman

with spawn; Guadeloupe

juvenile; Roatan

Aphelodoris antillensis **Bergh, 1879** **brown-speckled aphelodoris**

Size: to 30 mm Dorididae

ID: Body elongate, width about one-third length, height about half width; head small. Mantle soft, smooth, rim narrow and wavy, with a few low soft rounded tubercles except near the margin. No tubercles on the rhinophoral or gill sheaths. Rhinophores elongate, club-like, typically with 9-14 oblique lamellae but more have been reported, retractable, sheaths slightly elevated; gill dorsal, posterior, 5-6 bipinnate thick and fleshy leaves in a circle surrounding the anal papilla. Foot narrow (about half body width), concealed by mantle, front and rear rounded, rear sometimes protrudes when animal slides forward. Oral tentacles stout, blunt, rolled. Ground color translucent pale cream. Mantle peppered with brown/white/yellow dots, edge yellow, margin with radial streaks/blotches of coalesced dark dots (a dark margin may have white streaks). The rest of the back is variable: (i) two orange/red/brown/blackish blotches often containing blue to purple rings (ii) reddish-brown to blackish with two white blotches (iii) random distributions of variably sized blackish spots and blotches. Rhinophores and gill leaves are cream to yellow with brown and white spotting and whitish tips. Inhabits rocky reef areas where sponges are available to feed on. Finds hiding places in and under rocks or among algae. If disturbed it can swim using dorso-ventral undulations of its body. Spawn ribbon is a 4.5-turn spiral coil consisting of two long parallel rows of eggs. Depth: 1-25 feet. Bermuda, east Florida, Bahamas, most of Caribbean.

Doris juanformelli Ortea & Espinosa, 2017

morningstar doris

Size: to ~12 mm Dorididae

Curacao

ID: Body oval, length twice width; mantle rounded at front; rear of foot does not protrude when crawling. Mantle with a row of up to 14 large tubercles from each rhinophore to the gill; several transverse rows of smaller ones, smallest along mantle margin; the bases of the tubercles have radiating spokes of thick spicules that interconnect with those of neighbors to form a spiculose mantle; tubercles have apices with projecting spicules, giving them the appearance of a "morgenstern" – a medieval weapon with a spiky clubhead, but the largest ones are conical with sides reinforced by spicules that converge apically. Rhinophores are elongate with ~16 lamellae and an elongate apical mucron; the rhinophoral pit opening is flanked by two protective tubercles each with a rough and convex outer profile and an inner smooth and flattish one. Gill with ~7 bipinnate leaves encircled by a "crown" of 7 long vertical tubercles along the edge of the gill pit opening, their outer profile rough and convex, the inner one flat to convex. Mantle yellowish medianly and marginally, greenish-yellow in between. Mantle tubercles lime-green to bluish-green apically; foot pale yellow; rhinophores yellow, mucron tip white, protective tubercles green-spotted; gill leaves greenish-yellow. Associated with sponges like *Halichondria* spp. in relatively shallow water. Depth: to 10 feet. Cuba, Martinique, Curacao. Rare.

all photos from Bonaire

Doris acerico Ortea & Espinosa, 2017

pincushion doris

Size: to ~10 mm Dorididae

ID: Body elongate oval; mantle rounded at front; rear of foot protrudes when crawling; mantle with transverse rows of club-like variously sized tubercles, up to 12 tubercles per row; each tubercle with up to 8 radially projecting spicules at base that interconnect with neighboring tubercles to create a loose network of spicules within the mantle's surface, especially near the dorsal midline; larger tubercles with up to 30 internal spicules that break through their surface and create a thorny-looking club ("acerico" is Spanish for pincushion). Rhinophores are conical distally with ~10 lamellae and an elongate apical mucron; a vertical half-pipe tubercle on each side of the rhinophoral pit folds over the opening when the rhinophore is retracted. Gill posterior, 9 leaves, frontmost 7 with 9-10 pinnae, rear 2 smaller, front 1-3 vertical and rest inclined to rear; gill pit opening with 3 tubercles anteriorly – a central spiky club with a flattish one on each side that resembles those bordering the opening of the rhinophoral pit. Ground color is translucent yellow, the intensity increasing with the size of the animal. Dorsal tubercles are white to yellowish, their apices with dark pinkish spots and granules. Viscera appear pinkish; rhinophores yellow, mucron white; gill leaves yellowish pink with pink dots and spots, darker in larger animals. It has been found on hard substrates or under rocks, once in a *Thalassia testudinum* meadow. Depth: 3-45 feet. Cuba, Guadeloupe, Bonaire, Curacao, Costa Rica, Roatan. Rare.

Florida Keys

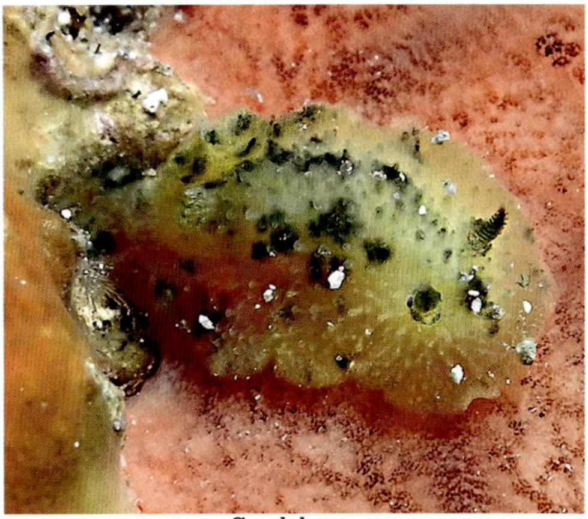

Guadeloupe

Bermuda

Bermuda

spawn; Florida Keys

***Doris bovena* Er. Marcus, 1955**

Size: to 20 mm Dorididae

ID: Body elongate oval, length twice width. Foot broad anteriorly, bilobed, sides almost straight, posterior pointed, sole narrower than mantle. Mantle covered with variably sized and randomly distributed low rounded warts. Rhinophores conical, 9-11 lamellae, sheaths low, encircled by 8-10 warts. Gill posterior, 8-12 unipinnate leaves, border of branchial cavity smooth but with a ring of subcutaneous glands. Ground color grayish yellow. Mantle with numerous irregularly distributed dark brown to purplish-black blotches of various sizes. Rhinophoral lamellae with brown shading. Gill leaves range from dark brown to mostly ground color. Has been found among mangrove roots and on rocky bottoms in shallow protected areas. Spawn is a gelatinous ribbon containing yellow eggs deposited on a hard substrate as a counterclockwise spiral of 2.5 turns. Depth: to 7 feet. Bermuda, Florida Keys, Cuba, Guadeloupe, Curacao, Aruba, Venezuela, Yucatan (Mexico), Brazil (Sao Paulo). Widely distributed but rarely seen.

Doris ilo **(Er. Marcus, 1955)**

Size: to 25 mm Dorididae

ID: Body elongate oval, its width about half its length, somewhat flattened; the mantle crowded with variously sized rounded tubercles containing thick bundles of linear internal spicules. Rhinophores spiculose, club-like, ~15 lamellae, the perimeter of the rhinophoral pit opening appears rough due to being lined with tubercles. Gill a circlet of **22 unipinnate leaves** where the frontmost two median ones are united at their bases and the rearmost six median ones loop inward toward the anal papilla at the center of the circlet; edge of gill cavity opening has very small tubercles but appears smooth. The mantle, rhinophores, and gill leaves are brownish-gray. It has been found under rocks and among sponges attached to the roots of mangroves in the intertidal zone. A rarely encountered species. Trinidad and Tobago, Brazil, Florida. Rare.

Key Largo, Florida

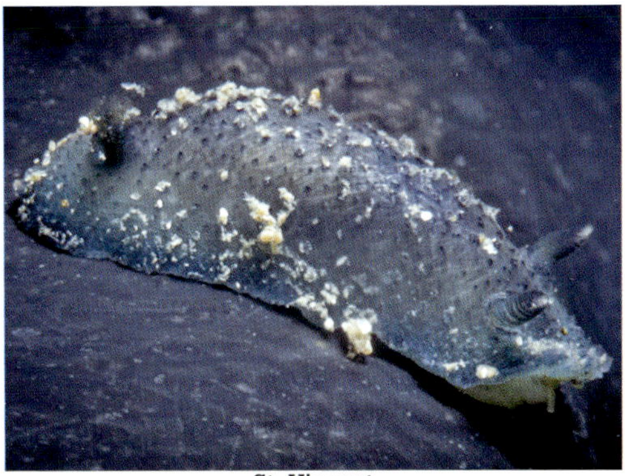

St. Vincent

Martinique

Panama

Roatan

Doris kyolis **(Ev. Marcus & Er. Marcus, 1967)**

Size: to 30 mm Dorididae

ID: Body elongate oval. Foot narrower than mantle, pointed rear may protrude slightly when the animal is moving. **Mantle covered with low, pointed, variously sized, spiculated tubercles** that progressively shorten (and disappear) toward the margin; small spicules project radially and may project above the mantle, large ones embedded into the mantle's top layer to stiffen it. Rhinophores club-like, up to 15 oblique lamellae, sheaths close-fitting, elevated, rim lobed and tuberculate. Gill dorsal, posterior, up to 11 unipinnate leaves sometimes flattened outwards onto the mantle with tips curled inwards (shaped like a double volute); edge of gill pocket slightly elevated, wavy, tuberculate; leaves partly encircle the anal papilla. Ground color **yellowish/greenish-gray to blue or black**. Rhinophores and gills usually ground color with white tips but can be much darker, as can the tubercles. The dorsal region above the visceral mass is usually darker. Inhabits shallow areas of mixed sand, rocks, mud, mangrove roots, detritus, seagrass meadows, often on sponges or under rocks and rubble. Feeds cryptically on sponges such as *Haliclona* sp., *Plakina* sp., *Dysidea etheria*, *Lissodendoryx isodictyalis*, and (off Mexico) *Halichondria melanadocia*; its color and texture tending to match that of its prey. Spawn is a yellow egg mass consisting of a flat ribbon in the form of a spiral coil with 2.5-3.5 turns, deposited on a hard substrate. Depth: to 25 feet. Southeast Florida, Cuba, Bahamas, Lesser Antilles, Brazil, Venezuela to Bay of Campeche. Uncommonly seen.

Brazil

Doris januarii **(Bergh, 1878)**

Size: to 40 mm

Dorididae

ID: Elongate oval, width about half length, somewhat elevated. Mantle covered with rounded tubercles, small ones anywhere but largest and longest near middle of back, size decreasing toward margin. **Foot slightly narrower than mantle**, pointed tail projects slightly. Rhinophores retractile, club-like, **~10 oblique lamellae**, a small apical knob; edge of rhinophoral pit usually with a large guard-tubercle on the left side and right side and a small papilla at the front and rear. Gill dorsal, posterior, 15-16 unipinnate leaves encircle the anal papilla, rim of sheath has 8 equidistant stalked tubercles. Ground color yellowish, green, brown, or gray. Body and appendages ground color with some darker shading mid-dorsally from between the rhinophores to the gill. Rhinophores and gill leaves are white-tipped; tubercles with some white frosting. Has been found under rocks in subtidal areas near estuaries and mangroves, especially those with an abundance of the encrusting sponge *Hymeniacidon heliphila*. Depth: to 40 feet. Brazil. Described originally from Brazil, it has long been considered a synonym of the East Atlantic and Mediterranean species *D. verrucosa* Linnaeus, 1758. However, based on detailed anatomical studies, Lima & Simone (2015) re-established *D. januarii* as a distinct species and suggested that purported Caribbean specimens of *D. verrucosa* be re-analyzed for confirmation.

Kice Island, Florida

Doris verrucosa **Linnaeus, 1758**

sponge slug

Size: to 70 mm

Dorididae

ID: Elongate oval, width about half length, somewhat elevated. Mantle covered with rounded tubercles, the small ones anywhere but the largest ones near the middle of the back; the largest are also more elongate than rounded, size decreasing toward the margin. **Foot considerably narrower than mantle**, pointed tail projects slightly. Rhinophores retractile, club-like, **11-16 oblique lamellae**, a small apical knob; edge of rhinophoral pit usually with a large guard-tubercle on the left side and right side and a small papilla at the front and rear. Gill dorsal, posterior, 13-18 unipinnate leaves encircle the anal papilla, rim of sheath with a ring of 8-12 equidistant stalked tubercles. Ground color yellow, brownish-yellow, greenish-yellow, yellowish-gray. Body and appendages ground color, darker shades mid-dorsally between the rhinophores and gill. Has been found under rocks in intertidal areas and in tide pools. In the East Atlantic it is known to feed on sponges in the genera *Halicondria* and *Hymeniacidon*. For defensive purposes it produces chemicals toxic to fishes. Spawn is a spiral ribbon with crowded whitish to yellowish egg capsules. Depth: 0-50 feet. Amphiatlantic and Mediterranean. Massachusetts to Florida, Tobago, Costa Rica. The very similar Brazilian *D. januarii* has fewer rhinophoral lamellae, a proportionally broader foot, more rounded and crowded tubercles, and a smaller maximum length (40 mm).

Guadeloupe

***Doris* sp. 1**

Size: ~20 mm Dorididae

ID: Body elongate, oval; mantle spiculose, covered with large wart-like rounded tubercles with spicules at their tips and radially at their bases; rhinophores with ~7 oblique lamellae, tip with small round knob. Ground color brownish-yellow. Rhinophores yellow, tips white. Tubercle papillae white. Found in coral rubble. Depth: ~27 feet.

Roatan

***Doris* sp. 2**

Size: to 15 mm Dorididae

ID: The body is elongate oval; the mantle is covered with wart-like rounded tubercles having knobby apices; rhinophores have ~7 oblique lamellae and a tip with a small round knob; gill posterior, with several leaves. Uniformly yellow. Rhinophores white-tipped. Tops of tubercles with white "frosting", some with small brown spots. Found in coral rubble at ~10 feet. Southeast Florida, Roatan. It looks very much like the rarely reported Mediterranean and East Atlantic species *Doris ocelligera* (Bergh, 1881).

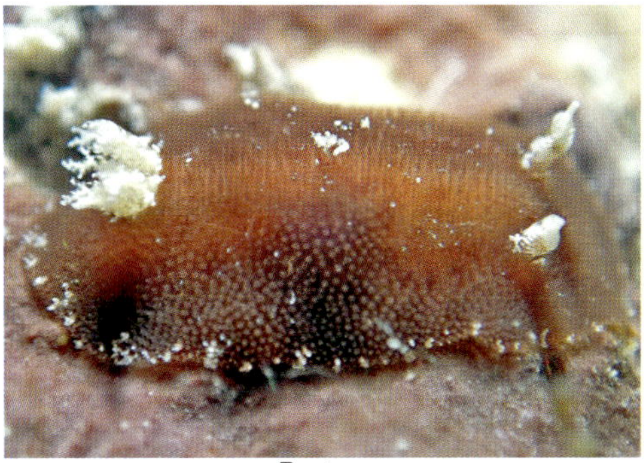

Roatan

***Doris* sp. 3**

Size: ~10 mm Dorididae

ID: Elongate oval, convex; mantle crowded with small rounded tubercles that appear to have laterally projecting spicules; there is a line of well-separated but prominent papillae along the mantle margin; rhinophores with ~9 lamellae and a cylindrical apical mucron; gill posterior with several multi-pinnate leaves. Brownish-orange; tubercles pale; blackish patches near mantle margins. Rhinophores, gill, and marginal papillae opaque white. Found on a hard substrate at ~20 feet.

Riviera Beach, Florida

***Doris* sp. 4**

Size: ~10 mm Dorididae

ID: Body oval, convex; mantle covered with small knob-tipped tubercles; rhinophores with ~9 lamellae and a cylindrical apical mucron; gill posterior with several multi-pinnate leaves. The ground color is yellowish-orange with a few scattered darker patches; the tubercles and gill leaves are white; the rhinophores are brownish-orange proximally and white distally. Found on a hard substrate at ~15 feet. Morphologically similar to *Doris* sp. 3, possibly a variant.

Curacao

Pompano Beach, Florida

Dorididae sp. 1
Size: ~25 mm Dorididae
ID: Body elongate oval, low; dorsum densely covered with low rounded tubercles, smallest near mantle margins. Rhinophores well separated, sheaths low and tuberculate, ~13 thin lamellae between sheath rim and tip, small cylindrical apical mucron. Ground color grayish-white, tubercles white; rhinophores yellowish. Found crawling on encrusting sponges. Depth: to ~15 feet.

Dorididae sp. 2
Size: ~10 mm Dorididae
ID: Body elongate oval, elevated. The mantle is crowded with small low pointed tubercles and some large conical ones; they appear to be spiculated. There is a longitudinal row of 3 very large rounded tubercles along each upper side between the rhinophore and gill. The rhinophores are club-like with ~6 oblique lamellae and have a prominent cylindrical apical mucron. The gill is posterior, with several tall narrow pointed leaves. Ground color gray. The smallest mantle tubercles are whitish gray; 3 large tubercles in each dorso-lateral row are brownish, white-tipped, and centered on broad round white spots. The rhinophoral tubercles are white-tipped, the mucron white, and an anterior blotch just below the apex is brown. The mantle edge is white-spotted.

Pompano Beach, Florida

Guadeloupe

Dorididae sp. 3
Size: ~10 mm Dorididae
ID: The external morphology and coloration are very similar to that of Dorididae sp. 2 (see above), but it differs from that species by being more densely covered by long and thin papillae as opposed to tubercles, by not having a longitudinal dorso-lateral row of three large orange tubercles on each side of the back, by lacking an orange spot at the top front of each rhinophore, and by having shorter gill leaves.

Guadeloupe

Dorididae sp. 4
Dorididae
Size: ~10 mm
ID: The body and head are crowded with short thin papillae; there are some long conical white ones on the back. The rhinophores are dark at the base, the distal lamellate club and cylindrical apical mucron are bright white. The gill has a few brownish leaves having bright white tips. The head is whitish.

Family GONIODORIDIDAE H. Adams & A. Adams, 1854

This family has about 200 species of small dorids in a dozen genera distributed worldwide in temperate to tropical seas. Nearly half of the genera are represented in the tropical West Atlantic.

Much morphological variability across taxa. Body oval to elongate to broad and flat; foot usually broad, tapers to a sharp or rounded point; mantle margin may be absent, reduced to a brim, or developed and broad; when margin is present its edges may be serrated, papillate, or spiculose; oral tentacles long and digitiform to veil-shaped lobes merged with the foot; rhinophores long and tapered, smooth to lamellate (sometimes sparsely); gill dorsal, several simple to tripinnate leaves, partially or totally encircling the anus; variously adorned with dorsal and lateral appendages: none, several, diversely shaped and distributed, commonly extra-branchial and/or extra-rhinophoral; morphology and distribution of radular teeth variable, rachidian tooth often absent.

They feed on entoprocts that attach to bryozoans, hydrozoans, colonial tunicates, sponges, and gorgonians, the various species having their own dietary preferences. Many have color and shape adaptations that provide them with camouflage while on their prey. Body lengths are typically less than 30 mm.

Ancula* cf. *evelinae **Er. Marcus, 1961**

Size: not specified Goniodorididae

ID: Body smooth, robust, elongate, width one-sixth total length, widest at mid-length, tapers to a laterally compressed tail with a dorsal keel; mantle merges smoothly into sides; front of head rounded, a short paddle-like protuberance on each side of the mouth; rhinophores long, stalk robust, club with ~7 lamellae, apical mucron; a pair of shorter and thinner divergent conical appendages share a base in front of each rhinophore; gill at mid-length with 3 leaves and a long, stout, backward-curving, finger-like tentacle on each side. Ground color translucent gray. **Head and body whitish, more yellowish-white around the midbody. Several chestnut to blackish markings**: (i) a median chain of 3-4 broken and somewhat triangular ones that increase in size from rhinophores to gill (ii) a broad netlike blotch behind the gill that straddles the tail's dorsal keel and has a rearward prolongation on each side (iii) a long lateral stripe posteriorly on each side of the tail (iv) a blotch between the rhinophoral bases extends forward from each base onto the adjacent mouth-flanking "paddle". The rhinophores are chestnut proximally, white distally; adjacent appendages translucent gray proximally, whitish with yellow band distally. **Gill leaves white with chestnut stalks; adjacent tentacles translucent gray with white tip and bright yellow basal spot. Pair of yellow spots on head**. No ecological information available. Depth:< 20 feet. Marcus described *Ancula evelinae* based on 2 preserved specimens collected in 1955 at Beaufort, North Carolina. His larger specimen was 3 mm long but neither its life stage nor its amount of contraction due to preservative, so in life this species could be much larger. No other specimens have been examined, nor have any been reported or photographed since 1961. The pictured individual from Northeast Florida is likely *A. evelinae* because of its proximity to that species' type locality as well as being a reasonably good match to Marcus' description.

Matanzas River, Florida

Riviera Beach, Florida

Bermudella* cf. *polycerelloides **Ortea & Bouchet, 1983**

Size: to 10 mm Goniodorididae

ID: Body elongate, smooth, without spicules, middle third ~50% higher; head squarish; foot wider than body, tapers to a sharply pointed tail. Rhinophores long, nonretractile, 5-9 cup-shaped lamellae on rear face. Gill on rear third of body, 7-9 bipinnate leaves. Body has up to 22 short cylindrical papillae, tips rounded, bases slightly wider: 2-3 ahead of rhinophores, 4-6 along each mantle margin, 6-7 dorsal between rhinophores and gill; **a single median dorsal papilla behind gill.** Oral tentacles cylindrical, smooth, about one third the rhinophore length. Ground color translucent whitish. The body is mottled with small, irregular, light/dark brown spots and irregular clusters of white dots and rays; sometimes bluish pigment. Rhinophores, gills, and papillae have the body coloration but the brown spots are more proximal – stems of gill leaves, bases of papillae, bottom half of rhinophores. Clusters of opaque white rays are more prominent at the rhinophoral tips. *B. polycerelloides* has been found living, feeding, and spawning on colonies of the bryozoans *Amathia vidovici* and *Amathia verticillata*. Spawn is a 4-10 mm long gelatinous cylindrical mass of crowded white eggs and embryos deposited on a bryozoan branch. Depth: 1-40 feet. Amphiatlantic. Brazil, Venezuela, East Florida. Sales et al (2019) suggest that many tropical West Atlantic records of *B. zoobotryon* may have been misidentifications of *B. polycerelloides*, and note that the unpaired median papilla behind the gill in *polycerelloides* is the most reliable external feature for distinguishing between the two. Statistically, *B. zoobotryon* has a smaller mean number of dorsal papillae (4) than *B. polycerelloides* (7) but their number of papillae ranges overlap considerably, making papilla counts unreliable for identification purposes.

St. Augustine, Florida

all photos from Riviera Beach, Florida

Bermudella zoobotryon **(Smallwood, 1910)**

Smallwood's okenia

Size: to 8 mm Goniodorididae

ID: Body elongate, narrow, smooth, width about one-third length, elevated, widest/thickest at midbody just in front of gill, from which it slopes down abruptly to rear; mantle spiculose; head squarish; tail narrow, thin, sharply pointed, about one-third body length. **Rhinophores long**, cylindrical, nonretractile, **top two-thirds of rear face with 3-6 cup-shaped equidistant lamellae**. Gill behind midbody, 4-6 variously branched leaves. **Several short papillae on mantle: singly on dorsal midline or pairs straddling it: a pair in front of rhinophores, a single behind; a single in front of gill preceded by a pair; a pair behind gill. Also, short papillae along lateral margins: 4-6 per side, paired laterally**. Papillae cylindrical to club-like, tips slightly swollen and bluntly rounded. Oral tentacles cylindrical, tapering, one-fourth length of rhinophores. Ground color translucent whitish. **Entire body is mottled with small light brown and dark purple spots and blotches**; back and sides with some larger white spots. Gill and oral tentacles body color; rhinophores white above, body color below. Originally found on the colonial bryozoan *Zoobotryon pellucidus* but known to live and feed on *Amathia verticillata*. Papillae are constantly in motion when the animal is crawling. The spawn is a transparent gelatinous tube of densely packed white eggs deposited on a bryozoan branch. Depth: to 20 feet. Amphiatlantic. Bermuda, Southeast Florida, Cuba, Brazil.

all photos from Riviera Beach, Florida

Bermudella **cf.** ***zoobotryon*** **(Smallwood, 1910)**

Size: to 8 mm Goniodorididae

ID: It was noted (Pola, 2014) that most records of *B. zoobotryon* in the West Atlantic should be considered tentative pending additional detailed anatomical studies based on new material, and that a complex of cryptic species being detected via new morphological and molecular analyses is very plausible. Note that the animals in the accompanying photos look very much like *B. zoobotryon*, but the distribution of papillae on the dorsum does not match the original description by Smallwood. This is a case where they "could" be intraspecific variants of *B. zoobotryon* or possibly distinct species in an as yet unknown "zoobotryon cryptic complex" of animals with subtle external and internal differences.

Brazil

Bermudella evelinae (Er.Marcus, 1957) Eveline's okenia

Size: to 8 mm Goniodorididae

ID: Body broadly elongate, width about one-third length, elevated at midbody, slopes down abruptly at rear. Head squarish, bluntly rounded at front; tail tapers to a sharp point, about one-third body length. Rhinophores are long, cylindrical, nonretractile, rear face with up to 8 irregular cup-shaped lamellae. Gill behind midbody, up to 7 short unipinnate and bipinnate leaves. **Several papillae – two along the dorsal midline and one on each side of it in front of the gill, 5-6 short conical blunt ones along each edge of mantle, a front-pointing one in front of each rhinophore.** Foot narrow, front corners small and blunt. Ground color translucent white. Back with opaque white spots; back and sides with a few brownish/purplish blotches; internal organ tissues yellow. Rhinophores are purple (the front sometimes whitish), the tip white; the tentacles and foot margins are white. It has been encountered in shallow water living on bryozoans, e.g. *Amathia distans*. It presumably preys on it. Depth: to ~15 feet. Georgia, Southeast Florida, Venezuela, Brazil. Rare.

Riviera Beach, Florida

Cuba

Goniodoridella miramarae (Ortea & Espinosa, 2000)

Size: to 5 mm Goniodorididae

ID: Body broadly elongate, mantle sharp-edged and cantilevered; foot as broad as body, tapers to a triangular tail that is about half the body length (elongates as the animal advances). Rhinophores long, ~8 lamellae on the posterior face, the bottom 3 large and cup-like with the rest tubercle-like and diminishing in size distally. Gill posterior, **four leaves in a somewhat transverse row** – outer two each with their own peduncle, middle two sharing a single thicker one, **each leaf unipinnate with 5-6 spaced lamellae** and a naked apex. The **cephalic veil has four conical lateral appendages**: 2 at the front and 1 at the outer side of each rhinophoral base. The body has a similar but vertical appendage in front of the gill and **2-3 other lateral ones along each flank, the one opposite the gill very long** (when directed backwards it is longer than the tail). Ground color white. Tips of all appendages yellow to bright orange. Sometimes a purplish pink spot behind the gill. It has been found crawling on hard substrates but has not yet been observed feeding. Probably eats small bryozoans like other *Goniodoridella* and *Okenia* species. It tends to contract when at rest but elongates when in motion. Spawn is a small mass of tiny white eggs, attached to the substrate. Depth: to 66 feet. Rare. First described from Cuba, later found in the Canary Islands. Canary Islands photos are shown for comparison.

Canary Islands

Riviera Beach, Florida

Saba

Goniodoridella picoensis (Paz-Sedano, Ortigosa & Pola, 2017)

Size: to 5 mm Goniodorididae

ID: Body elongate, high; foot tapers to a long slender tail. Mantle covered with variably sized spicules beside small bulges; **5 symmetrically distributed papillae per side – one in front of rhinophore, one behind gill, 3 in between**; papillae elongate, slender, cylindrical, tops abruptly tapered, diameter and length increase toward posterior; **one median mid-dorsal papilla**. Rhinophores elongate, conical, tips rounded, 7-9 lamellae on sides and rear but not near apex. Gill posterior, 4 papillae-like unipinnate branches form an arch around the anus, front 2 on a shared stalk. Ground color **translucent or opaque yellow to opaque white,** sometimes with a greenish cast. Yellow individuals often with a white dusting dorsally. In opaque yellow individuals the tips of the appendages and tail are orange; in opaque white individuals the tips are yellow, in translucent yellow individuals the tips may be ground color. Spicules are prominently bright. Has been found in areas of mixed algae. Depth: 20-100 feet. Amphiatlantic. Found originally at Pico Island in the Azores. Southeast and Northwest Florida, Bahamas, Virgin Islands, Saba. Rare. Referred to as *Okenia* sp. 1 in *Caribbean Sea Slugs* (Valdés et al, 2006).

with eggs; Destin, Florida

Riviera Beach, Florida

Cargoa impexa (Er. Marcus, 1957) disheveled cargoa

Size: to 7 mm Goniodorididae

ID: Body elongate, front broad, tapers to sharply pointed tail (about one quarter the body length). Mantle narrow, smooth, each side margin with ridge from in front of rhinophores to behind gill; **several papillae – a median one on a low crest halfway between rhinophores and gill**, 6 along each ridge behind rhinophores, rearmost 2 with common base; papillae stiff, club-like – **thin stalk, swollen top, short sharp tip**. Four long, conical, curved, forward-pointing tentacles on head – front pair on oral veil and a tentacle in front of each rhinophore. Rhinophores long, thick, conical, contractile, 7-10 lamellae on rear face, smooth elsewhere. Gill behind midbody, 3-4 unipinnate contractile leaves. Ground color **translucent gray with many white dots**. Median **dark brown to blackish patch** extends forward from behind rhinophores, rest of **back and sides yellowish to pale orange with some brown blotching**. All **appendages whitish** with some brown dots, brown pigment may be especially intense basally on rhinophores, tips of papillae sometimes yellowish, front tentacles prominently white. Foot margins dark brown dorsally. Found in areas of sand, scattered grass, and on rocks with attached algae such as *Sargassum*, *Cladophora*, and *Halimeda*. Known to associate with *Amathia* sp. bryozoans and assumed to feed on ascidians and bryozoans as do its congeners. Laboratory studies have shown that spawn consists of a translucent cylindrical oval coil (3 by 2 mm, up to 1.3 whorls) containing 560 white eggs on average. Off South Carolina, spawn has been found on oyster shell rubble. Depth: 3-25 feet. Amphiatlantic. Virginia to Southeast Florida, Bahamas, Cuba, Puerto Rico, Guadeloupe, St. Lucia, Curacao, Brazil. Rare.

Curacao

Riviera Beach, Florida (Atlantic coast)

Seahorse Key, Florida (Gulf coast)

Okenia ascidicola **M. P. Morse, 1972** **yellowspot okenia**

Size: to 12 mm Goniodorididae

ID: Body thick, broadly elliptical, midback elevated; **mantle with ridge along each side** from behind gill to rhinophores and arcing around them to meet in front; oral veil broad, fleshy. **Rhinophores long, thick**, ~25 slightly concave lamellae on top three quarters of rear face. Gill posterior, tall, up to 11 partly retractile bipinnate leaves, tips pointed, encircle anus. **15 tentacles on body**: **2 similar ones at base of each rhinophore** but thinner, slightly shorter, and taper to a sharp point; **4 shorter ones per mantle ridge; a bifid one on each side of gill**, **and a small triangular one just in front of gill**. Foot with protruding lateral margins, tapers to an obtusely pointed tail. **Cream with irregular reddish-brown spots/streaks, densest on the back; sides grade to whitish toward foot;** some white spots on head and sides. **Lemon-yellow:** median patch from rhinophores to gill, a few side blotches between ridges and foot. **Darker yellow spots**: along foot edges, on tail, patches on sides of mid-dorsal region, proximally on front tentacles and elsewhere on others. **All tentacles whitish distally**. Rhinophores reddish-brown-spotted, a yellow spot near apex. Gill leaves mottled brown and white. Feeds on tunicates. During ~4 hours it bores a round hole through the tunic, climbs inside, and (except for its gill) disappears as it sucks out the soft contents. In a few days, when the tunic collapses, it withdraws and relocates. Since it is hidden most of the time it is rarely seen, but it deposits a corkscrew-shaped white egg mass out in the open. Depth: to 20 feet. Although known mainly from the cold habitats off the northeast coast of the United States (Massachusetts, New Hampshire), it has apparently established a presence in the warm waters on the Atlantic and Gulf coasts of Florida. The Atlantic animal shown above is a good match for northern specimens, but the Gulf coast animal is less so. Its identification remains tentative.

← Riviera Beach, Florida →

Okenia **sp. 1**

Size: ~10 mm Goniodorididae

ID: Prominent hump in front of gill, tapers to acutely pointed tail; rhinophores very long, tapering, 7 well separated lamellae on rear face; gill leaves tall, tips tuberculate. Several papillae, thin stalks and swollen tips – 2 stubby ones in front of rhinophores, ~6 others along ridge on each mantle edge, some more in front of gill. Ground color translucent white with brown dusting. Back and sides mottled with brown and greenish dots/spots/streaks, sides with some white spots, sometimes a lateral pale orange wash with black blotches. Appendages ground color. In areas of mixed algae and bryozoans. Spawn is a cylindrical transparent gelatinous matrix with crowded white eggs. Depth: to ~15 feet. Bermuda, Southeast Florida, Bahamas, Jamaica, Puerto Rico, Martinique, Mexico.

Riviera Beach, Florida

Okenia sp. 2

Size: ~2 mm Goniodorididae

ID: Body small, thick. Rhinophores long, conical, ~12 rounded lamellae on proximal two thirds of rear face; 2 forward-directed oral tentacles, shorter and thinner than rhinophores. Gill posterior, several long reclining leaves; dorsum with a wavy median ridge from rhinophores to gill, terminating there with a long stout erect tentacle; mantle edge with ~8 tentacles per side, rear-most longest. Ground color translucent gray. Median dorsal ridge yellow-crested with reddish-brown terminal tentacle; ridge flanked on each side by a wide white stripe ending at the tentacle with a reddish-brown curved blotch that partially wraps around it. Oral tentacles reddish-brown dorsally, yellow ventrally. Lateral tentacles mostly yellow with reddish-brown tips, the series bordered below by a long broad irregular reddish-brown streak. Rhinophores whitish with a small reddish-brown spot in front of the base. Gill leaves ground color with cream stalks. Burrows into sand. Depth: ~15 feet. Southeast Florida. Rare.

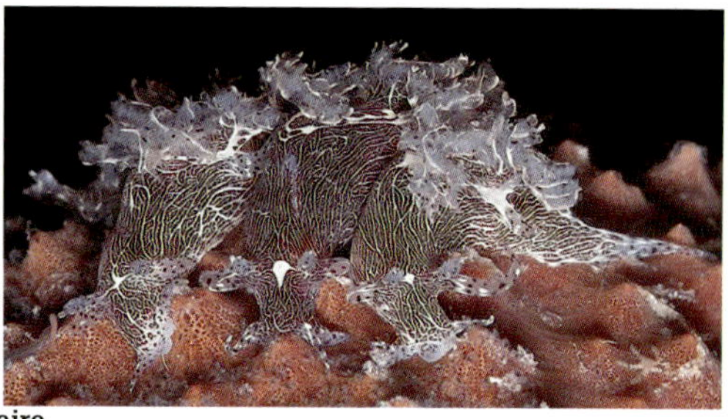

Bonaire

Grand Cayman

with eggs; Bonaire

Trapania bonellenae Á. Valdés, 2009

Ellen's trapania

Size: to 25 mm Goniodorididae

ID: Body elongate, narrow; mantle slopes directly to foot without distinct boundary. Rhinophores robust, cylindrical, distal third club-like with 5 lamellae, the rest smooth; a short cylindrical apical mucron. Oral tentacles short, thin, conical. Two pairs of long, conical, thick, recurved appendages – one flanking the gill, another the rhinophores. Gill posterior, bushy, 5 large tripinnate leaves. Body in shades of brown to blackish overlain with a complex network of crowded white striations: some short, variously oriented, randomly branched; most unbroken and longitudinal; some thicker and more prominent. A few irregular white patches: a transverse one in front of and branching onto the gill leaves, one between or in front of the rhinophores, a smaller one at the base of each oral tentacle. All appendages are grayish with black and white spotting. Found crawling on sponges. Apparently seasonal at Bonaire, observed for several years only between April and August. Depth: to 60 feet. So far, reported only from Bonaire and Grand Cayman.

St. Vincent

Bahamas

Grand Cayman

Bahamas

with spawn; Roatan

Trapania dalva Ev. Marcus, 1972 **white-V trapania**

Size: to 12 mm Goniodorididae

ID: Body elongate, width one-third length, height almost half the length; mantle slopes directly to the foot without a distinct boundary, with many small variably sized rounded tubercles. Rhinophores stout, middle two-thirds with 5-6 broad lamellae on each side intersected by smooth vertical strips – a broad anterior one and a narrow posterior one, apex with a narrow cylindrical projection. Oral are tentacles short, slender, and conical. Two pairs of conical, thick, usually backward-curving papillae: one flanking the gill, the other flanking the rhinophores. Gill behind midbody, tall, bushy, with 3 large bipinnate nonretractile leaves. The foot is narrow, front corners tentacular and recurved, tail triangular, half a body length. Ground color **dark brown**. Most dorsal tubercles are white but many (especially larger ones) are yellowish-orange, giving the appearance of **yellowish-orange spots**. There is a pattern of opaque **creamy-white body pigmentation, i.e a ragged-looking "V" formed by lines from midback to the outside base of the rhinophores**. Large **blotches around gill and rhinophores** and medianly on front of head; small spots/blotches on lower sides. All appendages are translucent gray with variable brown and white spotting/mottling, e.g. rhinophores can vary from mostly brown to mostly white. It has been found crawling on sponges, especially at night. Feeds on ascidians, bryozoans, and entoprocts. Depth: to ~40 feet. Bermuda, Southeast Florida, Bahamas, Virgin Islands, St. Vincent, Cozumel, Grand Cayman, Jamaica.

Destin, Florida

***Trapania* cf. *dalva* Ev. Marcus, 1972**

Size: to 12 mm Goniodorididae

ID: External morphology is like that of *Trapania dalva*. The color pattern differs by having less brown/white contrast due to extensive white "dusting", lacking yellowish-orange spots, lacking a distinct white "V" between the rhinophores and gill, and having more uniformly colored oral tentacles. On algae. Depth: ~40 feet. Destin, Florida. Perhaps a color variant of *T. dalva*.

Guadeloupe

***Trapania maringa* Er. Marcus, 1957**

Size: to 10 mm Goniodorididae

ID: Body slender, elongate, smooth, width one-fifth length, height one-third length. Mantle lacks a distinct edge, slopes directly to the foot. The foot is narrow, deeply furrowed medianly, its front corners with strong recurved tentacles and its rear tapers to a sharply pointed tail that protrudes by one-fifth body length. The oral tentacles are stout, elongate, and conical. The rhinophores are long and club-like with a robust stalk, a club with 6-7 lamellae, and apex with a narrow cylindrical mucron. The gill is dorsal, posterior, and has 3 nonretractile bipinnate leaves. There are **2 pairs of conical, thick, usually backward-curving papillae – one pair flanking the gill, the other flanking the rhinophores**. Ground color yellowish-cream. The body and appendages have irregular dense distributions of brown to reddish-brown pigment and some white spots. Rhinophores are white-tipped. It has been found with algae in shallow water. Probably feeds on encrusting ascidians, bryozoans, and entoprocts. Depth: to ~17 feet. Grenada, Guadeloupe, Brazil. Rare.

***Trapania* sp. 1**

Size: ~6 mm Goniodorididae

ID: Body elongate, broadens to the rear; head narrower than body, front rounded. Foot narrow; tail longer than half a body length, broad, pointed. Oral and propodial tentacles long and conical. Rhinophores have a broad lamellate club on a thick stalk, an apical cylindrical mucron. Gill posterior, three-leaved; a long, thick, recurved appendage beside each rhinophore and each side of gill. Ground color translucent gray. Back white with patches of crowded brown dots and some orange-yellow ones. Head white with antero-median orange-yellow stripe. Oral tentacles white with orange-yellow basal blotch. Rhinophoral stalk brown-dotted, club and apex white with some brown spotting. Extra-rhinophoral and extra-branchial appendages brown-dotted proximally, white distally, with some brown and orange-yellow dots. Tail white with a median orange-yellow stripe that bifurcates near the tip. Found on a sponge. Depth: ~70 feet.

Destin, Florida

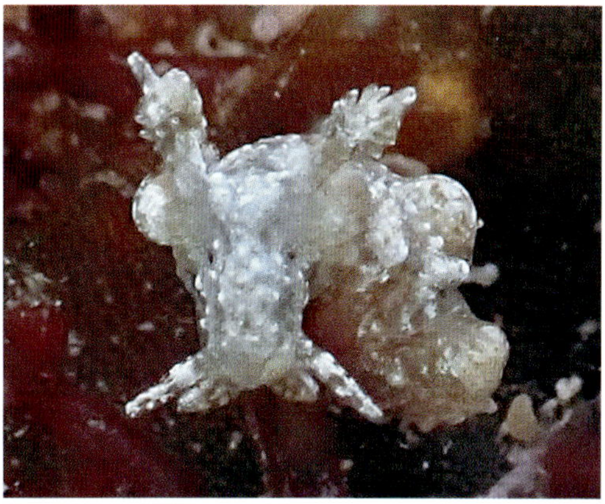

Pompano Beach, Florida

***Trapania* sp. 2**

Size: 3-5 mm Goniodorididae

ID: Body elongate, thick, tuberculate, humped behind midbody. Tail broadly triangular. Head long, broad, squarish. Oral tentacles short, tuberculate; propodial tentacles similar but shorter. Rhinophores with a smooth stout stalk, a bulbous club with ~6 lamellae; apex with a long thin cylindrical mucron; each rhinophore with a single long, thick, recurved appendage. Gill behind hump, several high bushy leaves, at least one thin extra-branchial appendage per side. Mostly opaque white; may have some small dark brown patches on the rhinophoral stalks and flanks, and some scattered brown dots. Depth: ~10 feet. Southeast Florida.

Guadeloupe

***Trapania* sp. 3**

Size: to 10 mm Goniodorididae

ID: Body elongate, thick, humped near midbody, covered with small rounded tubercles. Head long, squarish. Oral tentacles simple, cylindrical, short; propodial tentacles similar but shorter. Rhinophores with a thick stalk, broad club with ~10 lamellae, apical cylindrical mucron. Each rhinophore with a single long, thick, recurved appendage at the base. Gill at the dorsal hump; several high leaves with small bulbous tips; at least one thin extra-branchial appendage per side. Pale orange to whitish, tubercles white; blackish patches at the bases of the rhinophores and the oral tentacles, at the edge of the mantle near the gill, on the flanks, and on the sides of the tail. The tips of the rhinophores, oral tentacles, and gill leaves are whitish. Depth: ~10 feet.

Family HEXABRANCHIDAE Bergh, 1889

A small family of large phanerobranch dorids in the single genus *Hexabranchus*, distributed circumglobally in tropical regions. Until 2023 the Hexabranchidae consisted of only two species, the Caribbean *H. morsomus* and the Indo-Pacific *H. sanguineus*, when it was determined that *H. sanguineus* was actually a complex of 5 species. A member of this complex, *H. giganteus*, is one of the largest known dorids, attaining a length of more than 500 mm.

Body wide, flattened, soft; mantle smooth, broadly expanded, lateral margins rolled upwards to form folds and frills over the sides of the back; oral tentacles large, fleshy, flattened, lobate; rhinophores lamellate, inclined slightly back; gill complex consists of a circlet of individually inserted multipinnate branchial leaves that are contractile but not retractile, the number of "branchs" often more than "hex"; anus atop an elevated papilla that is central or sub-central within the gill circlet; radula multiseriate with numerous simple hamate teeth.

They feed on various sponges, usually in relatively shallow areas (< 60 feet) with coral, rocks, rubble, and sand, but some species are found as deep as 170 feet. Often referred to as "Spanish dancers" in reference to their flamenco-dance-like defensive reaction when disturbed – wave-like undulations of the mantle skirt to prominently flash the normally concealed bright colors of its dorsum, combined with vigorous dorso-ventral body flexions to facilitate escape by swimming.

St. Vincent

Puerto Rico

Roatan

Hexabranchus morsomus **Ev. Marcus & Er. Marcus, 1962**
Caribbean Spanish dancer
Size: to 400 mm Hexabranchidae
ID: Body elongate oval, the width about half the length. The mantle margin is thin, broad, curled over the mantle in places, the edge with undulations; its back with many stout conical tubercles. The foot is narrower than the mantle and tapers to a short bluntly-pointed tail. The rhinophores are long with a smooth stalk, a club on the distal half with up to 30 lamellae; the sheaths are low and smooth-rimmed. The gill is dorsal, posterior, with 6-7 tripinnate contractile leaves encircling a prominent anal papilla. **The ground color is orange to red to purplish. The mantle is "frosted" with white, usually concentrated near the mantle margins**; it may also have pale orange and yellow spots and patches; the edge usually has a thin orange line, the margin ground color; the curl of the margin conceals a white submarginal band that is visible only when it is swimming. Tubercles are ground color, often dusted with white except near their tops. The rhinophoral stalk is white/yellowish, club reddish/orange, tip white. Gill leaves are white to yellow, tips ground color. **Juveniles: mantle mostly whitish, edge orange, tubercles pinkish with red tips.** Usually under rocks and rubble around living reefs, especially near sponges on which it feeds. Defensive behavior is repeated uncurling and curling of its mantle margin to flash its normally hidden bright white submargin and then "dancing away" as it swims – flapping its margin and flexing its body up and down. It also secretes distasteful chemicals from its sponge prey. Spawn is a transparent flat ribbon standing on edge, deposited on a hard substrate as a spiral coil crowded with small red eggs. Depth: intertidal to 120 feet. Endemic to the Caribbean. Florida, Antilles, Venezuela to Bay of Campeche.

with egg mass; Roatan

Family Onchidorididae Gray, 1827

About 50 species in 7 genera distributed worldwide primarily in northern temperate seas (Mediterranean, north Atlantic coasts, northeast Pacific). The family's only presence in the tropical West Atlantic is via the genera *Onchidoris* and *Idaliadoris* – one record from southeast Brazil, another from Southwest Florida, and three from The Bahamas. Original Bermuda records from around 1900 are considered questionable.

Onchidoris and *Idaliadoris*. Body small, round to oval, wider than high; mantle soft or hard with a dense network of integumentary spicules, usually with variously shaped spiculated tubercles; foot broad, front not bilobed, slightly tapers to the rear; a broad semicircular oral veil surrounds the mouth, no labial tentacles; rhinophores are lamellate; sheaths are low to high, smooth, contractile, their rims poorly defined and bordered by several tubercles; the gill is postero-dorsal, has unipinnate contractile leaves arranged in a partial to complete circle around the anus on the dorsal midline; no gill pits or cavity; a large buccal pump (a muscular sucking crop) is on a long narrow stalk; the radula is with or without a rachidian tooth, there are 1-3 laterals.

Most are specialized suctorial feeders on encrusting bryozoans, where they live well camouflaged and deposit their spawn. Their typical depth range is from the intertidal zone to about 25 feet. Animal lengths are up to 11 mm.

Brazil

Idaliadoris brasiliensis
Alvim, Padula & Pimenta, 2011

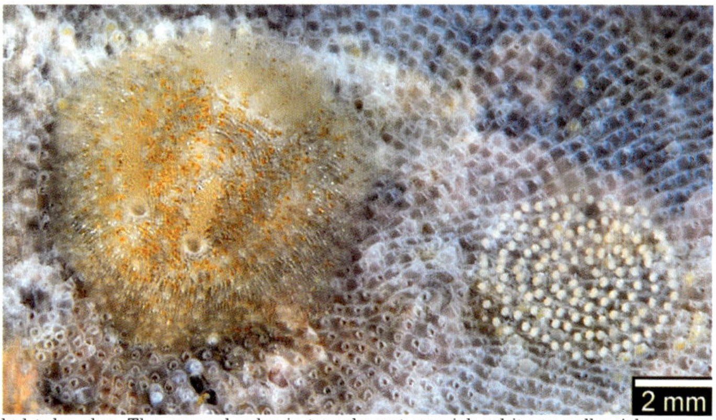

Size: to 8 mm Onchidorididae

ID: The body is elliptical, slightly depressed, front and rear edges equally rounded. The mantle is stiffened by spicules that are oriented transversely in the center, radially at the edges, and obliquely in between; **the mantle is also covered with elongate evenly-spaced tubercles**, larger in the middle and smaller/thinner at the edges. The rhinophores are long and slender, have 6-10 oblique lamellae, and are retractile into low sheaths; the sheath rims are smooth or with 1-3 tubercles. The gill is at the rear, with 11-13 nonretractile unipinnate leaves arranged as a broad closed circle around the anus, the front leaves largest and the spaces between the bases of the leaves with short rounded tubercles. The ground color is translucent grayish-white to yellowish-orange with a scattering of dark brown (sometimes orange) dots. The spicules are transparent, tubercles white. Feeds on and deposits spawn on the encrusting bryozoan *Parasmittina protecta* (Thornely, 1905). Spawn is a gelatinous ribbon with a single row of large egg capsules containing one cream to rosy egg per capsule, deposited as a counterclockwise spiral coil of 5-6 whorls. Depth: to ~26 feet. Found on the southeastern Brazilian coast, off northern Rio de Janeiro (Cabo Frio and Arraial do Cabo) and the northern region of the State of Sao Paulo (Ilhabela).

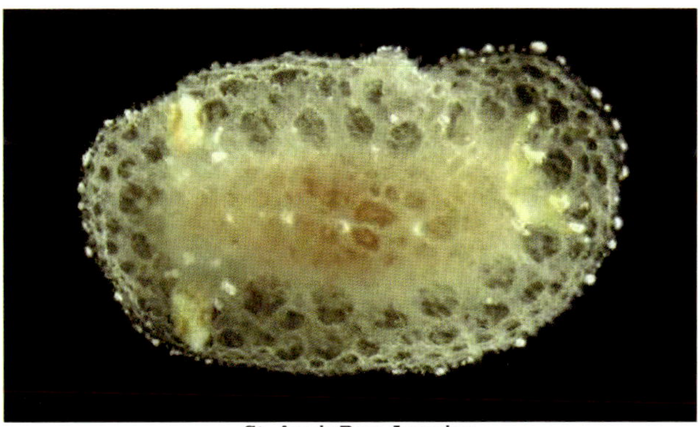

St. Ann's Bay, Jamaica

***Onchidoris* sp.**

Size: ~4 mm Onchidorididae

ID: The body is elongate oval. Ground color grayish-white with a mesh-like pattern created by a dense distribution of dark rounded patches that diminish in size toward the mantle edges. The visible internal organs give the mid-dorsal region a pale orange cast with variably sized brownish patches. The rhinophores are well separated and appear to be distally annulate with a pointed tip; yellowish white with some brown edging. The gill is well back on the body, consists of several (4-5) leaves, and is yellowish with white tips. There are numerous white tubercles: a prominent row along the mantle's perimeter, a median row of four between the gill and the rhinophores, and several elongate ones encircling the base of each rhinophore. Found under a rock at a depth of ~3 feet. Associated with bryozoans. So far, it has been sighted only in Jamaica.

Family Phyllidiidae Rafinesque, 1814

About 80 species of dorids in 5 genera distributed worldwide in tropical and subtropical seas, mostly in the Indo-Pacific regions but also with some presence in the Mediterranean Sea and Atlantic Ocean. Currently the family is represented in the tropical West Atlantic by one species in each of the two genera *Ceratophyllidia* and *Phyllidiella*.

Body oval, sometimes elongate, somewhat flattened; mantle wide, margin covers foot, usually toughened by integumentary spicules, surface covered with stiffened spiculated tubercles; the shapes, colors, and distributions of the tubercles vary with the species; the body wall and foot are also spiculose; the front of the foot is simple; the oral tentacles are short, usually grooved ventrolaterally, can be fused; the rhinophores are lamellate and retractile into pits; instead of a dorsal gill like other dorids they have a line of flat triangular gill leaflets oriented transversely along the groove between the mantle and the foot and concealed by the mantle margin; the gill series is interrupted by only the mouth and the reproductive apertures on the right side; anus usually dorsal and postero-median; penis armed; jaws and radula absent, replaced by oral glands on a suitably adapted buccal bulb.

Most species have prominent aposematic colorations typical of repulsive smelling/tasting animals. Indeed, phyllidids feed suctorially on sponges from which they derive not only nutrients but also sesquiterpenes that they store in mantle glands for producing and releasing characteristic malodorous and highly toxic chemicals to deter potential predators. Typically found by day in relatively shallow waters (to 100 feet) but have been recorded deeper than 2000 feet. Their range of lengths is 15-115 mm.

Utila

Ceratophyllidia papilligera **(Bergh, 1890)**

black-spotted nudibranch

Size: to 30 mm — Phyllidiidae

ID: Body oval, rigid, spiculose; mantle covered with scattered stalked papillae stiffened by spicules; papillae either separate or in broad-based fused clusters. Foot narrow (one third the mantle width). Rhinophores retractable, clubs with up to 20 lamellae on the distal half, protrude from low sheaths. About 100 gill leaves on the underside of the mantle along each side of the foot. Ground color white. Mantle with variably sized, rounded, dark to light gray spots with black margins; often a midline row of medium-sized spots flanked by 2-3 large ones, and a scattering of small ones elsewhere. Papillae that fall within the dark spots are dark with white tips, the large ones usually with several white tips of different sizes; other papillae are either entirely white or white with black spots. Rhinophoral stalks and tips are white, lamellae are black. Inhabits rocky areas of living reefs where it hides under rocks and crawls over coral heads in search of sponges to extract nutrients from. Depth: 6-600 feet. West Florida, Bahamas, Jamaica, Cayman Islands, Cuba, Dominican Republic, Virgin Islands, Guadeloupe, Martinique, Colombia, Costa Rica, Honduras, Belize, Mexico, northern Brazil.

Roatan

Bocas del Toro, Panama

Phyllidiella molaensis **(K. B. Meyer, 1977)**

Size: to 45 mm — Phyllidiidae

ID: Body oval, length twice the width. Mantle stiff, covered with papillae – some short and conical, some with 2-3 points, larger mid-dorsally. Rhinophores short and stout, sheaths low. Small gill leaves below the edge of the mantle at the junction with the foot, between head and posterior, increasingly better developed toward the rear. Covered with closely-packed broad groups of slightly irregular concentric rings of similar thickness, each group consisting of a central black spot encircled by a white band, a black band, and another white band, each white band containing a thin central gray "circle". Non-banded areas have random black and white lines and blotches. Rhinophores are black with white tips. Inhabits rocky areas. Depth: 10-50 feet. Panama, Costa Rica. Rare.

Family Polyceridae Alder & Hancock, 1845

A large family of phanerobranch dorids consisting of more than 25 genera in several subfamilies, totaling over 200 species distributed circumglobally at nearly all latitudes. About one third of the genera are represented in the tropical West Atlantic.

Body elongate and thick. Mantle margin is considerably reduced, sometimes to a narrow dorso-lateral ridge on each side of the body that tapers to a tail; tail usually somewhat thick and narrowly triangular but in some species has a paddle-like median crest to enable swimming by lateral flexions of the body; a rounded cephalic veil is usually a remnant of the reduced margin; the dorsum, margin and veil may have papillae that are either simple or branched; the rhinophores are lamellate, usually retractable into pits, sometimes into a sheath; the gills are dorsal, contractile but not retractable into a protective cavity, configurations varying from posteriorly situated circles/arcs/rows of simply multipinnate leaves to prominently elevated mid-dorsal clusters.

Prey preferences are species-dependent – bryozoans, colonial ascidians, even other polycerids. In a few species (e.g. *Plocamopherus* spp.) some lateral papillae have bioluminescent bulbous processes at their tips that flash when the animal is disturbed. The body length range is 10-80 mm.

Grand Cayman

20 mm; Roatan

Grand Cayman

Cozumel

spawn; Roatan

3 mm juvenile; Roatan

Kankelibranchus alhenae **Ortea, Espinosa & Moro, 2009** **Alhena's polycerid**

Size: to 25 mm Polyceridae

ID: Body elongate, thick, peaks near midbody, tapers to a pointed tail that is about half the body length. There is no apparent mantle edge. There are smooth conical papillae medianly on the head, back, tail, and along the sides, which also have many small branched appendages; there are 4-8 large, highly-branched, well-separated appendages along the front margin of the head. The rhinophores have a sturdy broad-based stalk, a distal club with 10-12 oblique lamellae, and a narrow cylindrical apical mucron. The gill is at the dorsal peak; has 3-5 tall leaves, largest at front; flanked by branched club-shaped appendages. Ground color translucent whitish. Body with a **network of irregular longitudinal orange lines**; some incomplete, some branching. **Yellow to white papillae** usually centered in a short longitudinal white streak, but most prominently along a median white line from near the front of the head to well back on the dorsum; the line sometimes intersected by a transverse cluster of papillae to form an **irregular white cross**. Branched appendages orange with white spots. Gill leaves and rhinophores ground color with orange spots/blotches; rhinophoral mucron orange with white tip. Found on algae, sand/rubble bottoms, on the hydroid *Symplectoscyphus filiformis*, and on orange bryozoans such as *Bugulina simplex* and *Canda* spp. that provide good camouflage. Spawn is a ribbon crowded with small orange eggs, deposited on a bryozoan as a spiral coil of ~2 whorls. Depth: to 100 feet. Cuba, Saba, Roatan, Cozumel, Belize, Grand Cayman. Rare.

Kankelibranchus incognitus
Ortea, Espinosa & Caballer, 2005

Kankel's polycerid

Size: to 7 mm Polyceridae

ID: Body elongate, broadly rounded at front, pinched in slightly at midbody; foot tapers abruptly from gill to a long **sharp tail** (two thirds the body length); a broad frontal veil begins well back of the rhinophores, edged with **5-6 large bushy appendages**, a few smaller ones, and some small pointed papillae on top. The nonretractile rhinophores have a robust stalk, **a broadly conical club** with 11-12 oblique lamellae, a small cylindrical apical mucron, and no sheath. The posterior gill is a semi-circle of 5 unipinnate tall nonretractile leaves. There are several variously sized **branching appendages along the notal ridge, mid-sides, midback, beside the gill, and on the front of the tail**, the largest beside the gill; some papillae are along the notal ridge, in front of and behind the gill. Ground color **bright orange** with **black dots** that sometimes coalesce into small black spots in front of the gill. Rhinophores are bright orange, mucron white; gill and all other appendages pale yellow to whitish. Found on orange bryozoans and *Galaxaura* red algae attached to rocks. Feeds on tunicates. Spawn is a low rounded orange mass in the form of an open circle. Depth: 8-60 feet. Bahamas, Cuba, Barbados, Grand Cayman. Rare.

Grand Cayman

Destin, Florida

Limacia janssi **(Bertsch & A. J. Ferreira, 1974)**

Size: to 8 mm Polyceridae

ID: Body oval, entire dorso-lateral perimeter with two rows of papillae; the papillae are elongate to oval, often a large globose top supported by a thin stalk, variably sized – larger and more globose typically at the rear and shorter/slimmer at the front; dorsum smooth, but sometimes with a median row of tubercles between rhinophores and gill. Rhinophores retractile, stalks short, clubs long, 10–13 lamellae. Gill posterior, with 3 bipinnate leaves encircling the anus. Rear of foot projects as a triangular tail. Ground color opaque white. Dorsum with orange spots and patches. Papillae translucent white, the long internal axis opaque white, becoming orange toward the apex, larger ones also with a subapical opaque white sphere. Rhinophores with translucent white stalks and orange clubs; gill leaves translucent white with orange blotching; tail white; viscera pinkish. Feeds on bryozoans. Spawn is a flat spiral coil of peach-colored eggs. Depth: intertidal to at least 20 feet. Known from the northern Gulf of California to Panama. The pictured animal from Florida may be the first record of this species in the Western Atlantic.

Limacia **sp.**

Size: ~10 mm Polyceridae

ID: Body elongate oval; dorso-lateral perimeter with rows of variably sized club-like papillae; the smaller more ventral papillae are slim, the more dorsal ones have a thin stalk supporting a globose top; dorsum smooth with a median row of low tubercles between the rhinophores and gill. Rhinophores conical, stout, smooth proximally, the rest with ~11 oblique lamellae. The gill is posterior, with 4 multipinnate leaves. The dorsum is whitish with yellow patches. Papillae are translucent white with a yellowish-orange internal axis that broadens into the bulbous tops, the apices bright yellow. Rhinophores and gill leaves are yellowish-orange; the rhinophores whitish basally. It is found on hard substrates, like rocks. Feeds on bryozoans. Depth: ~7 feet.

Guadeloupe

Boynton Beach, Florida

Martadoris divae **(Er. Marcus, 1958)**

Size: to 12 mm Polyceridae

ID: Body elongate, narrow, thick, broadest at gill, slightly constricted behind head; mantle smooth to slightly wrinkled, reinforced with small spicules; head rounded with slightly bilobed frontal veil; oral tentacles short; front of foot blunt with prolonged corners, tail long and pointed; an indistinct low notal ridge extends back from outer side of each rhinophore, ridges uniting behind gill to form a median crest that extends to tip of tail. Rhinophores retractile, stout, conical, ~17 crowded oblique lamellae, low apical knob, sheaths very low. Gill posterior, 3 branched leaves – tripinnate, non-retractile, broad-based, equal in size, forming a semi-circle around anus. Uniformly orange body, head, and tail, a few scattered white dots on mantle. Rhinophores orange with a white apical knob. Gill leaves orange basally, white at mid-height, translucent gray distally. On and under rocks. Depth: intertidal to 20 feet. Only 3 animals have ever been found: 2 in Brazil (1957, 2012), one in Florida (2021). Formerly *Nembrotha divae* Marcus, 1958 and then *Tambja divae* (Marcus, 1958).

Southeast Florida

Martadoris oliva **(K. B. Meyer, 1977)**

Size: to 20 mm Polyceridae

ID: Body elongate, smooth, laterally compressed, tapers to pointed tail; head rounded/truncate; oral veil reduced to 2 broad, shallow, cuplike processes. Gills at midback, 4-5 pinnate leaves. Rhinophores club-like, tips pointed, 14-16 oblique lamellae; sheaths low, rim smooth. Ground color olive to olive-yellow to light green, often with tiny flecks of yellow-blue or turquoise. May have a corrugated appearance due to many parallel longitudinal darker lines. Exterior surfaces of gill leaves ground color, interior surfaces white, tips purple. Rhinophores purple to translucent bluish-green with purple tips, may have white spotting; oral tentacles purple to bluish-green; front margin of notum dark green. Little is known about its ecology. Has been found on hard substrates. Depth: to at least 80 feet. Costa Rica (both coasts), Panama, Southeast Florida, Bahamas.

Destin, Florida

spawn; Riviera Beach, Florida

Riviera Beach, Florida

mating; Riviera Beach, Florida

Riviera Beach, Florida

juvenile; Riviera Beach, Florida

Plocamopherus lucayensis **Hamann & Farmer, 1988** **Lucayan plocamopherus**

Size: to 40 mm Polyceridae

ID: Body elongate, cylindrical, tapers to blunt tail, the tail is two thirds the body length; mantle slopes smoothly to foot, crowded with variably sized conical papillae (to 1 mm), less dense near foot; front of foot bilobed, corners with short rolled propodial tentacles; frontal veil with 8-10 irregularly-spaced appendages with branched tips, as long as tentacles. Rhinophores conical, ~33 lamellae from top almost to sheath, tip blunt; sheath low, rim papillate. Gill posterior; 6-7 tall bi- and tripinnate leaves with papillate stalks; each front side of gill structure has an elongate club-like appendage with a 2 mm wide oval cap (**light-producing organ**). Ground color **red to reddish-orange to reddish-brown**. All **papillae purple with white tips**; frontal veil appendages purple below, then opaque white, followed sometimes by a dark purple to black ring, tips translucent white. Rhinophores mostly ground color; small white tips; whitish near sheath rim; sheath rims purple. Gill leaves ground color, branch tips purple below and white above; gill's side appendages ground color below, followed by a broad to narrow purple band and bright white apical cap. A primarily nocturnally active species that has been observed feeding on ascidians and bryozoans (e.g. Bugulidae). When disturbed it can emit flashes of light from the light organs flanking its gills. Several other species of *Plocamopherus* have been observed swimming by laterally flexing the body and paddling with the tail, but they all have crested tails that assist in that function. This species, which lacks a crested tail, has not yet been seen swimming. Spawn is a transparent gelatinous ribbon crowded with orange egg capsules, deposited on its prey's branches. Depth: to at least 70 feet. Southeast Florida, Bahamas. Rare.

Destin, Florida

Plocamopherus pilatectus Hamann & Farmer, 1988

Size: to 28 mm Polyceridae

ID: Body elongate, cylindrical; foot slightly wider than body, front rounded, tapers to sharply pointed tail, the tail has uneven median crest; frontal veil broad (about one-third body length), front edge with up to 8 large arborescent processes and many smaller ones; rhinophores club-like, 11-12 lamellae, club and stalk of equal length, apex with small cylindrical knob; gill posterior, three finely branched tripinnate leaves that conceal anus; notal brim has 3 pairs of arborescent processes – two between rhinophores and gill and one behind gill, posterior pair each with a well-hidden cylindrical structure (probably a light-producing organ); 4-5 papillae along mid-side; eyes small, well behind rhinophores, relatively close together. Ground color yellow to yellow-ocher to orange with some scattered white specks. Arborescent processes and papillae white. Several white bands: one along front margin of veil continues (sometimes only partly) along notal ridges, a short one medianly behind the rhinophores, and one along median ridge of tail. Rhinophores ground color with white and brown dusting, tips white. Gill whitish. Eyes black. Has been observed at night feeding on the bryozoan *Amathia verticillata*. Swims with a lateral head to tail motion and its tail flattened vertically. When disturbed it can emit greenish-white light from the white parts of its body. Spawn is a convoluted flat ribbon of densely-packed white egg capsules deposited on a bryozoan branch. Depth: to 70 feet. Northwest Florida, Guadeloupe, Bequia, Grand Cayman. Rare. The original description made no mention of black dorsal spots or of the scattered white specks on the body being atop dark scrawls, as in some of the pictured individuals. Those are assumed to be intra-specific color variations.

with spawn; Destin, Florida

Riviera Beach, Florida

mating; Riviera Beach, Florida

Plocamopherus tilesii Bergh, 1877

Size: to 54 mm Polyceridae

ID: Body elongate oval, humped behind mid-length; head with **broad, flat, hemispherical oral veil**; numerous short, variably sized, somewhat flattened **appendages along veil's margin**; each upper side of body usually with a **longitudinal row of 3 papillate rounded tubercles** (transversely paired) ending just behind gill, rearmost one largest and having a **prominent globular apex** (light emitting organ); lower sides with a few smaller tubercles. Rhinophores club-like, lamellate, sheaths low. Gill behind hump, 3 non-retractable tripinnate leaves. Foot broad; folds to shape a **vertical "keel" at rear of body**; continues back as a short pointed tail. Ground color translucent whitish/yellowish. Body and head with large diffuse pale purplish-brown regions and crowded with mostly similar-sized **dark purplish-brown spots** and a scattering of irregular **orange patches**. Margin of oral veil **orange, appendages dark purplish-brown**. Rhinophoral lamellae dark-speckled, sheath rims orange with large blackish spots. Gill leaves whitish to brownish. **Tubercles orange with white papillae**, large pair flanking rear of gill have **bright white apical light organs**. Usually on or near sandy bottoms in relatively shallow water. Can swim forward by energetic lateral "paddling" of its posterior keel while making depth adjustments with its oral veil. Probably feeds on bryozoans, but has been observed burrowing into the substrate, possibly using its oral veil for sensing other kinds of prey. When disturbed it begins to swim away while emitting bright flashes from its light organs for a short period of time. Since the flashing subsides rather quickly it is suspected that the animal stores just enough light emitting product to temporarily distract a potential predator and enable escape. Depth: typically to 20 feet, but Pacific specimens have been dredged from 224 feet. Long known only from the western Pacific and eastern Mediterranean, until recently photographed off eastern Florida. Probably introduced into the West Atlantic via ship ballast water.

South Carolina

egg ribbon; South Carolina

Polycera chilluna **Er. Marcus, 1961**

harlequin polycera

Size: to 25 mm

Polyceridae

ID: A species known from Georgia and the Carolinas, and recently recorded in Venezuela. Body elongate, robust, width and height half body length, tapers to **sharply pointed tail**; stretched tail almost a body length; mantle skirt reduced to a low ridge along each side of back; rhinophores with broad conical shaft on lower third, lamellate club above with ~25 lamellae, small knob at apex; front of head broadly rounded, with **6 elongate conical appendages**; gill posterior, 5-6 simple tall, broad, pointed, non-retractile leaves; **large sharply-pointed appendage posteriorly on each flank of gill**; foot narrow. Ground color **grayish-black**. Several continuous and broken **yellow stripes along back and sides** from front of head to end of tail; orange stripes on lower sides and along edge of foot. **Frontal appendages yellow or orange**; other appendages mostly ground color with additional pigmentation; rhinophores with yellow tip and subapical black band; **gill leaves black-tipped, stalks yellow to orange; gill's lateral appendages with orange tip and subapical yellow band**. Found on its prey – bryozoan colonies. Depth: 20-110 feet.

Polycera hedgpethi **Er. Marcus, 1964**

Hedgpeth's dorid

Size: to 25 mm

Polyceridae

California

ID: Body elongate, narrow, foot tapers to long pointed tail; gill near mid-length; 7-9 tripinnate leaves. Each side of mantle with a low tuberculate ridge extending back from rear of frontal veil, uniting behind gill, and continuing as a median **caudal crest**. Front of head with **2-3 conical pointed velar appendages per side**, most lateral at front of ridge; **2-3 larger and blunter extra-branchial appendages on each side of gill**. Rhinophores club-like, 8-14 lamellae, elongate cylindrical apical mucron. Back, head, sides of body, and tail's dorsum with crowded dark brown to orange-brown to blackish irregular spots. Ridges and caudal crest grayish-white with yellow-spotted tubercles. Velar and extra-branchial appendages usually grayish near base and tip, yellow elsewhere, often a black band above the yellow, and scattered blackish dots throughout. Rhinophoral shaft has head color; club whitish below, blackish band at mid-length; then a subapical yellow band and white mucron. Gill leaves grayish-white to blackish, ribs white, tips yellow, blackish dots. Tip of tail usually yellow. Usually found among *Bugula* species of fouling/encrusting bryozoans. Depth: Intertidal to at least 80 feet. First reported from California, then Baha to Panama, later New Zealand, Australia, Japan, South and West Africa, Italy, Spain, **Caribbean coast of Costa Rica**. Probably introduced via shipping.

Riviera Beach, Florida

Polycera manzanilloensis

Ortea, Espinosa & Camacho, 1999

Manzanillo polycera

Size: to 6 mm

Polyceridae

ID: Body elongate, rounded at front, pinched-in behind rhinophores, broadens to gill, and then tapers to a point; edge of cephalic veil with **up to 14 short finger-like appendages**, 2 central ones twice as large as rest. Rhinophores robust, club-like, 6 thick lamellae, cylindrical apical mucron. Body with a few wide-based, low, conical, nipple-tipped papillae, e.g.: one median in front of rhinophores, 2 median just behind rhinophores, row of 4 evenly-spaced ones along each edge of back from behind rhinophores to just behind gill, a similar row of 4 along each flank but offset so that each papilla lines up with the middle of the gap in the row above it, and a **triangle of three larger thicker papillae behind the gill** – front 2 straddling midline, rear one on the midline. A raised area in front of the gill has a few much smaller papillae. Gill posterior with three bipinnate leaves. **Ground color olive green**. Papillae white; velar appendages white to yellowish with central brown ring near mid-length. Rhinophores green, lamellae may have reddish-brown spots. The tip of tail white. Gill leaves whitish, **middle with reddish-brown spotting. Large black blotch within triangle of papillae behind gill**. Specimens have been found on calcified red algae (*Galaxaura* sp. and *Bryothamium triquetum*). Depth: to at least 10 feet. Cuba, Southeast Florida. Rare.

all photos from Guadeloupe

Polycera herthae Ev. Marcus & Er. Marcus, 1963
Hertha's polycera

Size: to 9 mm
Polyceridae

ID: Body elongate, rounded at front, **pinched-in behind rhinophores**, broadens to gill, tapers to a point; tail about half the body length, tuberculate; mantle reduced to a **tuberculate ridge along each side of body** and an **oral veil with 8-10 short finger-like appendages**; gill posterior, **on a midbody hump**, at least 5 contractile leaves; back with 2 groups of tubercles – one behind rhinophores, one in front of gill; papillae on sides of body small, conical, with rounded tips; rhinophores retractile, short, thick, club-like on upper two thirds (at least 6 lamellae), tip cylindrical. Ground color shades of gray to brown, or pale green, with black and white specks. Rhinophores and gill leaves gray-brown with white to yellow tips. Rhinophores may have reddish-brown spots. Side papillae translucent with yellow tips. Oral veil white and yellow. Specimens have been found on seagrass beds among bushes of red algae. Depth: 10-80 feet. Southern Florida, Bahamas, Cuba, Antigua, Guadeloupe, Curacao, Costa Rica. Rare.

Destin, Florida

South Carolina

Mexico Beach, Florida

Polycera hummi Abbott, 1952
Humm's polycera

Size: to 25 mm
Polyceridae

ID: Body elongate, peaks at midbody, slopes forward to a bluntly rounded head; foot tapers back to a slender pointed tail with a raised crest at its tip; mantle skirt is a low fleshy ridge along each side of back. Body with small conical papillae, largest on ridges and tail's crest; **2 pointed conical appendages per side of head**, front one usually shorter. Rhinophores non-retractile; stalk cylindrical, club slightly longer, 13-14 oblique lamellae, vertical anterior groove; small cylindrical apical mucron. Gill just behind dorsal peak, **7-9 tall branched leaves**, anterior largest; **3-4 club-like postero-lateral appendages on each side of gill**, frontmost smallest. Ground color translucent grayish-white to cream with extensive brown or black mottling. Head, back, sides, tail, rhinophore/gill stalks mottled with patches of brown to black. Rhinophoral lamellae and gill leaves mottled with brown and black; tips white. Each front appendage has a black-mottled grayish band proximally, then a bright blue band, a yellow band near mid-length, a paler blue band above it, and a translucent gray apex. The proximal two thirds of each rear tentacle is ground color to pale brown, followed by **bands of bright blue, yellow, and paler blue; tip is translucent gray**. Tip of tail has a small blue horseshoe in front of (around) a yellow crest. Has been found on floating mats of arborescent bryozoans of the genus *Bugula*, especially on the brown bryozoan *Bugula neritina*. Probably feeds on the bryozoans and/or their associated organisms. Spawn is a transparent gelatinous ribbon crowded with small white eggs, deposited as a spiral coil of 1.5 turns. Depth: to at least 10 feet. Virginia to South Florida, Bahamas. Rare.

Riviera Beach, Florida

↑ Destin, Florida Key Largo, Florida ↓

***Polycera* cf. *hummi* Abbott, 1952**

Size: to 25 mm Polyceridae

ID: Based on scientific literature *Polycera hummi* has 3-4 conical appendages on each side of the gill. Although the coloration of the pictured animals is very similar to that of *Polycera hummi,* they differ by having **only two extra-branchial appendages.** It could be a variant that species. Alternatively, it could also be a variant of *P. rycia* (see below). Molecular analyses would be helpful. Found with bryozoans. Depth: ~10 feet. Florida. Rare.

all images from Riviera Beach, Florida

***Polycera rycia* Er. Marcus and Ev. Marcus, 1970**

papillose polycera

Size: to 9 mm Polyceridae

ID: Body elongate, peaks moderately mid-length, slopes gently forward and back, tapers to a slender pointed tail the tail is two thirds the body length; mantle skirt a low ridge along each side of back. Rhinophores with short robust stalk, club on distal two thirds with up to 9 broad oblique lamellae and thin cylindrical apical mucron. Gill at dorsal peak, a circle of 5-7 tall tripinnate leaves. About 5 short conical papillae on each ridge between rhinophores and gill, 8 more on each ridge extension to end of tail; 4-8 smooth pointed elongate appendages along front margin of head from one rhinophore to the other, lengths decreasing progressively from frontmost to rearmost. **Two elongate extra-branchial appendages on each side of gill**; usually papillose, either both conical, or a small knob-tipped conical one in front of an inflated looking bumpy larger one that sometimes has a thick ventral extension. Ground color grayish-white. Head, back, sides, tail, rhinophore and gill stalks mottled with patches of brown to black, patches composed of concentrations of dots. Papillae white; those on tail may be spotted with orange. Rhinophoral clubs and gill leaves white with some gray to brown pigmentation. Front appendages white to pale blue distally, but often with a sequence of colored bands from base to mid-length: white, pale blue, yellow/orange, white, **NARROW dark blue/purplish/black**. Extra-branchial appendages whitish; blue band above mid-length; sub-apical yellowish-orange band. Has been found on *Sargassum* (floating or attached to a substrate), on mud flats, and on algae with bryozoans. May feed on bryozoans. Spawn is a transparent gelatinous ribbon crowded with small white eggs, deposited as a spiral coil of 1.5 whorls. Depth: 3-25 feet. South Florida, Bahamas, Cuba, Grand Cayman.

Riviera Beach, Florida

Destin, Florida

Riviera Beach, Florida

Riviera Beach, Florida

6 mm juvenile; St. Vincent

Guadeloupe

Polycera odhneri **Er. Marcus, 1955**

Odhner's polycera

Size: to 25 mm Polyceridae

ID: Body elongate, peaks moderately at midbody, slopes gently forward and back, tapers to slender pointed tail; mantle skirt reduced to a low ridge along each side of back; ridge areas with a few large knob-tipped and irregularly-positioned conical tubercles, smaller ones lower on sides, some flank the midbody gill; two long slender tapering appendages in front of each rhinophore. Rhinophores non-retractile, shorter broad-based conical stalk, large club with ~15 oblique lamellae, vertical groove at front, prominent apical mucron. Gill with up to 7 tall (mostly) bipinnate leaves. Ground color light/dark brown to black; larger individuals usually darker. Back, sides, rhinophores, front appendages, top of tail, and gill leaves with white and orange/red spots; some scattered yellow dots. Antero-dorsally, that polka-dot pattern is reduced to a narrow median patch flanked by white or yellow stripes that originate behind the rhinophoral bases. Rhinophores and tail white-tipped, tubercles white. Front appendages may have a sub-apical dark band. Protruding foot margins translucent gray with orange or red spots. Has been encountered on various species of algae on reef flats and attached to boulders (e.g. *Galaxaura* sp.), on floating *Sargassum*, under rocks, and on bryozoans (e.g. *Zoobotryon verticillatum*). Depth: 3-80 feet. Bermuda, Southeast Florida, Bahamas, Cuba, Guadeloupe, St. Vincent, Barbados, Bonaire, Curacao, Venezuela, Yucatan Peninsula, Brazil. Rare.

juvenile and adult; Destin, Florida

Polycera **sp. 1**
Size: to 10 mm Polyceridae
ID: Body elongately oval, prominent pericardial hump near midbody; head narrower than body, front margin rounded, each front corner with a moderately long smooth tentacle; each front foot corner with a similar recurved tentacle; rhinophores stout, relatively short, lower half a smooth stalk, upper half a broad lamellate club, apex with a small cylindrical knob; a conical appendage, thicker and much longer than rhinophore, projects from inner front of each rhinophoral base and curves back around its outer edge onto the dorsum; gill posterior, low but almost a body width, with ~4-5 highly branched outwardly-inclined leaves; there are at least two extra-branchial appendages on each side of the gill base — one antero-lateral and similar to the rhinophoral appendage, and (at least) one much smaller one that is postero-lateral. Except for head and back, the ground color of all other visible parts is gray. Head and back white; all visible surfaces (including rhinophores, gill, appendages) densely covered with small clusters of irregular black spots and a few white scrawls. A transverse bright white band crosses in front of the gill, joining the two antero-lateral appendages. In the juvenile the dorsum has an irregular median white stripe that merges with the transverse band in front of the gill; this white patch has small orange spots but no black spots. Found on a hard substrate. Depth: ~70 feet. Destin (Florida).

Bermuda

Polycera **sp. 2**
Size: ~10 mm Polyceridae
ID: Body elongate, tapers to rear, peaks at mid-length; row of 8 short conical appendages along mantle margin anteriorly; rhinophores with ~10 oblique lamellae, short cylindrical apical knob; gill at mid-length with several tall branched leaves; thick, cylindrical, apically rounded appendage postero-laterally on each side of gill. Ground color cream; crowded with purple dots. Rhinophores with purple ring at base. White specks dorso-medianly, on blackish tail, and laterally over large blackish region (viscera?). Anterior head appendages and gill leaves white above, purple below. Bases of head appendages connected by dark purple arc. Depth: to 20 feet.

Riviera Beach, Florida

Polycera **sp. 3**
Size: to 10 mm Polyceridae
ID: A population of slugs off Southeast Florida that is most similar to *P.* cf. *hummi* (q.v.) but the 2 conical appendages on each side of the head are **much longer** and their **distal dark band is much wider**. The length may not be significant since those appendages can contract and expand. However, the width of the band could be significant. More studies are needed for clarification.

all photos from Brazil ↓ aggregation ↓

Tambja brasiliensis **Pola, Padula, Gosliner & Cervera, 2014**

Size: to 35 mm Polyceridae

ID: Body elongate, surface wrinkled; rear of foot extends as long pointed tail; head rounded; rhinophores conical, stout, ~20 closely spaced lamellae, retractile into well-developed sheaths; oral tentacles short, flattened; gill posterior, 5 non-retractile tripinnate leaves arranged in semicircle around tubular anal papilla; small lateral slot on each side between rhinophore and adjacent oral tentacle. Ground color yellow to pale orange to light green. Mantle and sides of body with scattering of **small bright blue spots**. Rhinophores dark green on bottom half, blue above. Externally the gill leaves are ground color proximally, blue distally, whitish between; internally the stalk and leaves are opaque white with blue apices. Foot edge light green to dark blue, foot posterior dark blue. Oral tentacles greenish near mouth, rest dark blue. Eyes behind rhinophores, prominent dusky translucent patch behind each eye. Found on vertical walls covered with bushy colonies of *Bugula* arborescent bryozoans. Spawn is a two-whorl spiral ribbon containing ~25 long rows of small orange eggs. With development the embryos turn white. It extracts chemicals from its prey (*Bugula dentata*) that show promising toxicity against some human cancer cells. Depth: 18-100 feet. So far, known only from Brazil.

photos from Brazil rear view

Tambja stegosauriformis **Pola, Cervera & Gosliner, 2005**

Size: to 100 mm Polyceridae

ID: Body elongate, tapers posteriorly; tail long, flat; mantle margin reduced, edges indistinct; front of head with rounded extension, broad and elevated front margin continues back as an elevated lobe ("lapel") to partially wrap around each rhinophore; oral tentacles short, wide, thick; gill at midbody, 4-5 non-retractile large tripinnate leaves form half-circle around tubular anal papilla; scattering of mostly ovate raised and variably sized **patch-like tubercles on back**, largest uniting to form a **median "crest" from behind gill to end of tail**, reminiscent of the upright plates along the back of a stegosaurus; rhinophores retractile, club-like, ~35 lamellae, sheaths broad. Ground color yellowish-orange or light brown. **Tubercle patches blue with darker edge and greenish middle**. Front of head, side "lapels", and posterior median crest prominently green. Small blue spots and two larger tubercles between rhinophores and gill. Rhinophores blue near base, then green with dark tip. Gill branches bright blue below, bluish green above, stalks bright blue along inner edge. Foot with wide bright blue margin. Probably feeds on arborescent bryozoans, but more study is needed. Depth: to 50 feet. So far, known only from Brazil.

continued on next page

↑ being eaten by *Tyrannodoris ernsti* ↑ photos from Brazil

***Tambja* sp. 1** **orangegill tambja**

Size: ~50 mm Polyceridae

ID: Body elongate, thick, tapers to a short rounded tail; mantle margin reduced, edges indistinct; front of head with flap having elevated sides, front margin of head continues back as a rounded elevated lobe ("lapel") to partially wrap around each rhinophore; Gill behind midbody; 4-5 pinnate leaves in a half-circle; many elongate, somewhat elevated, variably sized dorsal "tubercles" on the back; those in irregular rows along mantle margins merge in front of tail to form a pronounced crest. Rhinophores club-like, lamellate. Ground color chocolate brown. Dorsal tubercles yellowish-green with turquoise margins. Gill stalks orange, pinnae blackish with pale stems. Rhinophores ground color proximally, yellowish-green distally, tips purple. Front head flap yellowish-green with black edges. Visible foot edge blue. *Tambja* species feed on bryozoans. Depth: to 40 feet. Los Testigos, Venezuela. Rare. Its external morphology closely resembles that of *Tambja stegosauriformis* Pola, Cervera & Gosliner, 2005, but the coloration is quite different. Perhaps a color variant.

Los Testigos, Venezuela

Northwest Florida

***Tambja* sp. 2**

Size: to 15 mm Polyceridae

ID: Body elongate, smooth, laterally compressed, tapers to a pointed tail; head rounded; oral veil reduced to two broad, shallow, cuplike processes; gills at midback, 4-5 pinnate leaves; rhinophores club-like, tips pointed, ~15 oblique lamellae, sheaths low, rim smooth. Ground color yellowish orange with a corrugated appearance due to many longitudinal brownish to greenish lines. Exterior surfaces of gill leaves ground color, interior surfaces white, tips purple; rhinophores pale orange anteriorly, white posteriorly, tips purple; oral tentacles purple; front margin of notum dark green; tail purple tipped. Found on arborescent bryozoans. Depth: ~70 feet. Destin, Florida. Possibly a color variant of *Martadoris oliva* (q.v.). Similar-looking animals from Japan and Hawaii have been attributed to the species *Tambja amakusana* Baba 1987, but the identifications were not confirmed by physical analyses.

mating pair at 98 feet; Stetson Bank, Texas

Thecacera pacifica (Bergh, 1883)

Pacific thecacera

Size: to 30 mm Polyceridae

ID: Body elongate, narrow, elevated, width and height greatest near midbody, smooth, tapers to long broadly rounded tail; head bilobed at front, corners rounded; rhinophores club-like, lamellate; sheaths broad, elevated, rims flared and inflated at rear; gill posterior, ~5 leaves; two large blunt papillae flank the gill – one on each outer/posterior side, each papilla with a knob halfway up its inner face. Ground color reddish-brown. Rhinophores black-tipped; sheath rims black at front, blue at rear. Gill leaves whitish orange with black stalks. Apices of papillae, their side knobs, and posterior of tail have a blue-tipped black band. This benthic species is usually found in shallow intertidal areas with seagrasses, hydroids, sand, rubble, debris, on/under rocks, in the vicinity of bryozoans to feed on. Depth: 3-100 feet. Endemic to the Indo-Pacific, but there has been one report from the Gulf of Mexico off Texas.

Riviera Beach, Florida

feeding; Destin, Florida

Thecacera pennigera (Montagu, 1815)

feathered thecacera

Size: to 20 mm Polyceridae

ID: Body elongate, narrow, width and height greatest near midbody (height about one-fourth length), smooth, tapers to a long and pointed tail; head bilobed at front, each corner with a short conical projecting lobe; rhinophores club-like with ~8 oblique lamellae, tips knob-like; **each rhinophore with a pair of tall, stout, smooth, cylindrical, blunt tentacles on its outer side**; gill posterior, 5 tall branched leaves that partly enclose the anus, large knob-tipped appendage (sometimes bifid) on each rear flank. Ground color translucent/transparent grayish. All visible surfaces covered with irregular orange/yellow spots and smaller roundish black ones (sometimes a black "dusting"). Tips of rhinophores, their side appendages, and tips of gill side appendages are pale to prominently white. Viscera appear pale brownish orange. The oral tentacles are conical and grooved. Has been found in shallow lagoons among rock, sand, and algae, and on the arborescent bryozoan *Bugula turrita*. Feeds on bryozoans. Spawn is a wavy white ribbon. Depth: 2-65 feet. Worldwide. Massachusetts to Southeast Florida, Northwest Florida, Sao Paulo (Brazil). Rare.

Brazil

Tyrannodoris ernsti

(Pola, Padula, Gosliner & Cervera, 2014)

Size: to 80 mm Polyceridae

ID: Body elongate, surface smooth but strongly wrinkled; rear of foot extends as long pointed tail; head rounded. Rhinophores conical, 35-40 closely spaced lamellae, completely retractile into low sheaths. Oral tentacles short, well developed, dorso-laterally grooved along part of their length. Gill posterior, 5 nonretractile tripinnate leaves in a semicircle around the elevated anal papilla, frontmost 3 the largest. Ground color **dark green**. Body covered with **irregular longitudinal yellow stripes**; most crowded and broken on the sides (up to 14 per side); dorsal and lateral sets of stripes separated by a broad unstriped transition zone; marginal dorsal stripes broader anteriorly and wrap around the rhinophoral bases; rhinophores & gill are joined by 1-3 lines that branch into the gill leaves. Rhinophoral stalks and tips dark blue, lamellae dark blue-to-green. Gill leaves dark green with yellow body stripes proximally. Oral tentacles dark green proximally, dark blue distally. The foot margin and the patch behind each eye are dark blue. Found on vertical rocky walls, usually where *Tambja stegosauriformis* (its confirmed prey) and *Tambja brasiliensis* (its assumed prey) are abundant. Depth: to 25 feet. Eveline Marcus dissected a specimen in 1990 and labeled it "*Roboastra ernsti*" in honor of her late husband Ernst. She died before she could publish a formal description of this species. This is the second of only two *Tyrannodoris* species known in the Western Atlantic.

Destin, Florida

Tyrannodoris ricei **(Pola, Cervera & Gosliner, 2008)** **Rice's tyrannodoris**

Size: to 60 mm Polyceridae

ID: Body elongate, thick, almost cylindrical in cross section, tapers to a long sharply pointed tail; **surface very wrinkled**; the front of the head is rounded; the mantle lacks a distinct edge. The rhinophores are conical, retractile, have up to 50 closely packed lamellae, and low sheaths. The oral tentacles are short, stout, rolled, and **partially grooved longitudinally on the upper side**. The gill has 5 nonretractile similarly-sized tripinnate leaves in a semicircle around the anus. Eyes are in prominent patches behind and inboard of the rhinophores. Ground color **yellow to yellowish-orange**. The head, body, and tail have a **network of pale-yellow shallow grooves** and numerous variously sized **dark blue to black spots ringed with turquoise to white**. The oral tentacles, gill branches and stalks, and the rhinophores are yellow basally, then shade to turquoise/blue, becoming dark blue to purplish at the apex. The foot margins and the rhinophoral sheaths are similarly colored, but the sheaths are dark blue mainly at the rear. The eyespot patches are pale blue. It has been observed on sponges and hard substrates. Like other dorids in this genus, it is assumed to be a carnivore that feeds on other slugs. Depth: 50-85 feet. North Carolina, Northwest and Southeast Florida, Bahamas. Rare. Formerly *Roboastra ricei.*

Family AEOLIDIIDAE Gray, 1827

Consisting of nearly 100 species in 11 genera it is currently one of the 10 families of nudibranchs within the superfamily Aeolidioidea. Most members of this family are characterized by the combination of comb-like radular teeth and a diet of sea anemones. Their distribution is circumglobal at all latitudes.

Body elongate, usually broad but sometimes slender; foot somewhat to much wider than visceral part of body, sides taper gently and eventually converge sharply to the tip of a short tail; front corners of foot bluntly rounded to narrowly tentacular; oral tentacles usually quite long, stout basally but tapering to a rounded tip; rhinophores stout, either lamellate or papillate on all sides or just posteriorly, or simple and smooth everywhere. Cerata short to long, cylindrical and tapering, or compressed; often so numerous and crowded they cover the entire back; usually arranged dorso-laterally in vertical or oblique rows or arches, sometimes on shared low swellings; they are surrogate gills. Anus within frontmost post-cardiac ceratal cluster on the right side of the body wall.

Many species extract living photosynthetic dinoflagellates from the tissues of their anemone prey and convey them to specialized branches of the digestive gland where they are merged into "solar farms" that generate nutrients for the nudibranch. Such photosynthetic capability is usually indicated by brown reticulations within the body wall. In addition to providing a respiratory function the cerata can also deter predators – the stinging cells (nematocysts) acquired from anemones are transported through digestive gland branches to storage compartments (cnidosacs) at the ceratal tips. Lengths range from a few millimeters to a few centimeters.

Guadeloupe

Bermuda

Southeast Florida

Anteaeolidiella lurana **(Ev. Marcus & Er. Marcus, 1967)**

Size: to 12 mm Aeolidiidae

ID: Body elongate, tapers to the rear, pericardial hump prominent. Cerata have a long slender stalk, an inflated upper third, a conical tip, arranged in a line of crowded clusters of oblique rows along each side from just below the rhinophores to the rear of the body; there is a broad open space along the dorsal midline. The rhinophores and oral tentacles are conical and smooth, the tentacles slightly longer; the rhinophoral bases are close-set. Ground color translucent white. An orange stripe from the base of each rhinophore to the base of its adjacent oral tentacle. The back is orange with median opaque white patches – an elongate diamond just behind the rhinophores, a larger more oval one covering the hump, and progressively smaller diamonds/teardrops to the end of the body. Sometimes there is a diffuse longitudinal orange streak on each side of the head. Cerata have an orange to brown to purple digestive gland, white tips, sometimes a subapical white band. Rhinophores white on top third, pale to dark orange below. Oral tentacles ground color, white-tipped, sometimes an dorsal orange blush. Usually under rocks or on hard substrates in shallow water. Depth: to at least 3 feet. Eastern Pacific and amphiatlantic. Bermuda, Southeast Florida, Antilles, Panama, Yucatan, Brazil.

Fort Pierce, Florida

Anteaeolidiella **cf.** ***poshitra*** **Carmona, Bhave, Salunkhe, Pola, Gosliner & Cervera, 2014**

Size: to 10 mm Aeolidiidae

ID: The pictured animal has: a pale orange head loop that almost wraps around the base of each rhinophore/oral tentacle pair; white-tipped appendages (orange rhinophores, grayish oral tentacles, brownish cerata); an irregular median stripe of white spots and blotches from the rhinophores to the tail that broadens over the pericardium to give it a white-spotted appearance. It is a fairly good match for *Anteaeolidiella poshitra* Carmona et al, 2014, a species recently described from India. However, the morphology of this animal differs somewhat: its cerata are very slender instead of moderately inflated, and the rhinophores and oral tentacles are long and slender instead of stout and somewhat short. Nevertheless, it is probably *A. poshitra*. Ecological information is not available. Originally found under coral rubble.

Southeast Florida

Guadeloupe

Guadeloupe

Saba

Roatan

Martinique

Berghia creutzbergi **Er. Marcus & Ev. Marcus, 1970** **Creutzberg's spurilla**

Size: to 30 mm Aeolidiidae

ID: Body elongate, slender, tapers to a short narrow tail; front foot corners short, tentacular, recurved. Cerata cylindrical, wider above, nipple at tip, usually curled, in 5-6 well-separated clusters per side of back. Pericardium between 1st and 2nd clusters. Rhinophoral with adjacent bases, cylindrical with a distal club covered by large rounded papillae (may be sparse on the front face), and have an apical mucron. The oral tentacles are cylindrical, have rounded tips, are somewhat wrinkled transversely, and usually as long as the rhinophores but sometimes quite short. The back and head are usually crowded with uniform white to yellowish-white dots, sometimes coalesced into large spots. The cerata and rhinophores are brownish with white spots and white-to-cream tips; the cerata sometimes diffusely brown-and-white banded. Rhinophores may have a white/cream vertical stripe on the rear face. Oral tentacles are brownish on the proximal third, the rest being white/cream. Found on or under rocks in seagrass areas or crawling over sand and hard substrates. When moving forward it rocks its cerata from side to side; when at rest they contract to half their length and inflate. It feeds on anemones. Depth: 0-25 feet. Florida, Bahamas, Cuba, Lesser Antilles, southeast Brazil.

Florida Keys

adult and juvenile; Florida Keys

Grenada

mating pair; Florida Keys

***Berghia* cf. *stephanieae* (Á. Valdés, 2005)**

Size: to 20 mm Aeolidiidae

ID: Body slender, elongate, tapers to a pointed tail; the front foot corners are triangular and tentacular. The cerata are moderately flattened, their tips rounded, in 3-4 narrowly separated clusters of 5-7 uniformly spaced transverse rows of cerata per side of back from rhinophores to tail, some cerata even at the side of the pericardium; each ceras contains an extension of the digestive gland and an oval cnidosac at its tip. Rhinophores conical, tips rounded, with sparsely distributed small papillae or inconspicuous lamellae on all sides of the upper half. Oral tentacles conical, twice as long as rhinophores, taper to a rounded tip.

In the original paper (Valdés, 2005) describing this species the color pattern was described based on four small (4-7 mm) Florida specimens, as follows: ground color translucent gray; sides of body with some opaque white spots, back mostly opaque white, a triangular white patch broadens rearwards from between the rhinophores and ends behind the eyespots; **rhinophores and oral tentacles opaque white on their distal half, with some opaque white spots basally**; cerata grayish, distal third opaque white, the rest with diffuse white, and maybe a slight orange tinge. The paper also noted that the animal in the bottom right photo on this page is a good match for this species except for its brown cerata, the dark color probably due to recent feeding – a high concentration of zooxanthellae extracted from its anemone prey and stored in the digestive gland's epithelium.

Florida Keys

A more recent paper (Carmona et al, 2014) provided additional color information based on a specimen from Florida: beige or ocher lines along the inner edges of the ceratal insertions, a triangular ocher patch on each side of the head tapers to the base of the oral tentacle, the bottom two thirds of the ceratal branch of the digestive gland is beige to greenish, or brownish-ocher; cerata with a slight subapical yellow band. If that is indeed *B. stephanieae* it means that this species has a quite variable coloration. Based on these descriptions, the top left photo shows a pale animal that seems to be this species. The top right photo shows one just like it accompanied by a smaller one (juvenile?) with **blue cerata.** The top middle photo shows a pale one mating with a blue one, indicating that they are the same species. Clearly, additional studies (anatomical and molecular) are needed to confirm that animals like those shown on this page are *Berghia stephanieae*. So far, it has been reported only from south Florida, Grenada, and the Bay of Campeche (Mexico). It has been assumed that the "*Berghia verrucicornis*" being cultured in captivity for eliminating infestations of the sea anemone *Aiptasia pallida* from home saltwater aquariums is actually this species. However, it does not seem to have the color variations noted above.

Martinique

Colombia

Guadeloupe

Northwest Florida

Guadeloupe

Northwest Florida

Berghia rissodominguezi **Muniain & Ortea, 1999** **Risso's aeolid**

Size: to 52 mm Aeolidiidae

ID: Body slender, elongate, tapering; front foot corners tentacular, recurved, tail short. Cerata cylindrical, tips somewhat rounded, 6-10 clusters per side, first 2 most widely separated, a low pericardial hump between them. Rhinophores long, pointed, front smooth or wrinkled, rest densely "pearled" with rounded tubercles. Oral tentacles slender, up to twice as long as rhinophores. Diffuse orange back and head, front of head usually opaque white medianly. Orange oblique lines along inner edges of ceratal insertions; orange streak from rhinophore to adjacent oral tentacle; blotchy/ragged white median stripe from rhinophores to tail. Cerata translucent; digestive gland branches reddish-brown/blackish/greenish; apices white to yellow to orange, sometimes a subapical blue band. **Rhinophores cream to orange to brownish-red proximally, yellow to cream distally**. Distal third of oral tentacles white to yellow. Found under rocks or crawling on hard substrates in intertidal areas with anemones to feed on. When disturbed it can shed cerata as a distraction. Deposits its egg mass on a hard substrate as a spirally coiled gelatinous string of crowded capsules with 1-3 white eggs. North Carolina, Northwest Florida to Texas, Cuba, Lesser Antilles, Venezuela, Colombia, Panama, Jamaica, Brazil, Argentina.

Southwest Florida

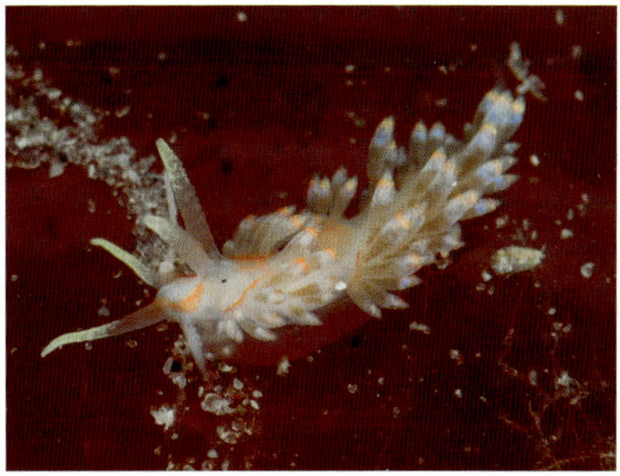

Berghia marcusi **M. Dominguez, Troncoso & F. J. García, 2008**

Size: to 17 mm Aeolidiidae

ID: Body slender, elongate, tapers to a pointed tail. Front foot corners tentacular and recurved. Cerata slender, nearly uniformly wide for most of their length, taper abruptly to a pointed tip; arranged in a row of 5-8 clusters along each side of the back; the bases of the five anterior-most clusters form an arch and the rest form rows; the number per cluster decreases posteriorly from ~12 to 2 or 3, the middle clusters having the largest cerata. The rhinophores are conical with rounded tips; densely covered with round papillae on the rear face, elongate erect papillae on the side faces; papillae may be in oblique rows on the front and sides. Oral tentacles are tapering, longer than the rhinophores. Ground color translucent white. A curved bright orange band from each rhinophore to the base of its adjacent oral tentacle; a median opaque-white undulating band (may be broken) from each oral tentacle to the tail; bright orange oblique lines at the borders of the ceratal insertions; cerata translucent with a clearly visible brown to orange digestive gland on their proximal two thirds, followed by a whitish or bluish band (sometimes indistinct), orange band, and a translucent white tip. Rhinophores and oral tentacles are usually translucent orange to reddish proximally, white to cream above. Under rocks in the intertidal zone and among mangroves where it probably feeds on anemones. Southeast Florida, Guadeloupe, Curacao, Jamaica, Brazil.

Roatan

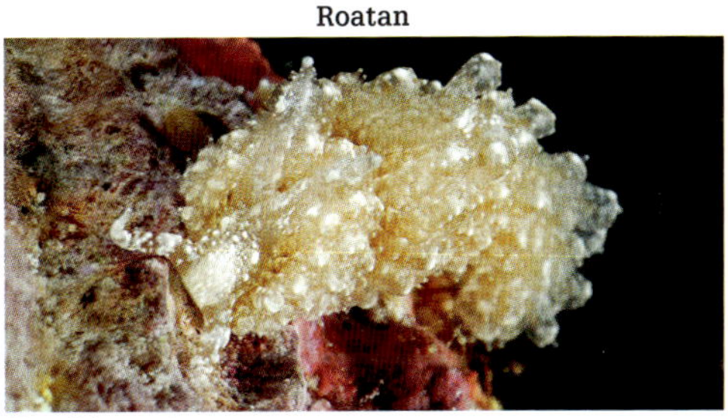

Berghia **sp. 1**

Size: ~15 mm Aeolidiidae

ID: Head and oral tentacles with low tubercles; tentacles short and conical; front foot corners robust and recurved. Rhinophores robust, conical; many papillae, denser and longer at the rear and short at front, mucron at the tip, bases fused. Cerata crowded, inflated, with a short apical mucron. Pale brown, papillae and tubercles white to white-tipped, the rhinophoral and ceratal apices white, the cerata translucent with a pale brown digestive gland branch, and what appears to be many whitish internal glandular structures. Found under coral rubble ~40 feet deep. Its crawling motion was jerky and spasmodic.

Anna Maria Island, Tampa Bay

Berghia **sp. 2**

Size: ~10 mm Aeolidiidae

ID: The body and head broad; front of head smoothly rounded; rhinophores and oral tentacles are simple, the oral tentacles longer; cerata long, thin, crowded. Ground color grayish-white. Head and appendages white; dorsum yellowish-orange; cerata translucent, proximal two thirds with a brown digestive gland branch followed by a diffuse yellowish-orange band and a whitish tip. The pictured specimen was found in the shallows off Anna Maria Island at the southern tip of Tampa Bay (Florida). It was near an artificial reef in an area with numerous *Exaiptasia diaphana* anemones, a common prey for Berghia species.

Key Largo, Florida

Destin, Florida

Riviera Beach, Florida

Islamorada, Florida

Key Largo, Florida

Destin, Florida

Berghia sp. **3**

Size: to 20 mm

Aeolidiidae

ID: The external morphology of the pictured animals is similar to that of *B. rissodominguezi*, *B. marcusi*, and *B. stephanieae* (q.v.), but the cerata of the former two species are considerably less inflated looking and the colors of those above are much more vibrant than those of *B. stephanieae*. The cerata are strikingly blue with tips that are either white, yellow, or pale orange above a yellowish band. The rhinophores and oral tentacles are pigmented like those of *B. rissodominguezi* and *B. marcusi*, and so are the orange markings on the head and at the ceratal insertions, albeit all of the orange pigmentation is much more reddish. One of the animals from Destin has a bright reddish-orange patch over the pericardium. These animals are typically found crawling on hard substrates in depths to ~10 feet. Like other *Berghia* species, they probably feed on anemones. The above-mentioned three species are very similar, and given the scarcity of information regarding their intraspecies variations, the reliable identification of animals like those shown above is quite difficult. More comparative analyses for these look-alikes would be very helpful. So far, these nudibranchs have been found only in Southeast and Northwest Florida and the Florida Keys.

St. Vincent

Guadeloupe

15 mm juvenile; Bonaire

Southeast Florida

Cerberilla potiguara **Padula & Delgado, 2010** **gold-fringed cerberilla**

Size: to 45 mm Aeolidiidae

ID: Body elongate, narrow; cerata conical, inclined/depressed rearwards, in ~15 transverse rows. Oral tentacles elongate, conical, smooth, angled outwards. Rhinophores much shorter, conical, smooth, bases adjacent. Foot wider than body, broadest anteriorly, front corners triangular and recurved. Ground color pale to dark blue; oral tentacles, rhinophores, rear edges of front foot corners are dark blue; yellow/orange stripe with black outside edge from rhinophore to oral tentacle, the stripes thread between the rhinophores to wrap around an elliptical orange spot; a yellow/orange stripe joins the oral tentacle bases. Dorsal cerata usually bluish (sometimes brownish) with an anterior distal yellow patch for the front-most cerata, the patch reducing to an apical spot for the rear-most cerata; the yellow usually grades to bright orange for lateral cerata. Inhabits shallow areas with grassy and sandy bottoms, where it burrows into the sand. Depth: 3-40 feet. Florida, Bahamas, Guadeloupe, St. Vincent, Bonaire, Grand Cayman, Mexico, Brazil (Natal). Rare.

Southeast Florida

Cerberilla tanna **Er. Marcus & Ev. Marcus, 1960** **tanna cerberilla**

Size: to 25 mm Aeolidiidae

ID: Body narrow, elongate. Cerata conical, inclined/depressed rearwards; in ~13 transverse rows. Oral tentacles elongate, conical, smooth, angled outwards; rhinophores much shorter, conical, smooth, bases adjacent. Foot wider than body, broadest centrally, front corners triangular and recurved. Ground color whitish to dusky yellow. Rhinophores brown, pale-tipped. Cerata pale brown proximally with apex translucent whitish/yellowish, an elongate dark brown patch on front surface (patch absent in juveniles), sometimes with an adjacent orange patch below it. There may be a broad, median, diffuse dark brown stripe behind the rhinophores. Oral tentacles, foot margins, and front foot corners bluish-white. Inhabits sandy bottoms in shallow water, usually buried in the substrate. A night feeder on sea anemones. Depth: to ~20 feet. Southeast and West Florida, Mississippi, Texas, St. Vincent, Bonaire.

St. Vincent

***Cerberilla* sp. 1**

Size: to 25 mm

Aeolidiidae

ID: The external morphology is similar to that of *Cerberilla tanna.* Ground color whitish to pale tan. Oral tentacles whitish/bluish; rhinophores orange but pale basally; triangular orange patch on head connects the bases of the oral tentacles, a patch on the tan specimen with a median bright red stripe; cerata translucent ground color with a thin irregular brown digestive gland branch ending in an elongate opaque white apex as a prominent reddish-brown arch-like spot edged below with orange; foot brown with pale margins. It was found at depths to ~20 feet crawling on sand and rubble substrates. Rare, known only from St. Vincent.

Bonaire

***Cerberilla* sp. 2** **Ellen's cerberilla**

Size: ~20 mm

Aeolidiidae

ID: Body oval. Foot wider than the body, broadest centrally, its front corners being short recurved tentacles. The oral tentacles are very long and laterally compressed. Rhinophores are short, smooth, cylindrical, and close together. The club-shaped cerata are in numerous rows along each side of the back, progressively less inflated and more crowded from front to rear; there is a median open area with low transverse folds on the front half of the back. Ground color tan. Rhinophores black on the rear face and apex; cerata orange with a black spot under the white tip; the front of the head has an orange triangular patch joining the oral tentacle bases; a black patch behind the rhinophores bifurcates forward and back: forward as fingers curving around the outside of each rhinophore, back to midbody as diffuse black stripes; oral tentacles black. A single specimen was found on a sandy bottom in relatively shallow water (< 20 feet). Both the external morphology and the color pattern are a close match for *Cerberilla asamusiensis* Baba, 1940, which is currently known only from Japan and Australia.

Bonaire

***Cerberilla* sp. 3**

Size: ~5 mm

Aeolidiidae

ID: Its external morphology is similar to that of *Cerberilla* sp. 2 but its rhinophores are relatively much longer. Its body, oral tentacles and rhinophores are pale tan; its cerata are orange and yellow with white tips. Found on sand shallower than 20 feet. One sighting only, at Bonaire. Maybe a *Cerberilla* sp. 2 juvenile.

Southeast Florida

Guadeloupe

Saba

Bulbaeolidia oasis Caballer & Ortea, 2015 oasis aeolid

Size: to 10 mm Aeolidiidae

ID: Body elongate, relatively broad, somewhat rectangular, flattened; head ovoid, wide; front foot corners short and rounded. Oral tentacles short, constriction at middle gives appearance of two bulbous swellings. Cerata are relatively short, slightly inflated looking, blunt-tipped, curve slightly inwards, and arranged as a row of ~14 clusters along each side from the rhinophores to rear of the body, each cluster a vertical row of 4-5 cerata, and the middle of the back exposed; the rearmost few clusters usually held erect like a peacock's tail feathers. The rhinophores are close together, long, with two bulbous swellings – one at mid-length and one subapical, the swellings sometimes discoidal, the apex a thick cylindrical knob. The eyes are at the outer sides of the rhinophoral bases. The ground color is creamy-white. The head and oral tentacles are opaque white; foot translucent whitish; cerata creamy-white, sometimes with a small reddish-brown spot at the tip; the posterior-most cerata may show brownish marbling; there is **a prominent median bright red spot behind the rhinophores.** Rhinophores are translucent brownish with a white to yellowish-white apex and brownish pigment extending down onto the sides of the head, sometimes with diffuse reddish transverse striations. Eyes black, set in a cream area with a reddish-brown perimeter. The brownish patch extending down from the outer base of each rhinophore has not been mentioned in the available literature for this species. On several occasions it was misidentified as the Indo-Pacific *Bulbaeolidia alba*, which differs by having a red rectangular band around the bases of the rhinophores, no reddish spots at the tips of the cerata, and no reddish striations on the rhinophores. *B. oasis* has been found on mixed algae, sandy rubble, and on/under rocks in relatively shallow water (to ~30 feet). It feeds primarily on sea anemones but has been recorded feeding on egg masses of *Elysia* species. When it crawls it moves forward in a jerky fashion with its rhinophores and rearmost cerata rocking back and forth rhythmically in unison. So far, it has been reported from Cuba, Southeast Florida, Bahamas, Saba, Guadeloupe, Barbados, Venezuela, Cozumel, Bay of Campeche, and southern Brazil. Rare.

Southeast Florida

Grand Cayman

Dominica

Southeast Florida

Curacao

Guadeloupe

with a parasite; Southeast Florida

Limenandra nodosa **Haefelfinger & Stamm, 1958** **warty baeolidia**

Size: to 15 mm Aeolidiidae

ID: Body elongate, slender, tapers to the rear; foot narrow, front rounded with recurved corner tentacles. Cerata long, papillose (warty), elongate at the apex, low and rounded elsewhere; up to 8 transverse rows per side from rhinophores to tail, 1-5 cerata per row, the first two rows very close together. Rhinophores elongate, conical, rear face with elongate papillae. Oral tentacles smooth, longer than the rhinophores. Ground color greenish-brown to grayish with small white dots, or whitish with dull olive-green dots. Medianly from head to tail there are several large opaque-white spots, each with up to three thin marginal rings–inner one pink, then yellow, then (sometimes) white; colors can be faint; the number of ringed spots varies with the size of the individual. Rhinophores, oral tentacles, and cerata with small white spots; rhinophoral apices white to cream; papillae white-tipped. Little is known about this very cryptic species. It has been found on the filamentous algae *Cladophora* and associated *Sargassum*. It probably feeds on anemones. Often moves with a snakelike winding motion. Depth: to at least 10 feet. Amphiatlantic. Southeast Florida, Bahamas, Saba, Guadeloupe, Dominica, St. Vincent, Bonaire, Curacao, Colombia, Costa Rica, Roatan, Belize, Bay of Campeche, Grand Cayman.

Roatan

Southeast Florida

Northwest Florida

with egg mass; Northwest Florida

Guadeloupe

Brazil

Spurilla braziliana MacFarland, 1909 **Brazilian spurilla**

Size: to 25 mm Aeolidiidae

ID: Body elongate. Ceratal morphology changes with age: short/thick very inward-curving in adults) but long/thin and slightly inward-curving in juveniles; a row of ≤9 arched clusters per side, a distinct gap at the pericardial hump, lengths decrease ventrally. Rhinophores lamellate; oral tentacles smooth, longer than the rhinophores. Ground color yellowish-white (**juveniles**), pale orange to red (**adults**). Head/back white-spotted, spots grouped into patches behind the rhinophores and on the pericardial hump. Zooxanthellae in the digestive system may produce a short dark line on each side of the head below the rhinophore and at the ceratal insertions. Cerata translucent dark brown (**juveniles**) shades of brown to bluish-gray to dark green (**adults**), white tips and some white spots. Rhinophores, oral tentacles, front foot corners ground color; rhinophores and tentacles white-tipped. In shallow areas (to 15 feet) with sea anemone prey. Deposits a white egg ribbon on a hard substrate as a convoluted spiral of several whorls. Circumtropical. Florida, Cuba, Antilles, Yucatan, Bay of Campeche (Mexico), Jamaica, Brazil, Argentina. Often misidentified as *Spurilla neopolitana*, which recent molecular studies have shown to be restricted to the Mediterranean and the eastern Atlantic.

Roatan

Southeast Florida

juvenile; Guadeloupe

with egg mass; Roatan

Spurilla dupontae **Carmona, Lei, Pola, Gosliner, Á. Valdés & Cervera, 2014** **Anne's spurilla**

Size: to 38 mm Aeolidiidae

ID: Body elongate, slender, tapers abruptly to the rear; the front corners of the foot are tentacle-like projections. The cerata are long, and cylindrical, their tips rounded and inward-curving, the diameter uniform for most of their length, arranged in up to 7 well-separated arches (vertex up) along each side of the back; the arch heights decrease rearwards, ceratal lengths decrease down each arch. The rhinophores are longer than the cerata, taper to a rounded tip, and have at least 16 lamellae. The oral tentacles are slender, tapering, and slightly longer than the rhinophores. Ground color pale brown. There are prominent pale orange, brownish, or brownish-green **reticulations on the body and the rhinophores due to the presence of symbiotic zooxanthellae living in its tissues**. The entire upper surface and cerata have yellowish-white spots, their size and density varying with the individual. Ceratal tips translucent white; rhinophores, oral tentacles, and front foot corners ground color; white tips on rhinophores, sometimes on oral tentacles. It has been found in sheltered areas on *Padina* algae and *Halophila* seagrasses. Feeds on sea anemones. Depth: 10-20 feet. Southeast Florida, Bahamas, Guadeloupe, Martinique, Grand Cayman, Roatan, a suspected occurrence off Venezuela. Rare.

Roatan

St. Vincent

Southeast Florida

with spawn; Guadeloupe

Spurilla sargassicola **Bergh, 1871** **sargassum spurilla**

Size: to 40 mm Aeolidiidae

ID: Body elongate, slender. Cerata short, thick, inward-curving, up to 7 arches per side of back. Rhinophores with 6-9 lamellae; oral tentacles smooth and slightly longer. Ground color translucent pale brown, body with pale orange, greenish, or brownish reticulations, most conspicuous near the head. Dorsal yellowish-white spots may coalesce into patches; sides dark orange/brown. Cerata white-tipped, covered with white to purple/ocher markings. The rhinophores are white-tipped, their distal third sometimes white or yellowish-white. The oral tentacles occasionally have yellowish-white markings. The only pelagic species of Aeolidiidae, its primary habitat being floating *Sargassum*, but is occasionally found on sand/rubble or stalks of seagrasses. It preys on sea anemones. Depth: 0-10 feet. Bermuda, South Carolina to Southeast Florida, Bahamas, Guadeloupe, St. Vincent, Roatan, Mexico. Uncommon.

Southeast Florida

Spurilla **sp. 1**

Size: ~20 mm Aeolidiidae

ID: Ground color translucent gray. White spots and blotches along the back and sides of the body and head; a median pink blotch at the front of the head. Rhinophores are white-spotted on the proximal half, the lamellae white. Cerata pink above, white-spotted, apex white, digestive gland reddish-brown. It could be a color variant of a known species. Depth: to ~10 feet.

Guadeloupe

Spurilla **sp. 2**

Size: ~20 mm Aeolidiidae

ID: The external morphology is similar to that of *S. braziliana* but the coloration is not a match. Ground color cream to pale orange; roundish white spots on the back and head; the distal halves of the rhinophores and oral tentacles are whitish; the cerata are dark brown with a bluish/greenish mottling; there are some turquoise blue spots; the tips white; Depth: to ~10 feet.

Family ARMINIDAE Iredale & O'Donoghue, 1923

Shell external, ovoid to fusiform, stocky; spire moderately elevated, usually conical, up to 8 whorls and a prominent protoconch; aperture narrow posteriorly, flares forward to an oval bounded by a long gently curved outer lip; columella with one to several strong folds; whorls rounded, sutures pronounced, body whorl sculptured with numerous punctate spiral grooves; periostracum thin; shell length usually less than 25 mm.

Body short and wide, fully retractable into shell, usually supplied with a chitinous operculum to seal the aperture following a defensive retraction. Cephalic shield with a pair of anterior lobes and a larger pair of anterior lobes that cover the front of the shell. No parapodia. Foot usually short, broad, front corners angular or tentacular, rear end shorter than shell.

They Inhabit both shallow and deep waters (to at least 10,000 feet). Prey on polychaete worms.

Bonaire

burrowing; Bonaire

feeding on *Renilla reniformis*; Guadeloupe

feeding on *Renilla reniformis*, Bonaire

***Armina* sp.** **smiling armina**

Size: to ~50 mm Arminidae

ID: Body flattened, elongate, wide, narrows toward the rear; back with many closely spaced longitudinal ridges, broad ones alternating with narrow ones, the broad ones mostly continuous between the front and rear margins of the mantle, the narrow ones that originate further back are often broken or short. Oral veil broad and semilunar. Retractile rhinophores behind the oral veil, short and thick, tips rounded, bases adjacent, and with many vertical lamellae. The front mantle margin is continuous, separated from and folding around the backs of the rhinophores, sometimes frilly-looking when contracted. There is a caruncle in front of the rhinophoral bases. The foot is as wide as mantle.

Ground color black; dorsal ridges white to cream to yellow, often narrowed at both midbody and rear to give the appearance of black transverse bands; mantle margin yellow. Rhinophores are black at the front, blue behind, orange above. The oral veil is pale blue to white with a **yellow margin**, a **submarginal black band** ("smile") and a **median black bar** ("nose") that extends to the rhinophores, the thickness/length of both variable. The caruncle is black, sometimes with just a transverse black bar. Usually found on sandy bottoms in subtidal areas near the sea pens and sea pansies (*Renilla reformis*) it feeds on. It is nocturnally active, typically buried in sand by day. Depth: 5-40 feet. So far, reported from Florida, Guadeloupe, Dominica, and Bonaire. Rare. For many years this slug has been known to exist in the Caribbean, but it will continue to lack a scientific name until specimens are collected, analyzed, and the results are formally published.

Southeast Florida

South Carolina

Armina wattla **Ev. Marcus & Er. Marcus, 1967**

Size: to 28 mm — Arminidae

ID: Body flattened, elongate, wide, narrows toward the rear. The back has up to 36 longitudinal ridges, broad ones alternating with narrow ones except centrally where the narrow ones are often missing; most begin at the front margin of the mantle and end at the rear margin but some originate further back. The oral veil is broad and semilunar. Rhinophores are short, thick, tips rounded, retractile, bases adjacent, protrude upwards at the rear of the oral veil, with numerous vertical lamellae. The front mantle margin is continuous, separated from and folding around the back of the rhinophores; there is a prominent caruncle at the front of the rhinophoral bases. The foot is as wide as the mantle. Eyes are visible through the epidermis at the outer base of each rhinophore. Ground color gray to black. Mantle edge and dorsal ridges white to yellow; rhinophores dark brown with white to yellow apex; oral veil and foot blackish with a white to yellow margin. Usually found on sandy bottoms in subtidal areas, near the sea pansies (*Renilla reformis*) it feeds on. Nocturnally active, typically buried in sand by day. If disturbed while exposed it will curl into a ball and float just above the substrate while tethered to it by a mucous thread attached to its foot. Extrudes a gelatinous wavy string of well-spaced creamy-white egg capsules (1 egg per capsule, 50 capsules per mm) into the water column, tethered to the substrate by a thin, tough, white fiber. Free-swimming lecithotrophic veligers emerge in 8-9 days, develop into juveniles in another 3 weeks, and reach sexual maturity in 2.5 months. Depth: 5-120 feet. North Carolina to Florida to Texas, Mexico. May be locally common but is rarely seen.

← Louisiana

Dauphin Island, Alabama →

Armina muelleri **(Ihering, 1886)** — **tiger armina**

Size: to 32 mm — Arminidae

ID: Body elongate, flattened, narrows posteriorly, rear tip pointed or with a small median notch. Oral veil semilunar with two large smooth-edged tentacular extensions (as wide as the mantle). Foot wider than the mantle. Mantle: up to 54 longitudinal ridges – wide continuous ones alternating with thinner broken ones, a continuous front margin, separated from and folding around the rear of the rhinophoral bases. Rhinophores retractile, short, thick, tips rounded, bases adjacent, behind a two-pointed caruncle at the rear edge of the oral veil, their ≤30 vertical lamellae converging at the tip Ground color red to black. Oral veil red or black. The front margin of the oral veil, dorsal ridges, mantle edges, and rhinophores are yellow. It has been found on shallow sandy bottoms near the sea pansies (*Renilla reformis*) that it feeds on, as well as on hard substrates near sea pens. Being nocturnally active, it is typically buried in sand during the day. When disturbed while crawling along the bottom it may curl up into a ball and float while remaining linked to a fixed object by a mucous thread attached to its foot. Its depth is typically 10-66 feet. North Carolina to Florida to Texas, Mexico, Dominica, Venezuela, Colombia, Suriname to Brazil. It may be locally common but it is rarely seen.

Family BABAKINIDAE Roller, 1973

One genus with four species, only one of which, *Babakina anadoni*, occurs in the tropical West Atlantic.

Body elongate, tapers to rear, entire length with a dorso-lateral ridge; foot narrower than body, front corners tentacular; oral tentacles well-developed, broad-based, tapering; rhinophores robust, joined basally and upwards for nearly half their length, bulbous-looking, lamellate on rear face, distally with a small cylindrical knob; cerata in crowded rows along both sides of body, but not distinctly clustered; radula uniseriate, rachidian tooth triangular with lateral denticles on each side of central cusp; masticatory border of jaws with 3–4 rows of denticles. Inhabit relatively shallow waters where they probably feed on hydroids.

Grand Cayman

Cozumel

Grand Cayman

Guadeloupe

Key Largo, Florida

Babakina anadoni **(Ortea, 1979)**

gaudy babakina

Babakinidae

Size: to 25 mm

ID: Body elongate, slender; front foot corners tentacle-like, recurved. Tail short and pointed. Cerata conical, crowded, ≤30 oblique rows per side, 2-4 cerata per row, longest dorsally; an open space behind the rhinophores. Rhinophores bulbous, fused proximally on a wide base, 20–29 broad lamellae, a short apical mucron. Oral tentacles narrowly conical, as long as the rhinophores. Ground color translucent gray, bluish-gray, pinkish. Head with a large median white to yellowish patch extending from the oral tentacles almost to the rhinophores; patch has a posterior "tail" and is edged with bright orange. A short median white stripe behind the rhinophores. Cerata purplish-blue below, then diffuse white, (pinkish-) orange, and a translucent white tip. Rhinophores brown to yellowish to orange to reddish, rear upper half white, a small mid-front white patch, white apex. Oral tentacles orange proximally, then blue, purple, white. Usually <15 feet deep near algae and gorgonians on hard substrates, under rocks in tide pools, and on its hydroid prey (e.g. *Myriothela* species). Amphiatlantic. South Florida, Bahamas, Guadeloupe, Grenada, Grand Cayman, Cozumel, Brazil. Rare.

Family BORNELLIDAE Bergh, 1874

One genus (*Bornella*) with about a dozen species that occur circumglobally in tropical to temperate latitudes. Only one species, *Bornella calcarata*, is known to inhabit the tropical West Atlantic.

Body elongate, narrow, soft, without a mantle margin; a row of several papillae along each side of dorsum with each papilla in one row directly opposite one in the other row; papillae sometimes branched but usually with 1-2 small gills attached to their inner base; head rounded; a short flattened oral tentacle on each side of mouth with 5-10 radiating finger-like lobes in one or two rows; rhinophores with lamellate clubs; rhinophoral sheaths elevated with rims usually vertically branched, resembling the dorso-lateral papillae. Each row of the radula has a prominent rhachidian tooth with a large denticulate median cusp, flanked by long and narrow lateral teeth (sometimes denticulate).

They appear to feed exclusively on hydroids. When disturbed or alarmed some can swim away by repeated flexions and extensions of their bodies.

Roatan

Guadeloupe

Utila

Guadeloupe

Puerto Rico

Bornella calcarata **Mörch, 1863** **tasseled bornella**

Size: to 80 mm Bornellidae

ID: Narrow body tapers to a pointed tail. A row of 6-9 well-spaced cerata per side of back; cerata short, simple, with single papillae at the rear, becoming taller, thicker-stalked, and more branched toward the front. Rhinophores small, conical, 10-25 lamellae; sheaths high, thick, with a tall central papilla and 1-5 smaller ones just below the rim on the rear surface. Oral lobes hand-like with 5-8 short blunt "fingers". Yellowish/whitish ground color usually obscured by reddish-orange surface pigmentation. Back with dark red reticulations (transverse most prominent); a median stripe of pale orange squares/rectangles from the rhinophores to the tail, each with 2 white bumps that align into rows. Sides with whitish oblong spots, more irregular and brighter near the foot's edge of the foot. Papillae tips white. Habitat variable: living reefs, rocky shelves, coral rubble, boulders, seagrass beds, sand bottoms, algae. Swims well using side-to-side body movements combined with flapping of its appendages. Feeds on hydroids. Depth: 10-165 feet. Bahamas, East Florida, Antilles, Costa Rica, Honduras, Cozumel, Quintana Roo (Mexico), Caribbean coast of South America and south to Brazil.

Family CORYPHELLIDAE Bergh, 1889

It consists of about two dozen species in the widely distributed genus *Coryphella*. Their biodiversity and abundance peak in arctic and cold temperate regions and then decline dramatically and progressively toward the tropics. *Coryphella verta* is one of the rare species with a wide distribution in warm tropical waters.

The species display a high variability in external and internal morphologies. Body wide to narrow; mantle edge continuous, discontinuous, or absent; cerata unstalked, in dorso-lateral continuous or broken transverse rows or clusters; rhinophores smooth to wrinkled to slightly tuberculate, sometimes lamellate or annulate; oral tentacles smooth, tapered, usually longer than the rhinophores; foot as wide as visceral part of body, front corners usually long and pointed; radula triseriate, teeth with variable shapes and number of denticles; anus pleuroproctic.

Most species prey on specific cnidarians, usually hydroids. Depths: from intertidal to several hundred feet. Lengths: 10-50 mm.

Southeast Florida

white-speckled variant; Southeast Florida

Bonaire

St. Vincent

spawning; Southeast Florida

spawning; Cozumel

***Coryphella verta* Ev. Marcus, 1970**

yellowline flabellina

Size: to 21 mm

Coryphellidae

ID: The body is elongate, slender, higher than wide. The foot is narrow with a flange as wide as the mantle, its front edge is bilobed, front corners are recurved tentacles half as long as the oral tentacles; the foot tapers to a long thin trailing tail. The mantle is reduced, its margin a sharp "brim". Cerata long, cylindrical, pointed, a row of 5-7 separated clusters per side of the back starting directly behind the rhinophores, the front two clusters prominently separated by a pericardial hump. The rhinophores are as long as the longest ceras; slender, smooth, cylindrical, tip pointed, bases close together. Oral tentacles are slender, as long as or longer than the rhinophores. Ground color translucent white to pale orange. There are three narrow-to-broad white or yellow stripes from the tip of the tail to the oral tentacles – one median and one along each side below the mantle's brim; the median one bifurcates between the rhinophores, each branch continuing onto an oral tentacle. Cerata are bright red to reddish-orange to brownish-orange, white-tipped, may be white-spotted, sometimes with a dark red subapical band. Each rhinophore (and sometimes each oral tentacle) white on its distal half. It feeds on orange *Eudendrium* hydroids and deposits a convoluted string of white eggs on hydroid branches. Depth: 3-100 feet. Florida, Virgin Islands, Saba, St. Vincent, Venezuela, Cozumel, Bay of Campeche, Brazil (Sao Paolo).

Riviera Beach, Florida

Coryphella sp. 1

Size: 15 mm Coryphellidae

ID: The external morphology is similar to that of *Coryphella verta* but the cerata are distinctly smaller and thinner. The color pattern is also similar to that of *C. verta* but there have not been any sightings of these nudibranchs with white-speckled cerata, and in one of the pictured individuals the cerata are dark purple. Depth: to 15 feet. Possibly a variant of *C. verta* or what has been referred to as *Flabellina* cf. *bandeli* in the book *Caribbean Sea Slugs* (Valdés et al, 2006).

Roatan

Coryphella sp. 2

Size: 8 mm Coryphellidae

ID: Its external morphology is similar to that of *Coryphella verta* but the cerata are distinctly smaller and thinner. The ground color is translucent gray. No white stripes. Cerata are translucent yellowish to greenish with an apical elongate white cnidosac and a subapical red band. Depth: 10-20 feet.

Roatan

Coryphella sp. 3

Size: 8 mm Coryphellidae

ID: External morphology similar to that of *Coryphella verta* but cerata are distinctly smaller/thinner. Ground color translucent gray. Narrow white striping like that of some *C. verta*, but the cerata have a bright yellow digestive gland branch that ends with a red band, followed by an elongate white cnidosac. It resembles *Coryphella* sp. 2 (above) which is from the same island, but that species has neither the white stripes nor the strikingly yellow cerata. Depth: ~10 feet.

Family CUMANOTIDAE Odhner, 1907

Consists of the single small genus *Cumanotus* that occurs in temperate waters of the northern hemisphere (Europe, northeast Pacific). Body elongate; head broad; oral tentacles very short to absent; rhinophores smooth, bases close together; cerata slender, some almost as long as the body, several transverse rows of cerata on each side of the back with up to 8 cerata per row, frontmost 2-3 rows ahead of the rhinophores, ceratal tip with a single long pointed cnidosac; foot very wide with two short propodial tentacles, tapers posteriorly to a short tail; radula triseriate, laterals denticulate.

Feed on hydroids. Length 8-30 mm.

Southeast Florida

Cumanotus beaumonti **(Eliot, 1906)**
Size: to 40 mm Cumanotidae
ID: Body broadly elongate; foot broad, the front foot corners are prominent tentacles; oral tentacles are short processes at the corners of the oral veil; rhinophores long, smooth, erect, bases close together; cerata conical, very long and slender, taper to a fine point with an elongate cnidosac, arranged in ~12 transverse rows of 6-9 cerata per row, 3 of the rows preceding the rhinophores. Ground color translucent white. Head and back with small clusters of white and gold speckles; digestive gland branches that enter the bases of the cerata are pink to purple. Cerata transparent, speckled with white and gold, their tips white. It has been encountered crawling along sandy bottoms or feeding on polyps near the tops of hydroid stalks (*Corymorpha nutans*, *Tubularia* sp.). It moves its cerata constantly, blending well with undulating hydroid tentacles. When disturbed it swims vigorously using a coordinated back and forth movement of its cerata. Depth: to ~50 feet. Known from the British Isles, Norway, France, and the Mediterranean. A single animal recently showed up off Southeast Florida. Rarely seen anywhere.

Family CUTHONIDAE Odhner, 1934

A family that currently consists of two genera encompassing nearly 50 described species, only one of which is not in the genus *Cuthona*. They occur circumglobally in frigid to tropical marine waters. Several of the species inhabit the tropical West Atlantic.

Body small, elongate, low, usually with a pericardial hump; foot normally wider than the body, its front simply rounded (including the corners), tapering to a long tail; oral tentacles smooth; rhinophores conical, smooth, slightly longer than the oral tentacles; cerata elongate, fusiform to club-shaped, in a longitudinal row of clusters along each side of the body, each cluster a distinct oblique row of many crowded cerata, with several pre-cardiac rows, the frontmost cerata range from next to the rhinophores to significantly behind them; ceratal tips usually have functional cnidosacs containing defensive glands; the anus is predominantly acleioproctic (high on the right side within the interhepatic space); renal pore near the anus; with or without a penial stylet; most with a penial gland; radula uniseriate, tooth broad, strong central cusp flanked by several denticles.

Most are found in relatively shallow subtidal areas on and around rocks, seagrasses, and algae and wherever else they can find their colonial hydroid prey, even on pilings, floating docks, and boat hulls. Lengths are typically less than 20 mm.

Guadeloupe

Cuthona barbadiana **Edmunds & Just, 1983**
Size: to 6 mm Cuthonidae
ID: Body elongate; oral veil semilunar; rhinophores and oral tentacles smooth, both about one-fourth body length. Cerata bowling-pin-shaped, slightly shorter than the rhinophores, in a line of up to five clusters per side, the frontmost two separated by a prominent gap. Foot narrower than the body, the front margin slightly indented and the corners rounded, it has a longitudinal groove, the rear projects as a triangular tail. Eyes behind the rhinophoral bases. Ground color translucent gray. A large white patch on the cheek tapers back partially toward the first ceratal cluster; narrow white stripes run back from the inner edges of the oral tentacles, merge between the rhinophores, and continue as a short median stripe to between the eyes, the stripe then bifurcating outwards and converging again to form the first of several similar oval "loops". Cerata are translucent brown with vertical white stripes (mostly parallel), a whitish apex with a clear tip, and a wine-red to dark purple base. Oral tentacles and rhinophores have a wine-red central band and white markings proximally, and are opaque white distally. Top of tail with a median white stripe. Eyes black. It has been found on algae such as attached *Sargassum* and *Padina* in shallow water. Assumed to feed on hydroids like other members of the genus. Depth: to ~20 feet. Bahamas, Barbados, Guadeloupe, Brazil (Ceará and Rio Grande do Norte). Rare.

St. Vincent

↑Southeast Florida↓

Cuthona herrerai Ortea, Moro & Caballer, 2002

blue cuthona

Size: to 12 mm Cuthonidae

ID: The body is elongate and narrow. The cerata are long and cylindrical with pointed tips, arranged in 13-14 rows that cross that cross the back and sides transversely; each row with 4-5 cerata. Oral tentacles are long, conical, smooth. The rhinophores are conical and smooth, as long as the oral tentacles, with their bases close together. The ground color is gray with irregular yellow spots. The midback has a dense distribution of yellow spots behind the second row of cerata. Head brightly colored: dark orange at the front, then a whitish-yellow cross-band, then a bright blue band that connects the bases of the oral tentacles, and finally a broad whitish-yellow band that passes between the rhinophores and encircles their bases. Cerata are opaque whitish-yellow with a blue band just below the tip, usually deeper yellow above the band. The rhinophores shade from gray to grayish-orange upwards, followed by a white band and an orange tip. Oral tentacles are opaque whitish-yellow with dark orange tips. Like other members of the genus, it has been found on hydroids, which it feeds on. Depth: 3-50 feet. It was first described in 2002 from Cape Verde in the East Atlantic. Similar-looking slugs found in Panama and Florida were later described as *Cuthona luciae* Valdés, Medrano & Bhave, 2016, a species that is currently synonymized with *C. herrerai* in WoRMS.

Saba

Cuthona iris Edmunds & Just, 1983

iris cuthona

Size: to 12 mm Cuthonidae

ID: Body elongate; oral veil semilunar. Cerata slightly inflated, tips pointed, up to one-fifth body length, in 7 transverse rows per side (3 in front & 4 behind the pericardial hump), 1-4 cerata per row, each row on a raised base. Rhinophores smooth, slender, conical, tips rounded, equal in length to the longest cerata. Oral tentacles are similar to the rhinophores. Foot slightly notched at the front, its corners rounded. Tail slender, moderately long. Eyes small, behind the outer rear of each rhinophoral base. Ground color translucent orange. A broad broken white stripe extends posteriorly from the front of the oral veil, between the rhinophores, along the dorsal midline to the tail's base, and continues onto the tail as a yellow median streak. Another similar white stripe along each flank starting with a large white patch on the side of the head. Cerata yellow with a central blue band, tips transparent. Rhinophores and oral tentacles orange basally, then yellow. Eyes black. On algae-covered rocky reef-flats and vertical walls with hydrozoans. Assumed to feed on hydroids like other members of the genus. Depth: 3-20 feet. Saba, Guadeloupe, St. Vincent, Barbados, Costa Rica. Rare.

Eastpoint, Florida

egg mass; Eastpoint, Florida

Southeast Florida

Cuthona perca **(Er. Marcus, 1958)** **lost cuthona**

Size: to 15 mm Cuthonidae

ID: Body elongate, slender, tapers to the rear; foot slightly wider than body, front rounded, tail up to one-fourth body length; head rounded, laterally widened. Oral tentacles and rhinophores slender and smooth, oral tentacles about one-tenth body length, rhinophores slightly longer. Cerata slender, fusiform, to one-fourth body length, cnidosacs at their rounded tips; arranged as a line of 8-10 clusters per side, the frontmost usually configured as an arch with up to 16 large and small cerata, the remaining clusters being slanted rows of cerata that decrease posteriorly in both number and size. Eyes are behind the rhinophoral bases. Ground color translucent gray. Head, body, oral tentacles, and rhinophores all white-flecked; head and body appear pale orange to yellowish, the color more intense on the sides of the head; front of head may be brown; internal organs partially visible as darker patches. Cerata with dark brown branches of the digestive gland, a narrow white subapical band, and a white tip. The foot is ground color. Eyes black. It inhabits coastal areas such as estuaries, lagoons, and harbors, on substrates ranging from rocks to piers and floating docks, to boat hulls. It tolerates a broad range of temperatures and salinities. Its prey consists of cnidarians such as anemones and hydroids (e.g. *Pennaria disticha* in South Carolina). When disturbed it inclines its cerata forward. Spawn is a kidney-shaped gelatinous mass of ~100 eggs attached to a substrate by a stalk. Numerous such masses are deposited close to each other. Eggs hatch in ~8 days at 23-25 Celsius. Depth: to ~4 feet. Cosmopolitan. Virginia, South Carolina, Florida, Colombia, Jamaica, Brazil. Rare. Since its first record from Brazil in 1958 it has been reported from several very distant locations, i.e. Jamaica, California, Florida, Hawaii, Ghana, Colombia, Venice, New Zealand, Japan, Black Sea, India. In most locations it is considered to be an invasive species brought from the West Atlantic via ships' ballast water. It is most frequently encountered in Lake Merritt, California.

Southeast Florida

Southeast Florida

Roatan

Les Saintes, Guadeloupe

Southeast Florida

Cuthona tina **(Er. Marcus, 1957)**
Size: to 10 mm Cuthonidae
ID: Body elongate, slender; head as wide as body; foot narrower than body, front slightly indented, corners rounded, tapers to a pointed tail. Rhinophores smooth, conical, tips bluntly rounded, bases close together, up to a third body length. Oral tentacles like the rhinophores but half as long. Cerata usually shorter than the rhinophores; a line of 4-5 clusters per side starting well back of rhinophores; front 2 clusters with 2-3 cerata each, the rest with 1-3 cerata each, a larger gap between the second and third clusters; each ceras a moderately thick spindle tipped with a prominent ovate cnidosac. A small eye at the rear of each rhinophoral base. Ground color translucent gray. Head and back extensively dotted and blotched with opaque white, with orange-brown hepatic tissues visible through parts of the dorsum; sides with scattered white spots; usually an orange-brown spot in front of the first ceratal cluster on each side. Rhinophores white with an orange-brown band below mid-length, sometimes with yellow or orange pigment at the rear of the base. Oral tentacles white distally, pale orange to transparent proximally. Cerata basally transparent to yellow-brown to orange-brown, then variably whitish with some irregular orange-brown blotching, and a white to transparent apex (cnidosac). Eyes black, in unpigmented areas. Orange coloration may be very pale to entirely absent in some individuals. Has been reported from shallow water on filamentous red algae, *Padina* sp., and *Sargassum* attached to a substrate. Also under rocks. Depth: to 15 feet. North Carolina, Florida, Bahamas, Lesser Antilles, Panama, Costa Rica, Jamaica, Brazil.

Cuthona **sp. 1**
pinkline cuthona
Size: ~16 mm Cuthonidae
ID: It has an elongate body that tapers to a rounded rear; the foot as wide as the body, each of its front corners a cylindrical tentacle; the tail is short and broadly rounded. The rhinophores and oral tentacles are long, cylindrical, smooth, with rounded tips. Behind and below each of the rhinophores there is a cluster of ~6 short cerata, and behind each cluster is a longitudinal row of ~6 well-spaced shorter cerata that extend to the rear of the body. Eyes small, at the outer rear edge of the bases of the rhinophores. Ground color cream to pale orange. Sides of body bright orange; three long narrow pink/purple stripes from front of head to rear of body – a dorsal median stripe and one on each side below the row of cerata. Rhinophores and oral tentacles translucent gray below, white above, and purple in between; foot whitish, foot corner tentacles whitish below and purplish above; eyes black. Discovered among algae and dead coral in shallow water. Depth: to ~15 feet. Known only from Riviera Beach (Florida).

Riviera Beach, Florida

Roatan

Cuthona sp. 2

Size: to 10 mm Cuthonidae

ID: Body elongate, tapers to rear; foot slightly wider than body, prominently visible anteriorly, tapers to a moderately long pointed tail; head elongate, front rounded; the cerata bowling-pin-like – cylindrical to slightly inflated-looking with a narrow elongate apical knob (cnidosac), grouped in a row of ~4 clusters per side, first cluster with up to 8 cerata and the rest with 1-2, front two clusters well separated by the pericardial hump. Rhinophores simple, smooth, conical, tips rounded, taller than cerata. Oral tentacles like rhinophores but thinner and half as long. Ground color translucent gray. Opaque white pigment covers most of the head and body. A narrow yellow median stripe from front of head to pericardial hump, a partly broken orange stripe from base of oral tentacle to base of adjacent rhinophore, curved orange stripe from rear of rhinophore to base of first ceratal cluster, irregular orange scrawls near bases of other clusters. Cerata opaque white with some dark spotting basally; apical knobs usually with a basal orange ring, white above. Rhinophores white with a dark band just below mid-length and sometimes another just below the tip. Oral tentacles with yellowish spots and blotches. Foot ground color with white spots. Among algae and hydroids in shallow water (to ~20 feet. Florida, Roatan, Guadeloupe.

Riviera Beach, Florida

Riviera Beach, Florida

Cuthona sp. 3

Size: ~10 mm Cuthonidae

ID: Morphology is similar to that of *Cuthona* sp. 2. Coloration is also like that of *Cuthona* sp. 2 but lacks the orange markings. Possibly a color variant. Discovered among algae and hydroids in shallow water. Depth: to ~20 feet.

Riviera Beach, Florida

Cuthona sp. 4

Size: ~8 mm Cuthonidae

ID: Body and head elongate; the cerata bowling-pin shaped; rhinophores long and slender; oral tentacles thinner and shorter. Translucent gray, head and back whitish. Cerata mottled with black, brown, orange, white, yellow; tips white. Rhinophores and oral tentacles white-speckled. Oral tentacles with central brown/orange band. Among algae and hydroids at ~20 feet.

↑Riviera Beach, Florida↓ ↑Cozumel↓

***Cuthona* sp. 5**

Size: ~10 mm Cuthonidae

ID: Body elongate, tapers to the rear; foot as wide as the body, tapers to a short and pointed tail; head squarish at the front. Cerata somewhat inflated and transversely flattened, mostly recumbent and crowded-looking, their tips rounded. Rhinophores simple, long, slender, conical, their tips rounded. Oral tentacles like the rhinophores but much thinner and shorter. Much yellow pigment on the head, back, and sides. Two broad purplish stripes originate from the base of each oral tentacle – one extending up to the base of the adjacent rhinophore, the other a horizontal one extending back across the cheek and continuing along the side to the posterior of the body. A purplish to bright blue transverse patch behind the rhinophores. Cerata purplish with a broad yellow vertical stripe on the front face, the stripe edged with dark purple, the apices translucent whitish. Rhinophores and oral tentacles are yellow, purplish basally. Found among algae and hydroids in shallow water. Depth: ~20 feet. Cozumel and Riviera Beach (Florida).

Riviera Beach, Florida

***Cuthona* sp. 6**

Size: ~5 mm Cuthonidae

ID: Body elongate; cerata fusiform, in a row of ~5 small clusters along each side; rhinophores simple, long, narrowly conical, taller than the cerata; oral tentacles similar to the rhinophores but smaller; tail long and pointed. Ground color translucent gray. The body and head white-mottled. Cerata whitish with orange-brown specks and a translucent whitish tip. Rhinophores and oral tentacles are white with an unpigmented band near the base and an orange band just beyond mid-length; the tail is ground color; eyes black in unpigmented patches. Found among algae at a depth of ~15 feet.

Grand Cayman

Cuthona sp. 7 **cayman cuthona**

Size: ~10 mm Cuthonidae

ID: The body is elongate, tapers to the rear; the foot is slightly wider than the body, prominently visible anteriorly, and tapers to a long pointed tail. The cerata are conical with pointed tips, grouped in a row of several clusters per side. Rhinophores are simple, smooth, conical, tips rounded, taller than the cerata. Oral tentacles like the rhinophores but thinner and half as long. Ground color translucent gray. Opaque white pigment covers most of the head and body; an orange stripe from the outer base of each oral tentacle to the base of its adjacent rhinophore; cerata orange-brown with white spots, tips white; the rhinophores and oral tentacles are white distally and diffuse orange proximally; tail whitish. Discovered among algae and hydroids in shallow water. Depth: ~20 feet.

Riviera Beach, Florida

Cuthona sp. 8

Size: ~10 mm Cuthonidae

ID: Cerata long, slender, conical, tips pointed, digestive gland ducts very narrow. Rhinophores conical, smooth, bases close together, almost as long as the longest ceras. Oral tentacles similar to the rhinophores. Eyes small, behind the rhinophoral bases. Ground color translucent gray. Head with a white patch at the front between the oral tentacles that narrows posteriorly and continues as a median stripe to between the rhinophores; superimposed on the white median stripe is a narrower orange stripe that tapers to a line ending somewhat behind the rhinophores. There is an orange line from the base of each oral tentacle to the outer base of the adjacent rhinophore. The back has some white patches. Cerata are translucent with an internal duct that is dark brown up to near the top where it is followed by a white streak. Rhinophores and oral tentacles are ground color on the proximal third, shading to orange on the middle third, followed by a white top. Eyes black. Found among algae and hydroids in shallow water. Depth: to ~20 feet.

Roatan

Cuthona sp. 9

Size: ~10 mm Cuthonidae

ID: Tail short and pointed; cerata short, inflated, inclined to the rear; rhinophores long, conical, with a few oblique lamellae; oral tentacles thin and long. Opaque white; a carmine patch at the rhinophoral base; cerata white with brownish-orange at the base; rhinophores purplish, white distally; oral tentacles white-speckled. Found on a hard substrate at ~20 feet.

Guadeloupe

Cuthona sp. 10

Size: ~10 mm Cuthonidae

ID: Orange with a white band between the first two ceratal clusters. Rhinophores conical with oblique lamellae; purple, tips white; oral tentacles long and white. Cerata moderately inflated, inward curving, orange with white apices. Tail long and thin. At ~20 feet.

Roatan

Cuthona sp. 11

Size: ~10 mm Cuthonidae

ID: The body is elongate and narrow, and tapers to the rear. Some of appendages are missing or regenerating. There is a row of ceratal clusters along each side from just behind the head to the end of the body, each cluster with several cerata sharing the small elevated pedestal, each ceras an inflated club atop a narrow stalk and with a small apical knob; the frontmost cerata are shortest, thin, cylindrical; rhinophores short, cylindrical, tips rounded. Ground color translucent gray; opaque white patch from behind the rhinophores broadens to cover most of the back; a small white blotch on the head; a short orange streak at the front outer base of the rhinophores. Cerata with brown stalks, white clubs and white apical knobs with a diffuse brown ring basally. Rhinophores are orange-tipped. Found on a mooring line at ~20 feet.

Grand Cayman

Cuthona sp. 12 chocolate cuthona

Size: ~10 mm Cuthonidae

ID: Body elongate, tapers to the rear. Cerata cylindrical, slender, grouped in a row of ~6 clusters per side, the first two well-separated; the number of cerata per cluster diminishing toward the rear. Rhinophores are simple, slender, conical, tips rounded, much taller than the cerata. Oral tentacles like the rhinophores but thinner and shorter. Ground color dark brown. Cerata dark brown basally, pale blue on the middle third, followed by a diffuse yellow/orange band and a translucent white tip. Rhinophores and oral tentacles are ground color below, white above. It was found on hydroids attached to *Sargassum* (anchored or floating), as well as crawling on a hard substrate. Depth: ~10 feet. Grand Cayman, Southeast Florida, Bahamas. This is the *Cuthona* sp. D in *Bahamian Seashells* (Redfern, 2013).

Roatan

Cuthona sp. 13

Size: ~5 mm Cuthonidae

ID: Body elongate, tapers to the rear; head elongate, front rounded. Cerata are slightly inflated looking with a narrow elongate apical knob; a row of ~4-5 sparse clusters per side. Rhinophores are simple, slender, smooth, conical, tips rounded, about as long as the longest ceras. Oral tentacles are like the rhinophores but thinner and shorter. The ground color is translucent gray. Opaque white pigment covers most of the body and variably on the front of the head. The cerata and the rhinophores are light brown below and white above. The rhinophores may be white near the base; they usually have an orange vertical stripe on the rear face. The oral tentacles are either all white or just on the distal half. Found on a mooring line at a depth of ~20 feet.

all photos from Destin, Florida

***Cuthona* sp. 14**
Size: ~10 mm Cuthonidae
ID: External morphology is similar to that of *Cuthona iris* (q.v.) but the oral tentacles are shorter and more slender. The color pattern is also similar to that of *Cuthona iris* but with some notable differences: (i) ground color pale orange to whitish (ii) cerata greenish proximally (iii) oral veil yellow with a median oval blue stripe. Found on hydroids. Depth: ~70 feet. It could be a variant of *Cuthona iris*.

Guadeloupe

***Cuthona* sp. 15**
Size: ~10 mm Cuthonidae
ID: Body elongate, slender; tail long, thin. Cerata bulbous below, tapering to a rounded tip, a row of 4 small clusters per side, the first cluster well back of the rhinophores. Rhinophores and oral tentacles are long, slender, conical, and smooth, their tips rounded. Oral tentacles are shorter. Ground color translucent bluish-gray. A reddish-brown elongate patch skirts the inner edge of each eye and extends partly up the base of its adjacent rhinophore. Rhinophores and oral tentacles whitish distally, darker proximally. Cerata pale orange below, shading to white above with a bluish apex. Found in seagrass. Depth: ~13 feet.

Riviera Beach, Florida

***Cuthona* sp. 16**
Size: ~6 mm Cuthonidae
ID: Body elongate. The cerata are fusiform with a rounded apex; a row of 5-6 clusters per side, clusters have 1-4 cerata. Rhinophores are longer than the cerata, smooth, cylindrical, their tips rounded; the oral tentacles are shorter. The front foot corners are triangular and recurved. Ground color translucent gray. Head blue, back whitish, the orange internal mass is visible beneath the rhinophores. Cerata are yellowish-orange with a white apex and a subapical dark orange/red band. Rhinophores are gray-blue with an orange tip. Photographed in a shallow area with dense algal growth. Depth: ~10 feet.

Roatan

***Cuthona* sp. 17**

Size: ~3 mm Cuthonidae

ID: Cerata fusiform, apex rounded; rhinophores shorter than the cerata, smooth, conical, tips rounded; oral tentacles shorter and thinner than the rhinophores. Ground color translucent gray. Head and back white, a white stripe along each cheek. Ceratal digestive gland gray with crowded black spots; apex white. Rhinophores white with a pale orange band at mid-length. Oral tentacles whitish with orange pigment dorsally on the base. Found crawling on a hard substrate. Depth: ~40 feet.

Riviera Beach, Florida

***Cuthona* sp. 18**

Size: ~10 mm Cuthonidae

ID: Body and tail are both long and slender. The cerata are like bowling pins; ~5 clusters per side; the first cluster is well back of the rhinophores. Rhinophores are long and conical. The oral tentacles are similar but smaller. A white patch between the rhinophores bifurcates onto the oral tentacle bases; there are some white spots between the rhinophores and the frontmost cerata; the rest of dorsum is white with orange mottling. A short red streak between the first ceras and rhinophore. Cerata white with a brown basal spot, blue apex, and orange subapical band. Rhinophores white, with a brown-orange band near mid-length. Found among algae and hydroids in shallow water (<10 feet).

Riviera Beach, Florida

***Cuthona* sp. 19**

Size: ~10 mm Cuthonidae

ID: Numerous cerata, slender to slightly inflated looking, tips rounded; rhinophores slender, conical, smooth, moderately long. Dorsum white. Cerata yellowish with a reddish blush, a bright red band just below the white apex. Rhinophores mottled with brown and white, the apical third bright red. Found among algae and hydroids in shallow water. Depth: ~10 feet.

Grand Cayman

***Cuthona* sp. 20**
Size: ~10 mm Cuthonidae
ID: Ground color translucent gray. Dorsum and head whitish, some white blotching. Rhinophores and oral tentacles are conical and smooth, the tentacles longer; both yellowish. Cerata in five clusters per side; pale orange, white distally. Tail short, triangular, with a white dorsal stripe. In shallow water off Grand Cayman.

Riviera Beach, Florida

***Cuthona* sp. 21**
Size: ~10 mm Cuthonidae
ID: Cerata and rhinophores are orange with white tips. Oral tentacles are pale orange proximally, dark orange distally. Front foot corners are recurved and white.

Riviera Beach, Florida

***Cuthona* sp. 22**
Size: ~10 mm Cuthonidae
ID: Body elongate, tapers to a sharply pointed tail. Rhinophores very long, slender, smooth, white distally, a short white stripe between their bases; oral tentacles are similar but shorter. There is a white stripe along each lateral row of (regenerating) cerata, the two converging to form a white stripe on top of the tail. There is a short red lateral stripe on the flank just behind each rhinophore.

Riviera Beach, Florida

***Cuthona* sp. 23**
Size: ~10 mm Cuthonidae
ID: Body and head are opaque white. Rhinophores are simply conical, white with a brown mid-length band. Cerata begin well back of the rhinophores, slightly inflated, brownish grading to white distally, the apex being a short and narrow translucent bluish cnidosac.

Family DOTIDAE Gray, 1853

A circumglobal family of small species found in cold and warm marine waters, most of them in the genus *Doto*.

Body elongate, slender; oral veil smooth, usually with a rounded lateral extension on each side of the head, the veil sometimes inconspicuous; rhinophores smooth, tapered, sheaths elevated, rims smooth or jagged and often flared or with a spout-like lip extending forward; stout cerata in a dorso-lateral row along each side of body, paired transversely, each ceras usually with a stack of circlets made of tubercles on its stalk; some species with variously shaped transparent pseudobranchs (gills) near the inner bases of the larger cerata; digestive gland branches extend into the cerata; radula narrow, uniseriate, teeth with a large median cusp flanked by a few denticles; jaws poorly developed to absent.

They feed on hydroids by boring through the stalk's outer protective layer of chitinous perisarc just below the polyps, and then sucking out the fluid from the caenosarc below. Slug lengths are typically less than 15 mm.

Attempts to reliably identify Caribbean species based solely on external features can be very frustrating because of intraspecies variability and species' appearance overlaps. In the words of a prominent expert: "they're all different yet they all look the same".

Cuba

Venezuela

Doto casandra **Ortea, 2013**

Size: to 10 mm Dotidae

ID: Body elongate, slender, tapers to the rear; tail short, bluntly rounded. Oral veil rounded, a longitudinal keel in front of each rhinophore; sides of body may have some conical papillae. A longitudinal row of up to 9 cerata per side of body, each ceras with a cluster of globose tubercles, usually arranged in 5-6 rings; a pseudobranch (gill structure) on the inner face of large cerata consists of a central axis with 3-4 branches on each side. Rhinophores are slender, smooth, bluntly-tapered, the sheath rim elongated anteriorly. Anus at the top of a papilla between the first two right cerata. Body brown to reddish-brown to blackish, sometimes with scattered white spots on the back and sides. **Ceratal tubercles are internally yellowish, orange, pale pink to strawberry red**, the apices orange to white; the central axis of a pseudobranch has a white or orange tip. Rhinophores are translucent pale orange with a **black internal axis** and white spots near the tip; sheaths are body-colored with an orange rim. Lateral papillae (if any) are white. Pre-rhinophoral keels often white. Anal papilla white or transparent. Found in relatively shallow water, often on mangrove roots with attached hydroids (e.g. *Eudendrium carneum, Dentitheca dendritica*), on which it feeds by boring holes into them and sucking out their fluids. Spawn is a gelatinous wavy string of white, orange, or pink eggs deposited on a hydroid's branches, egg color a function of the species of hydroid. Depth: to at least 30 feet. Southeast Florida, Bahamas, north and south Cuba, Puerto Rico, Curacao, Venezuela (Margarita), Costa Rica, Quintana Roo (Mexico). Very similar to *Doto chica*, with which it has been confused.

Doto chica **Ev. Marcus & Er. Marcus, 1960**

Size: to 8 mm Dotidae

ID: Body elongate, slender; 4-9 pairs of cerata per side of body; cerata up to one-fourth body length, up to 4 rings of tightly-packed round to oblong tubercles, up to 8 tubercles per ring, **apical tubercle prominently elongated**; pseudobranchs small, knob-like, ~3 on bottom inner side of large cerata. Rhinophores conical, long, thin, smooth; sheath low, rim smooth, no prominent forward expansion. A longitudinal ridge (keel) extends forward from the base of each rhinophore. Anal papilla between 1st and 2nd right cerata. Ground color translucent gray; random brown/blackish markings usually on back and sides, some scattered opaque white and pale blue spots, internal organs appear cream-white. Cerata brownish, white-spotted, tubercles have bluish and white dots, the apical tubercle usually lighter; dark brown blotches at the ceratal inner bases are not always conspicuous; pseudobranchs transparent. Rhinophores are transparent, have opaque white and/or bluish-white spots, sheath rims prominently white-dotted. Found among algae (e.g. *Thalassia testudinum*) in areas with muddy sand and mangrove roots. In The Bahamas, it has been reported as abundant on attached *Sargassum* in <3 feet and uncommon on floating *Sargassum* and mixed algae at 25 feet. Depth: 0-25 feet. Southeast Florida, Cuba, Puerto Rico, Bahamas, Curaçao, Margarita (Venezuela), Panama to Mexico, Brazil. Since the original description was based on a single preserved specimen, with unreliable color information ("general color is whitish, but no indication concerning the living animal is available"), the history of this species' sightings and redescriptions have been very confusing. The above color description is based primarily on the information in the books *Caribbean Sea Slugs* (Valdés et al, 2006) and *Bahamian Seashells* (Redfern, 2013).

Brazil

Riviera Beach, Florida

***Doto* cf. *chica* Ev. Marcus & Er. Marcus, 1960**

Size: to 8 mm Dotidae

ID: It differs from *Doto chica* (q.v.) by having rhinophoral sheath rims with a slight forward extension, a translucent gray to white to cream ground color, orange ceratal digestive gland branches having round yellow internal granules, pale orange rhinophoral sheaths, and prominently large apical tubercles that are more spherical than elongate. Nevertheless, it could be a variant of *Doto chica*. Found among algae near hydroids to ~10 feet in Southeast Florida.

with parasitic copepod egg sacs; Puerto Rico

***Doto* cf. *cabecar* Ortea, 2001**

Size: to 14 mm Dotidae

ID: Body elongate and slender. A row of 5-6 cerata per side of body, third ceras from the front is largest; cerata have ~4 rings of tightly-packed large rounded tubercles; the one apical tubercle may be somewhat larger, the subapical ring has ~6 tubercles; the pseudobranch on the bottom inner side of large cerata consists of 4 elongate narrowly-triangular tubercles. Rhinophores are slender, smooth, cylindrical, ~3 times the sheath length; the sheath rim is flared and slightly "beaked" forward. The back is dark brown/blackish with a few scattered orange spots and streaks. Cerata and tubercles are pale orange, tubercle apices with a pale blue internal sphere having a dark blue central spot ringed with white dots. Rhinophores are transparent with white spotting; sheaths transparent, rims white-spotted. Lives on or near hydroids, e.g. *Thyroscyphus marginatus*. Spawn is a string of white eggs deposited on a hydroid branch. Depth: to at least 60 feet. The pictured 8 mm individual differs from existing published descriptions by having pale orange instead of yellow cerata.

***Doto* cf. *divae* Ev. Marcus & Er. Marcus, 1960**

Size: to 10 mm Dotidae

ID: Elongate; foot narrow; oral veil wider than long, corners rounded; small rounded hump (keel) in front of each rhinophore. Cerata in a row of at least 5 per side, size decreases to rear, each ceras has many elongate tubercles with rounded tips, the single apical tubercle being significantly longer; each pseudobranch is a small lobed tubercle on the inner basal side of a large ceras. Rhinophores are thin and smooth; sheath about one-third length of rhinophore; sheath rim flared, slightly jagged, distended forward. Ground color translucent gray. Back and sides black with many white spots. Ceratal tubercles yellowish to pale orange, whitish tip with a large blackish/bluish internal spot. Rhinophores with a thin dark central axis along their entire length; sometimes with scattered white surface dots; sheath translucent, sparsely white-dotted, rim white. *D. divae* has been found on *Sargassum* and other algae near hydroids that it can suck nutrients from. Depth: to 20 feet. *D. divae* is known from Florida, Puerto Rico, Antigua, St. Kitts, St. Lucia, Barbados, Bonaire, Brazil. Rare. Identification is tentative. It has been repeatedly stressed in recent literature that additional studies of West Atlantic *Doto* species are required to remove inconsistencies and resolve potential synonymies, thereby allowing for more reliable identifications.

Indian River Lagoon, Florida

Bahamas

***Doto* cf. *caramella* Er. Marcus, 1957**

Size: to 8 mm Dotidae

ID: Body elongate, slender, tapers to rear; tail short and blunt; oral veil with a shallowly convex front and rounded lobes at its front corners. There are 5-6 pairs of well-spaced cerata; broadly ovate on a narrow basal peduncle; each ceras with 4 horizontal rings of ~6 rounded tubercles and a single larger moderately extended one at the apex; tubercles are well-separated both horizontally and vertically. Rhinophores are conical, long, slender, smooth; sheath low, rim flared and elongated anteriorly. The anal papilla is between the front two right cerata. Ground color translucent grayish-white. Back and sides of body variably pigmented with dark brownish-purple streaks, spots, and blotches. Rhinophores and sheaths both have some white dots; both also with brownish-purple markings – sheaths on sides, rhinophores as a broken internal axis. Ceratal digestive gland branches are dark brown to brownish-orange, tubercles opaque white. Found on or near green algae (*Caulerpa* and *Cladophora* spp.) in shallow water. Bahamas, Grand Cayman, Brazil. A rare species. Marcus described *Doto caramella* based on six Brazilian specimens, one in 1957 and five in 1967. All of them had pointed tubercles, which the pictured animal does not possess. Otherwise, the match to *D. caramella* is good.

all photos from Riviera Beach, Florida

***Doto* cf. *duao* Ortea, 2001**

Size: ~6 mm Dotidae

ID: Elongate; tail short, tapers abruptly; oral veil with rounded lobes; low hump (keel) in front of each rhinophore; a line of 5-6 tall narrow cerata along each side of back; each ceras with ~3 rings of rounded tubercles, topmost ring with 5-6 tubercles, the single apical tubercle significantly elongated. Rhinophores long, thin, smooth; sheath one-fourth to one- third length of the rhinophore, rim flared, distended forward. Finger-like pseudobranchs are on the inner basal sides of the cerata, 1 to 3 in number (small to large cerata, respectively). The anal papilla is between the 1st and 2nd cerata on the right hand side. Ground color translucent gray to cream. The back and sides are mottled with reddish-brown. The ceratal digestive gland is pale orange with off-white to yellowish granularity; a dark reddish-brown blotch at the inner base of each ceras. Tubercles have an internal blue/black spot and pale blue surface reflections. Pseudobranchs clear with white and/ or blue dots. Rhinophores and sheaths transparent and white-dotted – rhinophores distally, sheaths mainly on rim. Anal papilla white. On and near hydroids. Depth: ~10-20 feet. Southeast Florida. (*Doto duao* has been reported only from Costa Rica). Although the pictured animals agree reasonably well with the original description of *Doto duao*, they do possess more rings of tubercles and more tubercles per ring. The identification is therefore tentative.

Riviera Beach, Florida

Destin, Florida

Riviera Beach, Florida

egg string; Guadeloupe

***Doto* cf. *curere* Ortea, 2001** **blackstalk doto**

Size: to 7 mm Dotidae

ID: Elongate, slender, tapers to rear; tail short, blunt. Oral veil rounded at front, lateral lobes protrude slightly, an elongate keel in front of each rhinophore (top rounded). A row of 7-8 cerata per side of body, each ceras with many bluntly conical tubercles – the largest tubercle at the apex, up to 7 rings of 4-6 smaller ones below, the largest ceras is 3rd or 4th from the front; a pseudobranch (gill) on the inner face of large cerata extends half-way up from the base, consisting of a central axis with 3-4 rounded branches on each side. Rhinophores are slender, smooth, bluntly-tapered, retractable; the sheath is considerably shorter than half the length of the rhinophore, rim flared and distended forward/out/up. Anus between front two right cerata atop a very prominent papilla with a flared apex. Ground color translucent gray. Diffuse black smudges on the back and sides. Cerata white-dotted at the base; tubercles brownish-pink on the lower half, the upper half whitish with metallic blue surface reflections and an apical blue spot. Rhinophores with black internal axis and white dots on surface; sheath translucent with white-dotted rim. Keels and front edge of oral veil also white-dotted. Anal papilla white. Lives and feeds on hydroids; bores holes into them and sucks out their fluids. Spawn is a gelatinous wavy string of yellowish eggs deposited on a hydroid's branches. Depth: to at least 20 feet. Costa Rica, Southeast and Northwest Florida, Guadeloupe, Martinique. The original description of *Doto curere*, based on a single live specimen from the Caribbean coast of Costa Rica, includes a sketch of a ceras that shows the tubercles as **elongate conical**. The animals in the accompanying photos match the original description except for the shape of their tubercles. The identification is therefore tentative.

all photos from Riviera Beach, Florida

Doto fluctifraga Ortea & Pérez, 1982 **wave-breaking doto**

Size: to 8 mm Dotidae

ID: Body elongate, slender, small; tail short, broad, narrowly rounded. Oral veil rounded, broader than long, with a rounded keel in front of each rhinophore. One longitudinal row of 4-5 club-like cerata per side of body, the front two or three the tallest; each ceras has many hemispherical tubercles – the largest at the apex, 3 rings of 5-6 smaller ones below; a pseudobranch on the inner face of large cerata extends half-way up from the base as either a **single keel, a double keel, or two stacked inverted "U"s** (for 2 mm, 5 mm, and 8 mm animals respectively). Rhinophores are slender, smooth, bluntly-tapered, retractable; the sheath is up to half as long as the rhinophore, its rim broad and slightly distended forward. The anus is on top of a papilla between the first two right cerata. Ground color semi-translucent gray with visible gray to beige viscera. Body and head with irregular black/brown markings. The ceratal branches of the digestive gland are gray; the tubercles white with a black apical spot, which may be absent or inconspicuous in smaller/interior tubercles. Rhinophores transparent, black-spotted below and white-dotted above, the sheath rim white-dotted. The inner faces of the pre-rhinophoral keels are black-spotted. The anal papilla is white. The foot and tail are unpigmented. It lives on or near hydroids of the family Campanularidae, usually in turbulent shallow water (the Latin word "fluctifragus" means "wave-breaking"). It eats by boring holes into the hydroids and sucking out their fluids. Its spawn consists of a transparent gelatinous matrix containing white eggs, deposited near the hydroid's base. Depth: to 15 feet. An eastern Atlantic species (Azores to the Canary Islands) that was recently recorded in Southeast Florida, Bahamas, and Guadeloupe. Rare in the West Atlantic.

Roatan

Doto* cf. *fluctifraga
Ortea & Perez, 1981
Size: to 8 mm Dotidae
ID: The external morphology of the pictured animal differs from that of *Doto fluctifraga* as follows: the largest ceras of the pictured individual has five rings of tubercles instead of only three; the rims of the rhinophoral sheaths are flared much more. Its coloration differs as follows: the ground color is translucent yellowish instead of semi-translucent gray, the tubercle apices have some blue reflections, and the rhinophores lack Doto *fluctifraga's* typical black markings. Found on a hydroid. Depth: ~15 feet.

***Doto pita* Er. Marcus, 1955 pita doto**
Size: to 7 mm Dotidae
ID: Body elongate, slender, small; tail slender, pointed; oral veil rounded, inconspicuous; one longitudinal row of 4-6 cerata per side of body, the front three largest and similar in size; each ceras jagged-looking with a few small, irregularly-placed, rounded tubercles; rhinophores long, slender, smooth, bluntly-tapered, retractable; sheath narrow, 20-30% length of rhinophore, rim circular and slightly flared; anus on a papilla attached to rear of base of right front ceras. Ground color semi-transparent gray. Back with many white spots and some diffuse brown to orange-brown mottling, darkest anteriorly. Hermaphrodite gland prominently visible as a series of large white lobes extending back from between first and second cerata. Ceratal stalks and tubercles transparent, branches of digestive gland white, glands in tubercles sometimes spherical, tubercles with small white tips, stalks and tubercles may have reddish-orange to brownish surface spotting, inner base of cerata may have reddish-brown spot. Rhinophores and sheaths transparent and brownish internally; rhinophores white-dotted, especially on upper half, sometimes with orange spots (the *D. iugula* color variant); sheath rim white-dotted. Anal papilla white. Found on hydroids (e.g. *Lobophora variegata*) or on nearby algae (*Thalassia*, *Sargassum*) in shallow water. Feeds by boring holes into the hydroids and sucking out their fluids. Deposits a wavy string of white eggs on hydroid branches. Depth: 0-80 feet. Florida, Bahamas, Antilles, Costa Rica, Quintana Roo (Mexico), Brazil.

Destin, Florida

Riviera Beach, Florida

Cuba

***Doto pygmaea* Bergh, 1871**
Size: to 10 mm Dotidae
ID: Body elongate, dorsally convex; tail short, broadly triangular. Rhinophores are conical, retractable, the sheath rims even to lobate. Cerata are tall, bulbous-tipped, paired in a row of 4-7 per side (usually 5), each ceras concave on the lower half of the inner face; 15-30 tubercles on the outer face, 0-2 on the upper inner face, a ceratal apex with a larger "bulb"; no pseudobranchs. Ground color translucent gray to nearly transparent. The dorsum and flanks have a variable density of blackish spots and blotches. The ceratal surface is either nearly opaque due to variable pigmentation in shades of black, or transparent and revealing an orange-pink interior; the tips of the tubercles are white. The front of the head, tip of the tail, rhinophoral sheaths, and the proximal inner face of the cerata are unpigmented. It inhabits the hydrozoan *Obelia geniculata* that is found on floating algae (*Sargassum*, *Lepas*) or detritus. Amphiatlantic. Bermuda (Sargasso Sea), Georgia, Bahamas, southern Cuba.

Riviera Beach, Florida

Riviera Beach, Florida

Dominica

egg string; Riviera Beach, Florida

egg string; Martinique

Doto torrelavega **Ortea & Caballer, 2007**

Size: to 10 mm

Dotidae

ID: Elongate, slender, arched body with scattered conical papillae, most numerous, largest, and widely spaced on the flanks. Foot narrow, tapers to a short blunt tail. Oral veil rounded, wider than long. A stout vertical keel extends forward from each rhinophoral base, its front vertical and its oblique top angling down to the rhinophoral base. Cerata in a row of 6-7 per side, the 2nd or 3rd from the front largest; each ceras has many globose tubercles – one apical, at least 4 rings below it, a ceras' lower inner half often with one (sometimes 2) large pseudobranch shaped like an open hand with 4-6 fingers. Rhinophores are conical, smooth, bluntly-tapered, retractable; sheaths are about half the length of the rhinophore, sometimes tuberculate near the middle, rim jagged, front slightly extended. Anus atop a prominent tuberculate papilla between the inner sides of the front 2 right cerata. Ground color orange with iridescent gold-yellow dots. Ceratal stalks and tubercle interiors grayish-white to bluish-white; **tubercle surface with apical orange band**; pseudobranchs bluish-white. Rhinophores are yellowish-orange, white-speckled; sheath paler than the body. Anal papilla golden-orange with white dots. Found in areas with rocks and detritus on *Thyroscyphus* hydroids (e.g. *T. ramosus* and *T. marginatus).* Feeds on the hydroids by boring holes into their tissues and sucking out their fluids. The original description mentions the spawn as a short wavy ribbon, 8-10 eggs high and 2-3 eggs thick, pink in color. The accompanying photos show a white ribbon deposited on a hydroid branch. Depth: 12-80 feet. Southeast Florida, northern Cuba, Dominica, Martinique, St. Vincent. Rare.

Roatan

Roatan

Grand Cayman

Roatan

***Doto* cf. *uva* Er. Marcus, 1955**

Size: to 11 mm

Dotidae

ID: Body elongate, slender, tapers to the rear; tail moderately long and pointed. Cerata in a row of up to 6 per side of body, size decreasing rearwards, each ceras with ~5 rings of rounded tubercles and a single larger apical tubercle. Rhinophores thin, smooth, bluntly tapered; sheath smooth, rim flared, rim distended forward. Ground color white to grayish. Head, back, and sides with dark brown to blackish spots and patches. The ceratal digestive gland branches are pale orange to pinkish; the tubercles white to bluish. Rhinophores are transparent with a dark central axis, some white dots, a white tip; sheath translucent, white-spotted, the rim white. Inhabits areas with hydroids. Depth: to 20 feet. Roatan, Grand Cayman. *Doto uva* was first described by Marcus from a single preserved Brazilian specimen, the description updated based on 2 live Brazilian specimens (Marcus, 1957), and then again on one live specimen from Key Largo, Florida (Marcus, 1960). Although the external morphology for the animals was reasonably consistent, the coloration was not. The body color ranged from "white with light orange viscera that appeared pink in the cerata" (1955) to light pink or light yellow (1957), to "irregularly light brown" (1960). The head and dorsum had some black spots and the rhinophoral sheath had a blackish ring (1955); black pigment was not mentioned in 1957; melanophores were present on rhinophores and sheaths, the back and flanks, and the upper tubercles in 1960. A later description of Jamaican specimens by Thompson (1980) differed even more, both in color and external morphology, leading to a suggestion (Ortea, 2001) that the vagueness and inconsistencies of the early Brazilian descriptions had led to many misidentifications. In fact, Ortea contends that *D. uva* may not even exist in the Caribbean at all. The book *Caribbean Sea Slugs* (Valdés et al, 2006) includes photos from Puerto Rico and Costa Rica that add yet another variation to the mix – a dark to almost black background color, bluish tubercles, and rhinophoral sheath with a bluish rim.

all photos from Riviera Beach, Florida

Doto varaderoensis **Ortea, 2001** **Varadero doto**

Size: to 10 mm Dotidae

ID: Body elongate, slender; tail slender, rounded. Cerata in one longitudinal row of up to 7 per side of body; middle ceras largest, with up to 5 crowded rings of rounded tubercles; apical tubercle largest, topmost ring with up to 7 tubercles; pseudobranch with up to 5 upward-pointing "fingers" at the inner base of larger cerata, its middle finger largest (can be taller than the ceras). Rhinophores very long, slender, smooth, tips rounded, partially retractable; sheaths much shorter, the rim's edge flares forward; **pre-rhinophoral keels are** elongate, low, and rounded. Ground color translucent gray. The back and sides may have faint reddish-brown mottling and many white spots; the whitish/pinkish/orange mass of the digestive gland is clearly visible through the body walls. Cerata with white to cream digestive gland branches, stalk base reddish-brown, inter-ceratal spaces whitish; **tubercles translucent and faintly bluish, with an internal small blue/blackish sphere** (sometimes indistinct). Pseudobranch fingers clear, with small blue/black dots. Rhinophores and sheaths translucent, white-spotted, its black axis sometimes fragmented, **the sheath rims with 9-11 blue dots**. Keels white-spotted anteriorly. It has been found on *Thyroscyphus* sp. hydroids, where it bores holes into tissues and sucks out fluids. Deposits white eggs in a wavy gelatinous ribbon. Depth: 3-20 feet. Southeast Florida, Cuba, Puerto Rico, St. Vincent, Roatan.

Roatan

Doto wildei Er. Marcus & Ev. Marcus, 1970

Size: to 10 mm Dotidae

ID: Body elongate and slender. About 7 pairs of cerata in 2 rows along the sides of the back, those in each row almost touching, each ceras with 4-5 horizontal rings of densely-packed mostly rounded tubercles and a single larger one at the apex, the overall appearance resembling a bunch of grapes; however, some neighboring tubercles in a ring coalesce into extended horizontal structures and, if all of them coalesce, can even change the ring into a bumpy "donut". Rhinophores are slender and smooth; a sheath is less than half as long as the rhinophore and relatively snug; a sheath's rim is smooth, slightly flared, and distended forward; a pseudobranch is at the inner base of each ceras, and consists of 1-4 finger-like structures of variable height. The general appearance is orange with white spots and some dark brown markings on the back; head translucent gray. The tubercles are paler orange to yellowish. Rhinophores are clear, white-spotted; their sheaths translucent with a few white spots, the sheath rims densely white-spotted. Pseudobranchs are transparent. Found in shallow water on hydroids and on algae and sponges near hydroids. Spawn is a wavy gelatinous ribbon containing many round orange eggs, deposited on a hydroid branch. Depth: 3-25 feet. Roatan, Southeast Florida, Curacao.

St. Lucia

Roatan

Roatan

Riviera Beach, Florida

Doto cf. *wildei* Er. Marcus & Ev. Marcus, 1970

Size: to 10 mm Dotidae

ID: See notes for *D. wildei*. Coloration is like that of *D. wildei* but with one notable exception – the rhinophores have a long and prominently dark internal axis, a feature that has not been mentioned in the scientific papers describing that species. Their spawn is a wavy gelatinous ribbon containing many round yellowish eggs, deposited on a hydroid branch. Depth: 3-25 feet. The pictured slugs have been found in shallow water on hydroids and on algae/sponges near hydroids. St. Lucia, Roatan, Southeast Florida. The species *Doto awapa* Ortea 2001, described from the Caribbean side of Costa Rica, is similar enough to *D. wildei* to be regarded by some taxonomists as the same species, but **its rhinophores do have a dark internal axis**. Although the only description of *D. awapa* does contain some significant unique features, the possibility remains that the pictured slugs could be that species.

The following several pages contain numerous examples of *Doto* spp. with indeterminate identifications. Some may be color variants of previously described species, some may be new species, some may be variants of one another. Because of significant overlaps in the appearances of known species the uncertainty in the reliable identification of any of them is quite large. Further study is needed for this group.

Roatan

Doto sp. 1

Size: ~8 mm Dotidae

ID: Body elongate, slender, small. Cerata in longitudinal crowded row of 5-6 per side, each ceras with numerous rounded tubercles stacked as horizontal rings, the apical tubercle largest; large multi-lobed pseudobranchs are present low on the inner ceratal faces. Rhinophores are long, slender, smooth, conical; the sheaths are less than half the length of the rhinophores, their **rims smooth, broadly flared, crenulate**. Fleshy pre-rhinophoral keels are present. The oral veil is prominently extended laterally. Ground color tan. Head with dark brown mottling; pre-rhinophoral keels and front of oral veil white-dotted. The ceratal digestive gland is yellowish to pale orange. The **tubercle apices have a dark spot centered in an area with pale blue reflections**. Pseudobranchs are transparent with white internal structures. The rhinophores are white-dotted, lack a dark internal axis; the sheaths are white-dotted, especially on the rim. Found on hydroids. Depth: ~20 feet.

Riviera Beach, Florida

Doto sp. 2

Size: ~8 mm Dotidae

ID: Body elongate and slender. Cerata in one longitudinal row of 5-6 per side of body; the first and second cerata well separated, the second and third the tallest; each ceras with numerous rounded tubercles, the apical tubercle largest; there are large pseudobranchs low on the inner ceratal faces. Rhinophores are long, slender, smooth, conical; the sheaths less than half the length of the rhinophores, their rims smooth and flared. A pre-rhinophoral keel is present. The oral veil is prominently extended laterally large. The anal papilla is between the first and second cerata on the right side. Ground color tan. Head, back, and flanks have dark brown mottling (less so on the pericardial hump) and a scattering of white dots. The ceratal digestive gland is brownish; some tubercles are yellowish-white to pale orange and distributed irregularly; others are pale blue with a prominent apical black spot and arranged in rings. Rhinophores are variably white-dotted, with either no internal axis or a few small black fragments; sheaths white-dotted, especially on the rim; pre-rhinophoral keels white. Pseudobranchs are clear with some white dots. Anal papilla white. Found on hydroids. Depth: to 20 feet.

Riviera Beach, Florida

Doto sp. 3

Size: ~10 mm Dotidae

ID: Body elongate, slender, small. One longitudinal row of 8 cerata per side of body, each ceras with many tightly packed rounded tubercles. Rhinophores are long, slender, and smooth; the sheath is one-third the length of the rhinophore; its rim is broadly flared, smooth, and slightly distended forward. The oral veil is broad, its front rounded. The large anal papilla is in front of the second ceras on the right side. Ground color cream. The dorsum has a dark reddish-brown to blackish median stripe with scattered white dots; the stripe extends onto the oral veil. Ceratal stalks pale orange; tubercles white with bluish apices; pseudobranchs at lower inner sides of cerata translucent bluish. Rhinophores are transparent, a broad black central axis along the entire length, tip white, surface white-dotted; sheaths transparent, white-dotted, rim white. The anal papilla is grayish, its apical rim white dotted. Found on a hydroid. Presumably feeds on them. Spawn is a wavy ribbon of densely-packed white eggs deposited on a hydroid branch. Depth: to ~15 feet.

Destin, Florida

Doto sp. 4

Size: ~10 mm Dotidae

ID: Body elongate, slender. A row of 5-6 cerata per side; each ceras with ~2-3 rings of up to 5 small spherical non-touching tubercles per ring, apical tubercle largest. Rhinophores long, slender, smooth; sheath half the length of the rhinophore, the rim smooth and flared. Dorsum black, white-dotted. Ceratal digestive branches are yellowish and grainy; many tubercles are blue internally. Rhinophores black, white-dotted, tips white; sheath black, rim white. Found on a hydroid. Spawn is a U-shaped cylinder crowded with white eggs. Depth: ~70 feet.

Martinique

Doto sp. 5

Size: ~10 mm Dotidae

ID: Body elongate, slender. A row of ~5 well-separated cerata per side; each ceras with ~3 rings of up to 6 rounded tubercles (largest at apex). Rhinophores conical, long, slender, smooth; sheath one-third length the rhinophore, rim smooth, distended forward. Whitish; back with black blotches, sides dark-dotted; cerata with pale orange digestive gland branches, tubercles white to bluish, black dot at apex; rhinophores transparent, with a full-length blackish axis, distally white-dotted; sheaths clear, also white-dotted. Found on a hydroid. Depth: ~20 feet.

Doto sp. 6

Size: ~8 mm Dotidae

ID: Body elongate, slender. A row of ~5-6 crowded cerata per side; each ceras with ~4 rings, each ring with as many as 6 spherical tubercles, the largest being apical. Rhinophores are cylindrical, long, slender, smooth; the sheath somewhat short and with a flared rim. Cerata pale orange, apices whitish. Rhinophores blackish with white dots. Found on hydroids, which are probably their prey. Spawn is a long string of orange eggs within a clear gelatinous matrix that has been deposited on the stalk of a hydroid. The pictured animals were found at a depth of ~20 feet. *D. wildei* is similar but its rhinophores are not blackish. The species *Doto awapa* Ortea, 2001, which is regarded as a possible synonym of *D. wildei* in *Caribbean Sea Slugs* (Valdés et al, 2006), **does** have rhinophores with a black central axis, but its distinctive egg mass is quite different from that shown in the left photo.

Grand Cayman

Doto sp. 7

Size: ~10 mm Dotidae

ID: Body elongate, slender; tail broad, abruptly pointed. Cerata in a longitudinal row of 7 well-separated cerata per side of body, second pair tallest; each ceras with several rings of rounded tubercles, apical tubercle prominently longest; pseudobranch above the rear base of each ceras. Rhinophores long, slender, smooth; sheath about one-third the length of the rhinophore, rim smooth, flared; a broadly rounded keel in front of each sheath. Anal papilla prominent between the first two right cerata. Dorsum creamy-white with a median streak of brownish-orange blotches; a few white dots, especially along the inner ceratal bases. Head/tail translucent gray with brown mottling and a surface sprinkling of white dots. Cerata cream, the stalks brownish at the inner side of the basal peduncle, apical tubercles transparent and white-dotted; pseudobranchs translucent gray, white-dotted. Rhinophores transparent, surface white-dotted, a broken brown axis along the proximal two-thirds; sheath transparent, white-dotted, its rim greenish & white-dotted. Anal papilla and tops of pre-rhinophoral keels white. At ~50 feet on a hydroid. The pictured animal resembles the *Doto chica* on page 216 of *Caribbean Sea Slugs* (Valdés et al, 2006), but the pseudobranchs of that species are described as being at the **inner** bases of the cerata, not above the bases on the ceratal posteriors.

Martinique

Martinique

Doto sp. 8

Size: ~10 mm Dotidae

ID: Elongate, slender; a row of 4-5 cerata per side; each ceras with clusters of elongate tubercles. Grayish-white; back/flanks with some black spots; cerata white, tubercles with an apical black spot. Found on a *Halocordyle disticha* hydroid at ~7 feet.

Northwest Florida

Doto sp. 9

Size: to 8 mm Dotidae

ID: Elongate and slender. A row of crowded cerata per side; each ceras with many elongate tubercles having rounded tips. Rhinophores are long, slender, and smooth; their sheaths relatively low, with a flared rim and slightly distended forward; the large anal papilla is between the first and second cerata on the right side. Brownish with scattered white dots. Tubercles with yellow granules proximally, apex white-dotted and bluish. Rhinophores with a continuous black axis; the distal half and sheath rim white-dotted; pseudobranchs clear with some white dots; anal papilla opaque white. Found on a hydroid at ~20 feet.

all photos from St. Vincent

Doto sp. 10 St. Vincent doto

Size: ~10 mm Dotidae

ID: Body elongate and slender. A longitudinal row of 5 well-separated club-shaped cerata per side of body, each ceras with many tightly-packed globose tubercles – largest one at the apex; pseudobranchs low, multi-lobed, at the inner bases of the cerata. Rhinophores are long, conical, and smooth; their sheaths are about one-third the length of the rhinophore, flared, rim distended forward. The body appears bright orange due to clearly visible viscera; all surfaces are white-dotted. Cerata are penetrated by orange branches of the digestive system, tubercles are white to yellowish to orange, pseudobranchs are purplish. Rhinophores and sheaths are white-dotted (the rhinophores densely); the sheath rims are white. Found on a hydroid. Presumably feeds on it. Spawn consists of orange eggs in a transparent gelatinous string deposited on a hydroid branch. Depth: ~15 feet. St. Vincent, Grand Cayman. Resembles *Doto torrelavega*, but the rhinophores are too white and the ceratal tubercles have neither bluish internal reflections nor a prominent orange ring. More studies are needed to determine possible intraspecific variations in that species.

Doto sp. 11

Size: ~10 mm Dotidae

ID: The cerata have ~4 rings of small rounded tubercles; each ring with at least 12 tubercles, the single one at the ceratal tip much larger. Rhinophores cylindrical, thin, project from a wide sheath that is about half the length of the rhinophore; the sheath rim is somewhat jagged but not distended forward as a "beak". A prominent narrow keel in front of each rhinophore. Body and head mottled with reddish-brown; ceratal digestive gland branches orange, tubercles whitish with a pale blue internal sphere and a blackish apex; rhinophores densely white-spotted with some dark axial spotting, sheath rims black-spotted. The pre-rhinophoral keels are white. Several *Doto*'s have been described as having black pigment on their sheath rims, but in all cases some of their other distinguishing features exclude those species as a match for the pictured animal.

Puerto Rico

Guadeloupe

Doto sp. 12

Size: ~10 mm Dotidae

ID: Rhinophores long, thin, smooth; sheaths broad, length about one-third rhinophore length, rim smooth; low keel in front of each rhinophore; a row of 6-7 cerata per side, up to 7 rings of round crowded tubercles and a single apical one. Head and body with black/white mottling. Rhinophores translucent, white-dotted, with a black internal axis and white tip; sheaths gray, white-dotted; rims and keels white. Not a good match for any described West Atlantic species.

Riviera Beach, Florida

Doto sp. 13

Size: ~10 mm Dotidae

ID: Rhinophores long, thin, smooth; sheaths broad, about one-third the rhinophore length, rim smooth and slightly distended forward. There are about 6 cerata per side, each with many small elongate side tubercles; apical one large and conical, 2-3 pseudobranchs at its inner base. Mottled with black and pale orange. Rhinophores translucent, white-dotted, with a black internal axis and white tip; sheaths whitish. Cerata brownish-orange, tubercles white with a dark blue spot, white-tipped apical tubercle, pseudobranchs transparent and white-dotted.

with parasitic copepod egg sac; Puerto Rico

Doto sp. 14

Size: ~11 mm Dotidae

ID: Elongate, slender. A row of 5-6 cerata per side, each with ~5 rings of crowded rounded tubercles and one at the apex. Rhinophores are slender, smooth, cylindrical, and much longer than the sheath. Dorsum pale orange. The cerata and tubercles are yellow to pale orange, the tubercle apices hemispherical with faint blue reflections, scattered white dots, and a central blackish spot. Rhinophores opaque tan, tips white. Found on a hydroid at ~20 feet. Perhaps a variant of *Doto cabecar*, but that species has translucent rhinophores.

Guadeloupe

***Doto* sp. 15**

Size: ~10 mm Dotidae

ID: The body is elongate and slender. There is a row of 5 to 6 crowded cerata per side; each ceras is an ovoid cluster of small round tubercles arranged in about 4 rings. The rhinophores are long, slender, smooth; the sheaths about one-third the length of the rhinophore, flared, distended forward. The ceratal digestive gland is yellow and prominently granular with many small gold spheres. The tops of the tubercles are white with a central black dot, the apical tubercles have an orange-gold ring. Rhinophores are transparent with a dark brown and somewhat broken central axis, surface white-dotted, tip white to bluish; the sheath transparent and white-dotted. It was found on a hydroid. Presumably feeds on it. Depth: 6-80 feet. *D. torrelavega* has also been described as having a general golden coloration and an apical orange-gold ring at the top of some tubercles. However, the pictured individual differs as follows: (i) rhinophores have a dark internal axis (ii) rhinophoral sheaths do not seem to have a lacy/jagged rim (iii) tubercles have a black dot at the apex. Nevertheless, due to the rarity of *D. torrelavega*, it remains possible that this is a heretofore unknown color variant of that species.

Riviera Beach, Florida

***Doto* sp. 16**

Size: ~5 mm Dotidae

ID: Cerata globose, without distinct tubercles; rhinophores short and cylindrical; sheath half the length of the rhinophore, rim snug against it. A pale orange digestive gland, head mottled with black, flanks blackish above and white below. Cerata pale orange, apices whitish with black dots and some bluish highlights. Rhinophores clear, white-dotted, distally with black internal pigment; sheath rims white, black-spotted just below. Found on a hydroid. Depth: ~10 feet.

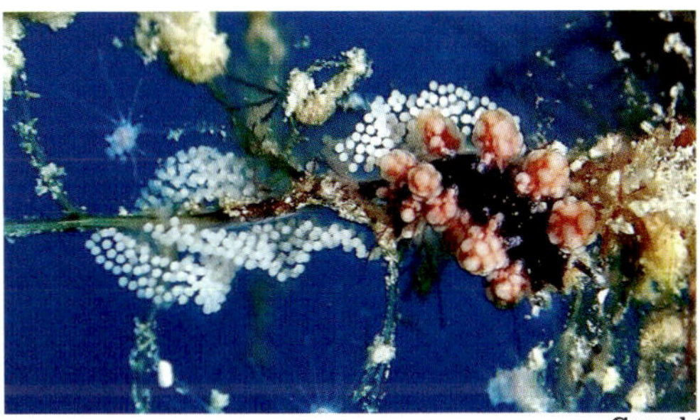

Grand Cayman

***Doto* sp. 17**

Size: ~5 mm Dotidae

ID: Body elongate; 5-7 pairs of well-spaced cerata; each ceras relatively narrow and cylindrical with ~5 rings of small grape-like tubercles. Rhinophores long, conical, smooth, sheaths slightly flared and distended forward. Ground color grayish, body and head blackish-brown. Ceratal digestive gland branches pink, tubercles whitish. Rhinophores white-dotted with a broad dark inner axis, tips and sheath rims white. From mixed algae. Depth: to ~3 feet. Bahamas, Grand Cayman. This animal is a good case study for demonstrating uncertainty in *Doto* identification. One specimen was collected by Colin Redfern and shown as image 736A in his book *Bahamian Seashells* (Redfern, 2001), tentatively identified as *Doto uva*. Ortea (2001) suggested that it couldn't be *Doto uva* because that species does not occur north of Brazil, and later that it was probably the *Doto chica* that he had just redescribed in that paper. In 2013 Ortea applied that redescription to define a new species – *Doto casandra* – and *Bahamian Seashells* (Redfern 2013) now lists this animal as *Doto* sp. B. So, is it *Doto uva*, *Doto chica*, *Doto casandra*, or something else?

Doto sp. 18
Size: ~10 mm Dotidae
ID: Body elongate, slender, tapers to the rear; tail short and blunt; 5-6 pairs of well-spaced cerata, narrowly ovate on a narrow basal peduncle; each ceras has 4 horizontal rings, each ring has ~6 rounded tubercles and a single larger moderately extended one at the apex; tubercles well-separated both horizontally and vertically. Rhinophores conical, long, slender, smooth; the sheath low, rim flared and elongated anteriorly. Ground color translucent grayish-white. Back and sides of body variably pigmented with dark brownish-purple streaks, spots, and blotches. The proximal two-thirds of the rhinophores with a brownish-purple fragmented internal axis; the distal third white-dotted. Ceratal digestive gland branches dark brown to orange-brown, tubercles opaque white. No ecological information available. Depth: ~10-20 feet. The closest match is *Doto caramella* Er. Marcus, 1957, but its cerata are broadly ovate with pointed tubercles.

Grand Cayman

Guadeloupe

Doto sp. 19
Size: ~10 mm Dotidae
ID: Body elongate and slender. Six pairs of cerata with rings of crowded rounded tubercles; pseudobranchs at inner bases of cerata. Rhinophores are cylindrical, slender; sheaths elevated, broad, rim flared and distended slightly forward. Body mottled reddish-brown; ceratal tubercles white with an apical black dot; pseudobranchs transparent with a black dot and white spots; rhinophores and sheaths translucent gray, white-dotted; rhinophores with proximal fragmented black axis.

Riviera Beach, Florida

Doto sp. 20
Size: ~10 mm Dotidae
ID: Body elongate. Rhinophores are cylindrical, slender, long, tapering; the sheaths are relatively low and snug around the rhinophores, the rim flared and slightly distended forward. There are several cerata per side, each a tall conical cluster of round yellowish tubercles and dark digestive gland branches inside the ceratal stalks. The body is mottled with reddish-brown and pale orange. Rhinophores and sheaths are translucent gray, the sheath rims are densely white-dotted.

Grand Cayman

Doto sp. 21
Size: ~10 mm Dotidae
ID: Rhinophores are cylindrical and slender; sheaths low, broad, rims broadly flared. Several cerata per side, each with rings of round tubercles and one much larger at the apex. Body mottled light/dark brown, covered with white dots. Rhinophores and sheaths dark brown, white-dotted; rhinophoral tips and sheath rims bluish-white. The tubercles are white with a dark blue apex and a large internal paler blue sphere giving an overall blue cast; digestive gland branches orange.

Grand Cayman

Doto sp. 22
Size: ~10 mm Dotidae
ID: Body elongate. Rhinophores cylindrical and slender; sheath low, rim flared, distended forward. About 6 cerata per side, middle one largest, each one tall and narrow with a short basal peduncle and several rings of round tubercles. Body mottled with brown/white/orange. The rhinophores and sheaths are translucent whitish with opaque white spots, the rhinophores with a thin dark internal axis. Cerata mostly orange with some of the distal tubercles prominently white.

Guadeloupe

Doto sp. 23

Size: ~10 mm Dotidae

ID: There are 9 pairs of cerata with rings of crowded rounded tubercles. Rhinophores cylindrical, slender, long. Body pale gray, mottled with black. Tubercles are pale yellow, whitish apically, digestive gland branches brown. Rhinophores translucent gray with white spots and white tips. Among hydroids.

Panacea, Florida

Doto sp. 24

Size: ~10 mm Dotidae

ID: Body elongate and slender, tapering to an acutely pointed tail. Eight pairs of tall slender cerata, many with elongately rounded tubercles. Rhinophores short, slender; sheath elevated, rim flared, distended forward. Body mottled with brown, white, yellow. Cerata brown with black spots. Rhinophores white.

Riviera Beach, Florida

Doto or *Eubranchus* sp.

Size: ~6 mm Dotidae/Eubranchidae

ID: Body elongate; foot broad, tapers to triangular tail; oral veil rectangular, longer than broad, front corners slightly rounded; a longitudinal row of four well-spaced cerata behind each rhinophore; each ceras a long finger-like central stalk surrounded proximally by a "sheath" rimmed with 4-6 short stout lobes; tip of stalk slightly inflated; rhinophores as long as cerata, conical, smooth, with **no sheaths** (a developmental defect?). Ground color white. Body and cerata with some brown mottling and dots. White digestive gland branches penetrate ceratal stalk and lobes. Rhinophores and front of head transparent; head peppered with dark spots; rhinophores white-spotted distally, with dark dots and smudges basally. From mixed algae and hydroids. Depth: ~12 feet.

Family EUBRANCHIDAE Odhner, 1934

This family currently consists of four genera with a total of about 60 described species (most of them in the genus *Eubranchus*), many more already awaiting description, and others likely to be soon discovered through molecular analyses of cryptic species complexes. They occur circumglobally at all latitudes. Several species inhabit the tropical West Atlantic.

Body elongate, low, short; the foot usually wider than the visceral part of the body, the front corners rounded to pointed; oral tentacles are short and smooth; rhinophores are long and smooth; cerata are few, swollen, sometimes with annular constrictions and rings of tubercles, arranged in simple or branched rows; the anus is acleioproctic on the right side just behind the renal pore; radula triseriate, lateral teeth broad and rectangular with a single cusp near the inner end of the posterior edge.

They feed on small colonial hydroids. Caribbean species are typically 3-6 mm long.

Grand Cayman

Riviera Beach, Florida

Riviera Beach, Florida

Eubranchus conicla **(Er. Marcus, 1958)** **cone eubranchus**

Size: to 4 mm Eubranchidae

ID: Body elongate, tapers posteriorly; foot narrower than body, tapers to a short pointed tail. Cerata are somewhat to highly inflated, tuberculate, with a sharp-tipped globose apical knob, the tallest ceras almost half a body length; cerata are arranged in 3 clusters per side – a precardiac one with 3-6 cerata along an arc that peaks inwards, and 2 postcardiac ones aligned transversely with 1-3 cerata each. Rhinophores are conical, smooth, long; oral tentacles are similar but thinner and shorter, usually bent down and recurved. The anal papilla is prominently elevated behind the pericardium right of the dorsal midline. Ground color translucent gray; variously dotted/mottled with opaque white; brown or greenish spots on the back and head, sometimes in patches on the body's sides and the rear of the foot. Cerata white to cream, sometimes with brown/greenish spots; tubercles white-dotted; apical knob translucent gray/whitish with a white tip and a dark subapical band. Rhinophores and oral tentacles ground color with 1-2 dark bands. Found on sand, grass, and algae (e.g. anchored *Sargassum*). Cerata detach easily. Depth: to ~27 feet. Uncommon in Southeast Florida, Cuba, Jamaica, Barbados, Tobago, Venezuela, Panama, Costa Rica, Honduras, Brazil. Common in Bahamas. Some of the pictured animals lack a dark subapical ceratal band, a feature that was present in Marcus' original description. Most subsequent records of this species have considered the band's absence to be an intraspecific variation, but that is an assumption that needs to be tested.

Riviera Beach, Florida

Eubranchus leopoldoi **Caballer, Ortea & Espinosa, 2001** **measled eubranchus**

Size: to 67 mm Eubranchidae

ID: Body elongate, slightly elevated anteriorly, tapers to a short pointed tail. Cerata inflated, tip nippled and bulbous; arranged in a line of 3 well-separated clusters per side, the frontmost cluster with 5 cerata on an arch, the remaining clusters (behind the pericardium) each with 2 cerata in a row that inclines forward and up. Pericardial region slightly elevated. Rhinophores and oral tentacles smooth, conical, tip rounded; rhinophores long, slender, twice the length of the longest ceras; oral tentacles shorter and thinner; front foot corners short, narrow, rounded, project postero-laterally. Ground color grayish-white, crowded with reddish dots. Back and sides with some scattered white spots. Cerata translucent with a narrow reddish-brown branch of the digestive tract clearly visible, the surface dotted with red/white/orange-brown, a yellow/gold band at the most inflated part, the topmost "bulb" yellow near the tip with a bluish ring below it; areas around ceratal insertions with white pigment. Rhinophore proximal half grayish with red dots and some white spots, distal half white with a broad yellow/gold/brownish-orange irregular band (sometimes broken); oral tentacles similar to rhinophores. Foot translucent pink. Reported from areas of mixed algae, attached and floating *Sargassum*. Depth: to ~33 feet. Amphiatlantic: Japan, Bali. West Atlantic: SE Florida, Bahamas, Isla Margarita (Venezuela), Costa Rica. Rare.

Roatan

Eubranchus **sp. 1**

Size: ~6 mm Eubranchidae

ID: Body is elongate and tapers to the rear. The cerata are highly inflated up to a constriction near the top which is followed by a short and less bulbous portion that tapers abruptly to a pointed apex; cerata are smooth, in ~5 small clusters per side of back, the frontmost cluster positioned considerably behind the rhinophores. The rhinophores and oral tentacles are smooth and conical, their tips rounded; rhinophores are longer than the tallest ceras, oral tentacles shorter and thinner. The ground color is translucent grayish; a thin yellow median stripe extends from between the rhinophores to the pericardium. There is a thinner orange line from the base of the first ceras to the base of the adjacent oral tentacle, the line interrupted by a round unpigmented patch that contains a small black eye; the head's front margin is orange. Below the constriction the cerata are mostly dark with white spots; the segment above the constriction is white, has a "ring" of orange spots at its base, a thin brownish band directly above it, and a sharply pointed transparent tip. The rhinophores are whitish with two irregular diffuse brownish bands. The oral tentacles are blotched with white and brown. The pictured animals were found among algae in shallow water. Depth: ~15 feet. Roatan and Riviera Beach, Florida.

Riviera Beach, Florida

Riviera Beach, Florida

Eubranchus sp. 2

Size: ~6 mm Eubranchidae

ID: Elongate body; cerata smooth, very inflated, small bulb at top, ~5 small clusters per side of back. Rhinophores smooth, slender, conical, tips rounded, longer than the cerata; oral tentacles are similar to the rhinophores but are shorter and thinner. Eyes small, each in an unpigmented patch at the postero-lateral base of each rhinophore. Ground color translucent grayish. Head and back with irregular white and brown patches/spots; an oblique orange streak behind each eye. Cerata opaque white above, brown below, apical bulb translucent white. Rhinophores whitish with a broad brownish-orange band below mid-length. Oral tentacles ground color with white tips. Eyes black. Found among algae in shallow water. Egg capsules large and white. Depth: ~15 feet.

Guadeloupe

Eubranchus sp. 3

Size: ~5 mm Eubranchidae

ID: Body elongate; configuration of cerata difficult to distinguish – appears to be an anterior and posterior cluster on each side, possibly with 1-2 missing or regenerating clusters in between, each pair consisting of a large one and a small one, each ceras with a narrow peduncle and a bulbous tip with a somewhat elongate pointed cnidosac. Rhinophores and oral tentacles are simple and smooth; the tail moderately long. Ground color translucent grayish. Internal organs appear as whitish to yellowish. There is a crimson blotch directly over the cardiac region. Cerata whitish with faintly yellowish digestive gland branches; some cerata with a reddish-brown vertical streak at the base. Rhinophores with diffuse pale orange color proximally. There is no banding on the rhinophores or oral tentacles. Two individuals were found with species of *Halophila* seagrasses. Depth: ~10 feet. Western Guadeloupe (Bouillante).

Riviera Beach, Florida

Eubranchus sp. 4

Size: ~5 mm Eubranchidae

ID: Body elongate, tapering. A row of 4 tuberculate cerata per side (one is missing in the pictured animal). Each ceras is supported by a narrow peduncle, broadens upwards and then narrows to a pointed apical bulb. The rhinophores and oral tentacles are simple and smooth. Ground color white. Flanks with a broad irregular greenish-brown stripe; cerata white with prominent pinkish digestive gland branching, apical bulb white; rhinophores and oral tentacles white with two dark bands. Found among hydroids at ~10 feet.

Family FACELINIDAE Bergh, 1889

A diverse family of aeolid nudibranchs consisting of well over a hundred species in about thirty genera distributed throughout all oceans. The approximately ten genera that are known to occur in the tropical West Atlantic usually have the following characteristics.

Body elongate, slender, tapers to the rear; foot broad, narrower than the body, tapers to a long and pointed tail; head rounded; front of foot rounded, bilobed with a median notch, front corners triangular/tentacular; oral tentacles long, thin, smooth, tapered; rhinophores may be smooth, annulate, tuberculate, papillate (often just on the posterior face), perfoliate, sometimes with 1-3 subapical swellings, the annuli occasionally irregular or incomplete, rhinophores shorter than the oral tentacles; cerata smooth, variably shaped – long, short, tubular, tapering, curved, tips sharp to rounded, in a row of separate clusters on each side of body, clusters in the form of arches and rows; the cerata in all genera except *Favorinus* have apices with functional cnidosacs for storing nematocysts; anus cleioproctic, on the right side behind the inter-hepatic space; radula uniseriate, teeth usually triangular with a prominent central cusp flanked by several short lateral denticles.

Most feed on hydroids but species of *Favorinus* prey on the eggs of other sea slugs. When alarmed, some species wave their cerata threateningly while pointing their cnidosac-armed tips at the aggressor. Slug lengths range from 10 to 30 mm; habitat depths from very shallow to 150 feet.

Riviera Beach, Florida

St. Croix

Riviera Beach, Florida

Guadeloupe

with spawn; Riviera Beach, Florida

Learchis evelinae **Edmunds & Just, 1983** — **Eveline's learchis**

Size: to 14 mm — Facelinidae

ID: Tail long, pointed. Cerata slender, pointed, long, a row of many clusters per side of back, a pronounced gap between clusters at the pericardium. The **rhinophores are smooth**, tapering, pointed, length almost that of the longest ceras; oral tentacles one third the body length or more. Foot narrower than body, front corners recurved and tentacle-like. Ground color translucent gray. Back with varying amounts of irregular opaque white patches. Head with a prominently **white transverse patch in front of rhinophores** extending back between them onto the back; sometimes a white median streak from front of head. **Orange streak/patch across cheek** from base of oral tentacle to below base of rhinophore. **Cerata translucent with a brown tube of the digestive gland**, tip white. **Rhinophores are pale orange below, white above**, the tip white but sometimes red; sometimes it is **three-banded** – white or pale-orange in the middle and bracketed by orange. Oral tentacles are orange near the base and whitish elsewhere. Tail white. Small black eyes in large unpigmented patches behind rhinophores. Feeds on hydroids. Creates a defensive weapon by recycling stinging cells from its prey into the cnidosacs near the tips of its cerata. Spawn is a convoluted, transparent, gelatinous string with round white eggs deposited on a hydroid branch. Depth: to ~20 feet. Bermuda, Florida, Bahamas, much of the Caribbean.

Riviera Beach, Florida

Guadeloupe

Cozumel

adult & juvenile; Puerto Rico

with egg string; Roatan

Riviera Beach, Florida

Learchis poica **Ev. Marcus & Er. Marcus, 1960** **Christmas tree hydroid learchis**

Size: to 25 mm Facelinidae

ID: Front of foot with extended corners; tail short, pointed. Cerata slender, cylindrical, nipple-tipped, a row of 7-8 clusters per side of back; clusters are oblique rows ahead of pericardium, arches behind; 2-12 cerata per cluster, number decreases rearwards; middle of back open. Rhinophores long, tapering, **up to 10 widely spaced annuli that are often irregular/incomplete, sometimes giving the rhinophores a tuberculate appearance**. Oral tentacles long, slender, pointed, smooth. Eyes small, each in an unpigmented patch at rear outer edge of rhinophoral base. Ground color translucent white to silvery-gray to pale orange, with white speckling. Opaque white patches on head and behind pericardial hump, occasionally forming a broad median white stripe that reaches the tail; each flank may have an irregular line of opaque white markings; often an orange line from below rhinophore to neighboring oral tentacle, sometimes broadening to form a cheek patch. Cerata translucent clear, surface with white markings; digestive gland duct pale orange, red, brown, blackish; tips translucent white, usually with a pale orange subapical ring. Rhinophores and oral tentacles translucent gray to brownish proximally, whitish distally; bases of oral tentacles sometimes with an orange wash and a white dorsal blotch. Eyes black. Inhabits reef areas harboring algae with attached hydroids that it feeds on (e.g. the Christmas tree hydroid, *Halocordyle disticha*). Sometimes in clusters on floating algal debris, hydroids, and *Sargassum* attached to hydroids or shoreline rocks. Feeds by opening its mouth to distend its buccal mass over a hydroid polyp, cuts it off with its jaws, and ingests it. Spawn is a sinuous string of white eggs deposited on a hydroid stalk. Depth: 1-60 feet. Carolinas, Bermuda, Florida, Bahamas, most of Caribbean.

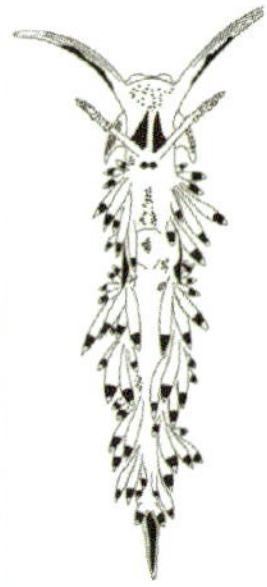

Bay of Charagato, Cubagua (Venezuela)

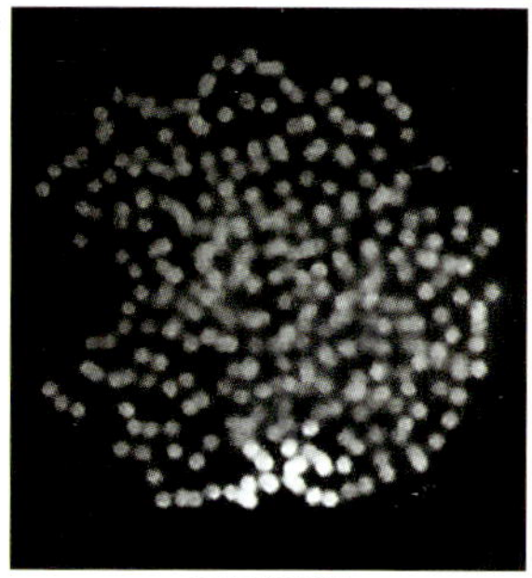

egg mass

Learchis ignis **Crescini, De Sisto & Villalba, 2013**

Size: to 13 mm Facelinidae

ID: Foot narrow, wavy, bifurcates at middle of body; tail short, pointed; rhinophores are smooth proximally, with irregular or incomplete rings distally, bases close together. Oral tentacles twice the length of the rhinophores. Cerata long, narrow, club-like, a row of 6 clusters per side from behind the rhinophores to the start of the tail; each cluster an oblique line of cerata that decrease in size ventrally, number of cerata per cluster decreasing posteriorly from ~13 to 1. Ground color cream, somewhat darker behind the second ceratal cluster due to viscera. Dorsum with scattered orange speckles; an orange triangle behind the base of each oral tentacle. Rhinophores and oral tentacles transparent proximally, then bright yellow with whitish tips. Cerata pale brown, apices orange, transparent cnidosac at the tip. Tail with a median orange streak. Jaws visible as a diffuse orange spot. Specimens were found on a shipwreck associated with the coral *Tubastrea coccinea*. Spawn was deposited as a flat, circular, and somewhat disorganized mass of round white egg capsules. Depth: 6-20 feet. So far, known only from Bay of Charagato, Cubagua (Venezuela).

Saba

Jupiter, Florida

Guadeloupe

Key Largo, Florida

Anetarca piutaensis **(Ortea, Caballer & Espinosa, 2003)**

Size: to 8 mm Facelinidae

ID: Foot wider than body, front corners with recurved short tentacles, tapers to a thin pointed tail. Cerata relatively short, spindle-shaped, on a short narrow basal peduncle; apex tapers abruptly to a short thin knob; cerata form a row of 6-7 clusters per side, the frontmost (pre-cardiac) cluster with ceratal bases along an arch, the other clusters on a vertical line; a pronounced separation between the first and second clusters; number of cerata per cluster decreases posteriorly; tallest ceras per cluster always the most dorsal. Rhinophores long, bases close together on a common elevated peduncle, up to 7 widely-spaced oblique lamellae, apex with a prominent cylindrical mucron. Oral tentacles long, taper to a rounded tip, curve upward slightly. Ground color translucent gray. Head and body with yellowish, white, and brown mottling. Head with a purplish spot below and in front of the rhinophoral base, a yellow patch occasionally behind the rhinophores, a yellow/orange blotch between the bases of the oral tentacles. Cerata bluish-white below with a dark brown base, followed by a diffuse red/orange band, a broader yellow band, and a white-ringed translucent gray apical knob. Rhinophores purplish, distal lamellae and tip yellow. Oral tentacles yellow. Foot translucent pale orange, front margin yellow-spotted, tail yellow. There is little ecological information available for this species. Found among algae, probably feeds on hydroids. Depth: to ~15 feet. South and Southeast Florida, Saba, Guadeloupe, Grenada, Costa Rica.

Mona Island, Puerto Rico

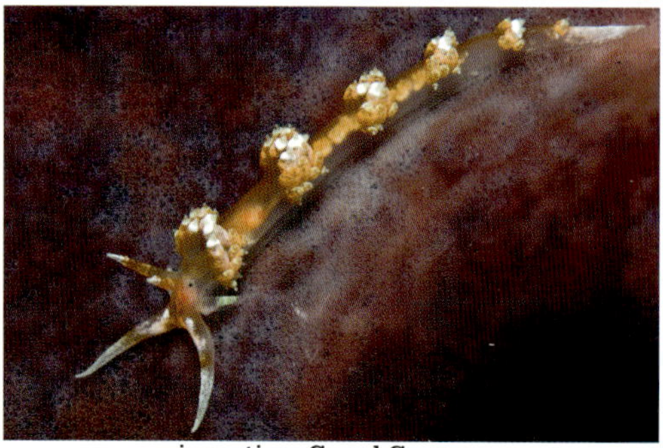

in motion; Grand Cayman

Bonaire

Austraeolis catina Ev. Marcus & Er. Marcus, 1967
longhorn austraeolis

Size: to 22 mm Facelinidae

ID: Long, smooth, tapers to a slender pointed tail; front foot corners tentacular, recurved. Oral tentacles and rhinophores long, slender, pointed; rhinophores half as long, ~12 annuli near the top. Cerata fusiform, smooth, up to 7 well-separated clusters per side; cerata per cluster decreases from up to 20 in the first cluster to 1 in the last; cluster cerata decrease in size laterally outwards, heights decrease from front to rear, front-most cerata as tall as the rhinophores. Ground color translucent pale brown to gray. Head and back with varying amounts of pale to deep orange patching, from almost none to almost totally covered. Cerata are usually orange to brownish with a white top and an unpigmented conical apex, the amount of white variable. Oral tentacles usually whitish with a central orange-brown band and a prominent basal dorsal white spot. Rhinophores orange-brown with white tips. Foot corners white-tipped. Usually found on reefs crawling over algae, sand, and especially brown/orange sponges which provide excellent camouflage. Like other aeolid nudibranchs, it probably feeds on cnidarians, such as hydroids. It has been reported feeding on upside-down jellyfishes. Depth: 20-100 feet. North Carolina to Florida, Bahamas, Turks and Caicos, much of the Caribbean.

all photos from Guadeloupe

Dicata odhneri Schmekel, 1967 **Odhner's dicata**

Size: to 13 mm Facelinidae

ID: Foot tapers to a short pointed tail, front corners with triangular extensions. Rhinophores are about one third the body length, smooth, conical, bluntly pointed; oral tentacles shorter, smooth, taper to a fine point. Cerata are in a line of 7-8 clusters per side, those in each of the first three clusters form an arch (vertex up), the rest in diagonal rows; cerata elongate, smooth, are inflated looking and constricted at the base, tips pointed. Translucent white. Body/head opaque velvety-white except over the pericardium; sometimes just opaque white speckles on ground color. Rhinophores and oral tentacles usually lemon yellow distally and white elsewhere (proportions quite variable), sometimes with an unpigmented basal region separating the white and yellow. Cerata white with unpigmented patches at the bases and tips. Foot white. Found on rocky substrates among algae, sponges, colonial hydrozoans, and sea squirts. Seems to be a nocturnal predator. Spawn is a spiral ribbon of ~2 whorls with spherical white eggs. Depth: 0-150 feet. A long-known (but rare) species in the western Mediterranean and the east Atlantic from the Canary Islands to the British Isles. Now also in Guadeloupe.

Riviera Beach, Florida

Riviera Beach, Florida

Jupiter, Florida

Riviera Beach, Florida

Jupiter, Florida

Cratena minor **Padula, A. K. Araújo, Matthews-Cascon & Schrödl, 2014** **orangespot cratena**

Size: to 25 mm Facelinidae

ID: Body elongate, slender, tapers to a pointed tail. Cerata long, slender, pointed, in 6 well-separated groups. Rhinophores and oral tentacles long, slender, smooth, pointed; oral tentacles longer than the rhinophores, often laterally projecting and recurved. Eyes behind rhinophores. Body, head, and foot translucent white. Cerata translucent gray with orange, bright red, or dark red digestive gland branches, many white specks, and white tips. Rhinophores translucent gray near base, pale orange at mid-section, white above. Oral tentacles are white except near the base, where they are translucent gray. There is a nearly quadrangular bright orange spot on each side of the head between the rhinophore and oral tentacle. The tail is white. Feeds on hydroids, such as the red *Eudendrium carneum*. It creates a defensive weapon by recycling stinging cells from its prey into the cnidosacs near the tips of its cerata. Its egg mass is a gelatinous transparent string with closely packed white eggs. Depth: to ~20 feet. Southeast Florida, Brazil. The very similar Mediterranean and Eastern Atlantic species *C. peregrina* differs as follows: it is considerably larger (the reason for "minor" in *Cratena minor*); the ceratal tips are bright blue; orange spots are smaller and rectangular; jaw and tooth differences.

Islamorada, Florida

St. Augustine, Florida

Cratena pilata (A. A. Gould, 1870) **striped cratena**

Size: to 24 mm Facelinidae

ID: Elongate, slender, tapers evenly to a pointed tail; foot wider than the body, front corners recurved and sharply pointed. Cerata slender, nearly cylindrical, taper to a point, the longest about half a body length, a row of 6 clusters per side – the 3 frontmost clusters a ventrally open arch, each of the others being an oblique row angling down and back; the number of cerata per cluster diminishes rearwards (e.g. from 12 to 2 in a 6 mm long specimen), the number increasing with the animal's size, and the uppermost cerata the longest. Rhinophores are long, tapering, distally smooth to slightly tuberculate. Oral tentacles are smooth and tapering, usually longer than rhinophores, often slightly shorter than the longest cerata. The ground color is translucent grayish-white. A broken and irregular orange-and-white median stripe from the oral tentacles to the tail is usually flanked by small white spots and may be crossed by orange bars. Cerata are transparent, have a broad brown to pale orange digestive gland axis and a whitish tip; they may appear "banded"– a transparent apical cnidosac, preceded by a white region, another transparent section, and another white band. Rhinophores are whitish with a brownish-orange central band. The oral tentacles are whitish with a pale orange proximal flush. In its northwest Atlantic habitat it feeds on hydroids such as *Pennaria tiarella* and *Tubularia crocea*. The spawn is usually deposited on a hydroid as a gelatinous spiral ribbon of closely packed white capsules, each containing one egg. Depth: to ~20 feet. Most common from Maine to North Carolina, occasionally further south: Bermuda, Florida, Saba, Louisiana, Texas, Brazil. Although the appearance of the pictured animals is largely in agreement with the descriptions of specimens from North Carolina to Maine, they do display some notable differences in color patterns (e.g. transverse orange dorsal bars). Accordingly, their identification remains tentative.

Saba

Fernando de Noronha, Brazil

Phidiana riosi Garcia and Troncoso, 2003 **Rios phidiana**

Size: to 18 mm Facelinidae

ID: Body elongate, slender; foot narrow, rounded anteriorly (no propodial tentacles); tail broadly triangular, moderately long. Oral tentacles long (one-fourth body length), conical, tapering. Rhinophores slightly shorter than the oral tentacles, club-like, 6-7 large annuli alternating with smaller ones. Cerata cylindrical with pointed apices; 12 clusters per side, each cluster a row angling down and back, 5 of the rows in front of the pericardial hump, ceratal size decreases ventrally in each row. Eyes behind rhinophoral bases. Ground color reddish. Body's back and sides with numerous white spots. Oral tentacles reddish with white spots on the proximal third, middle third orange, distal third translucent white. Rhinophores are orange with white tips. Cerata are translucent with clearly visible dark brown digestive gland branches on the proximal third, the rest whitish with an orange tip (cnidosac). Foot reddish. When disturbed it slightly curves its body, directs its rhinophores forward, and straightens its cerata into an organized pattern, with more dorsal cerata becoming erect and lower ones sticking out sideways. Depth: to 46 feet.

St. Vincent

Cuba

Grand Cayman

Grand Cayman

Utila

Phidiana lynceus **Bergh, 1867** **lynx phidiana**

Size: to 45 mm Facelinidae

ID: Elongate, slender, head small; foot protrudes somewhat laterally, tapers to a pointed tail (25-50% body length). Oral tentacles long, smooth, usually curved outwards with tips curled up and back. Cerata fusiform, smooth; 3 indistinctly separate clusters per side: front 2 with 6-9 oblique rows, number of cerata per row increasing rearwards from 2 to 9; rearmost cluster with 4-5 oblique rows, number of cerata per row decreases rearwards from 6 to 1; longest cerata most dorsal. Rhinophores longer than the longest ceras, proximal third a short smooth stalk supporting a club of slightly **oblique annuli** (sizes may vary), a vertical groove on the rear face, apex with a small cylindrical mucron. Ground color translucent gray. Head, body, and oral tentacles often **suffused with shades of orange.** A **blue or white median dorsal stripe from tip of tail to head, bifurcates between rhinophores, a branch continues onto proximal half of each oral tentacle**; stripe width variable, it may be absent, **a similar stripe along each flank**. Oral tentacles grayish-white to yellowish distally, may have a diffuse middle orange band. **Rhinophoral stalk unpigmented, club orange below, white to yellowish-white above**. Cerata translucent, broad branches of digestive gland prominent on lower half to two-thirds, orange to pale brown to dark brown to black; ceratal tips white to banded – white/brown/cream, white/brown/white, pink/white, orange/white, yellow/white. Inhabits shallow areas with rocks, reefs, mangroves. Feeds voraciously on polyps of the hydroid *Myrionema amboinense* from which it extracts and sequesters zooxanthellae and stinging nematocysts; the former for nutrient production by photosynthesis, the latter for storing in their cnidosac tips for deterring predators. **If hydroids are not available it resorts to cannibalism**. Spawn is a gelatinous cylindrical string usually attached to undersides of rocks as a spiral coil of 1.5-2 whorls and up to 5.5 mm across; sometimes attached to algae or ascidians among mangrove roots; has crowded rows of transparent, fluid filled, almost spherical capsules each with a small white egg. As embryos develop, they become transparent colorless in 6 days, at which point the veligers hatch. Depth: 0-10 feet. Circumtropical. Common throughout the tropical West Atlantic. Locally abundant.

Riviera Beach, Florida

Riviera Beach, Florida

Bocas del Toro, Panama

Guadeloupe

Roatan

preying on *Doto* eggs; Riviera Beach, Florida

Favorinus auritulus **Er. Marcus, 1955** **long-eared favorinus**

Size: to 13 mm Facelinidae

ID: Body elongate, foot tapers to long slender pointed tail. Cerata long, slender to moderately inflated, smooth, slightly curved, tips pointed; up to 6 crowded clusters per side, longest most dorsal, cluster size diminishes rearwards. **Rhinophores close together, long, with 1-3 (usually 2) bulbous swellings**. Oral tentacles smooth, cylindrical, tips pointed, nearly as long as but thinner than rhinophores. Ground color translucent gray to brownish. White patches on back and head; large transverse unpigmented region behind rhinophores. Cerata usually with white markings over orange/reddish/tan (diet-dependent) digestive gland branches; tips white, sometimes 1-2 subapical white bands; rarely all white. Rhinophores brown to tan, upper swelling and apex white. Oral tentacles usually white distally, area between bases sometimes bluish. Tail white. Inhabits shallow areas with various habitats but usually with plentiful algae, where it feeds primarily on the eggs of other sea slugs. Has been reported on and under rocks and floating with *Sargassum*. Spawn is a gelatinous string arranged as a spiral coil of 7-18 regularly spaced whorls, clockwise or anti-clockwise from the center, and deposited either inwards from the end of the outer whorl or outwards from the center. Each millimeter of string has ~30 white capsules, each capsule with a single egg. Veligers hatch in ~5 days. Depth: 0-25 feet. Bermuda, Southeast Florida, Bahamas, Brazil, most of the Caribbean. Another species of *Favorinus* in the West Atlantic might be the "white aeolis" *Favorinus branchialis* (Rathke, 1806), which is relatively common in the European Atlantic and Mediterranean. It also has variability in color pattern and number of rhinophoral swellings but reaches a greater maximum length (20 mm) and its egg coil is a tighter spiral. There have been scattered but unconfirmed reports of that species in the West Atlantic and photos from opposite sides of the Atlantic show animals (of similar size) that are virtually indistinguishable.

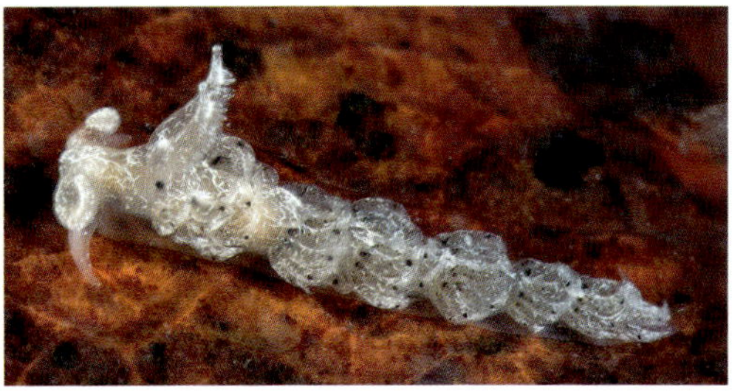

Roatan

Noumeaella kristenseni (Ev. Marcus & Er. Marcus,1963)
frosty noumeaella

Size: to 16 mm Facelinidae
ID: Body elongate, slender, tapers to a long thin pointed tail; head rectangular. Cerata long, slender, pointed, usually curled, a row of 6 clusters per side. Rhinophores long, smooth in front papillate behind. The oral tentacles are stout, moderately long; the front foot corners are recurved and just as long. Eyes are at the outside rear of the rhinophoral bases. Ground color translucent gray to pale yellowish-brown. All surfaces with a variably dense network of white to silvery-white lines, especially noticeable on the head, oral tentacles, front foot corners, and the cerata; sometimes also densely white-dotted, with dots often coalescing into dorsal white patches. Cerata usually have a few small orange-brown spots; sometimes with many scattered small black spots and a large one basally. Eyes black. Inhabits shallow algae-rich reef areas with abundant hydroids that it feeds on. In The Bahamas it is rather common on *Microdictyon marinum* and attached *Sargassum*. Like many members of this family, it can create defensive weapons by recycling stinging cells from its prey into the cnidosacs near the tips of its cerata. Depth: to ~10 feet. Southeast Florida, Bahamas, Turks and Caicos Islands, most of the Caribbean.

Roatan

↓Cozumel↑

Palisa papillata Edmunds, 1964 **nippled palisa**
Size: to 16 mm Facelinidae
ID: Elongate, narrow; foot tapers to a moderately long, pointed tail. Cerata inflated looking, apical cnidosac nipple-like, a row of up to 8 distinct clusters per side, total number of cerata increases with size of individual (~200 for largest). Pericardial hump between front 2 clusters. Oral tentacles long, pointed, tips often incurved. Rhinophores are about one third as long, relatively smooth on front face, **tuberculate on rear face**, tubercles in irregular oblique rows, increasing in number with animal's size (e.g. 3 mm animal has 7 rows, 8 mm animal 40 rows); a small cylindrical apical mucron. Front foot corners are long recurved tentacles, as long as the longest cerata. Ground color **translucent gray to pale brown**. Back, head, oral tentacles, and front foot corners **mottled with white**. **Cerata pale brownish with a yellow or orange cast**; variable white dusting; digestive gland duct **sometimes pale blue**; some brown to blackish spots near the ceratal base, just below the cnidosac, and sometimes in between. Rhinophores ground color, tubercles brownish, tips white. Margin of foot with 4-5 **transverse white blotches**. Eyes black. Found among algae in the vicinity of its probable prey – hydroids. Depth: to at least 40 feet. South and East Florida, Panama, Roatan, Cozumel, Grand Cayman, Jamaica. Rare.

↑↓ Riviera Beach, Florida ↓↑

Riviera Beach, Florida

with spawn; Destin, Florida

***Palisa* sp.**

Size: to 35 mm

Facelinidae

ID: Body elongate, narrow; foot tapers to a moderately long pointed tail, front foot corners are long recurved tentacles, as long as longest cerata. The cerata are inflated looking, a row of clusters per side of back, the pericardial hump between clusters 1 and 2. Oral tentacles very long and pointed; rhinophores one third as long, tuberculate on the rear face, with a small cylindrical apical mucron. Ground color translucent gray to whitish. The back and appendages variably white-mottled. The head and the proximal portions of the rhinophores and oral tentacles have a **dense network of reddish lines**. There are **several prominent orange stripes** – a longitudinal one in front of each rhinophore and several transverse ones that cross the foot margins and extend partly up the sides of the body. Cerata are pale brownish with white blotching and **blackish pigment near the base,** the tips white. Distally, the rhinophores and oral tentacles are usually whitish. It has been found with algae in the vicinity of its probable prey – hydroids. Its spawn is deposited on an algae leaf as a coiled flat ribbon, on edge, containing rows of small white eggs. Depth: to ~15 feet. The only known West Atlantic *Palisa* is the rare *P. papillata* Edmunds, 1964, which was originally described based on the examination of 24 specimens from Jamaica and one from Southern Florida. Neither that description, nor any subsequent one, has ever mentioned the presence of either a network of reddish lines or orange stripes. Although it is possible that these animals are color variants of *P. papillata*, that should not be assumed until specimen examinations are performed.

Riviera Beach, Florida

Grand Cayman

Pauleo jubatus Millen and Hamann, 1992

white-speckled pauleo

Size: to 46 mm Facelinidae

ID: Body slender, elongate, rounded; foot slender, front corners with short recurved tentacles; tail long, slender. Cerata smooth, taper to an inward-curving tip; a row of 8 separated clusters per side of back, each cluster projecting from an arch-like pedestal. Rhinophores are slender, narrowly tapering, **smooth or weakly lamellate**; oral tentacles are thicker, up to twice as long, slightly flattened, tapering, point antero-laterally, often recurved upwards. Head, back, and appendages **translucent yellow to pale orange**, suffused internally with a diet-derived color – orange, yellow, tan, pinkish-beige, blue-gray. Usually a variably wide **white median stripe** (or a series of irregular spots/patches) originating at the front of the head, passing between the rhinophores, and ending at the tail while narrowing and widening as it passes the clusters of cerata. A small white patch, sometimes very diffuse, in front of and just below each rhinophore. The ceratal internal color varies with the food ingested; **top two-thirds usually has white or bluish-white spots/mottling**; apical cnidosac white; bottom third unspotted. The rhinophoral tip and lamellae (if present) are white. Oral tentacles may have a dorsal white line extending (and narrowing) from the base. It inhabits coral reefs where it can feed on the sea rod *Plexaurella dichotoma*. When alarmed it curls the tips of its cerata forward in a defensive posture and can easily discard the cerata as a wiggling distraction while it escapes. The cerata regenerate. Depth: 10-100 feet. Southeast Florida, Bahamas, Turks and Caicos, Belize, Roatan, Grand Cayman, Jamaica.

West Caicos

Grand Cayman

Facelinidae sp. 1

Size: ~20 mm Facelinidae

ID: Body elongate, narrow, tapers to the rear. Oral tentacles thick, long, tips rounded and usually curled. The rhinophores are shorter than the oral tentacles, cylindrical, tuberculate on the upper rear face, the tip with a small cylindrical knob. The cerata are inflated looking, tips conical, a row of several clusters along each side of the back, the frontmost 2 clusters well-separated by the pericardial hump, the longest ceras half the rhinophore length. The foot is wider than the body, its front foot corners elongated into stout recurved tentacles (half the rhinophore length); tail moderately long and pointed. Eyes at outer rear of the rhinophoral bases. Ground color translucent pale orange. Back with two irregular broad white stripes – one along the inner edge of each row of ceratal clusters; some white spots below the stripe in the interspaces between clusters. Sides with gray mottling; ~6 white patches on the lower side along the foot margin. Cerata are white with some black dots; tips clear with a subapical orange ring. Oral tentacles, front foot corners, and head with large white patches. Rhinophores with a white apex and a diffuse subapical orange band. Eyes black. Found crawling on hard substrates and algae in near hydroids. Depth: to ~15 feet. Possibly a color variant of *Palisa papillata*.

Islamorada, Florida

Riviera Beach, Florida

***Facelinidae* sp. 2**

Size: ~20 mm Facelinidae

ID: Body elongate, narrow, tapers to rear; oral tentacles conical, long; rhinophores shorter than oral tentacles, conical, tuberculate on upper rear face; cerata spindle-like, tips abruptly pointed, a row of several clusters along each side of back, front foot corners are short recurved tentacles. Ground color translucent gray. Broad white stripe from front of head to at least the rhinophores; orange patch at front of each rhinophore. Cerata pale blue with white tips. Oral tentacles white; rhinophores whitish-orange below, white above. Found crawling on a hard substrate covered with algae. Depth: ~10 feet. Compare with *Berghia* on p. 159.

Cozumel

***Facelinidae* sp. 3**

Size: ~15 mm Facelinidae

ID: Body elongate, narrow; a row of 4-5 well-separated ceratal clusters along each side from just behind head to end of body; each ceras slender, conical, with elongate pointed apex; rhinophores long, annulate; the oral tentacles is long, conical; tail long. Ground color pale orange. Dorsum with large carmine blotches and darker spots/dots. Head with diffuse pale carmine stripe from front of rhinophoral base onto dorsal surface of adjacent oral tentacle; also, some narrow dark carmine lines – a median one from front to between rhinophores, and one along each cheek that extends forward from behind rhinophore and abruptly turns vertically about halfway to oral tentacle. Cerata blackish with white vertical stripes and white apex. Found on coral rubble. Depth: between 25 and 40 feet.

***Facelinidae* sp. 4**

Size: not available Facelinidae

ID: The cerata are crowded, broadly inflated, with a conical apex having a small knob at its tip. There are several very small tubercles scattered across the broad surface of each ceras, but similar tubercles are very crowded on its conical apex. Each rhinophore is quite elongate and very slender, with a small but prominent **apical spherical bulb** that has an extremely thin and trumpet-like distal projection; it also has several small rounded tubercles. The ground color yellowish-orange, the cerata are transparent. The rhinophores are orange with an opaque white bulb and distal projection. All of the tubercles are opaque white. It was found among algae at a depth of ~10 feet.

Riviera Beach, Florida

Family FLABELLINIDAE Bergh, 1889

Based on a combination of morphological and molecular analyses of many widely distributed putative specimens of Flabellinidae, the composition of this circumglobal family has been narrowed to about forty species in about six genera. The most speciose genera, *Coryphellina* and *Flabellina*, have been reported in the tropical West Atlantic.

Body usually elongate, narrow, tapering; notal ridge present, reduced, discontinuous, or completely absent; cerata in separate clusters, on elevated lobes or distinct stalks (which may be highly ramified); oral tentacles are usually elongate, tapered, with a broad triangular base; rhinophores smooth, annulate, or papillate, as long as or shorter than the oral tentacles; foot usually broad, front corners tentacular; anus opens below ceratal clusters (pleuroproctic); distinct oral glands are present; radula triseriate; rachidian teeth usually with a narrow compressed cusp and distinct denticles; lateral teeth denticulate, with attenuated bases.

They are brightly colored and feed on hydroids. The lengths of Caribbean species are typically 10-30 mm. Depths range from very shallow to 75 feet.

Saba

Riviera Beach, Florida

Bonaire

with spawn; Destin, Florida

Coryphellina hamanni **(Gosliner, 1994)** **Hamann's flabellina**

Size: to 32 mm Flabellinidae

ID: Body elongate, tapering; foot tapers gradually to a short narrow tail, its front corners tentacle-like and recurved. Cerata narrow at the base, widen upwards, most having an abruptly rounded apex with a narrow short sharply-pointed tip; cerata in a line of ~7 separated clusters on each side with the ceratal insertion pattern and number varying by cluster: front cluster is 4 obliquely vertical rows of 3-4 cerata with bases slightly raised above the mantle, rear cluster is one row of 2 cerata, each remaining cluster is an arch of 2-7 cerata. The rhinophores are slightly longer than the cerata, their bases close together, middle two-thirds of the rear face has up to 100 elongate papillae, apex conical and smooth. Oral tentacles are thin, elongate, taper to a sharp tip; shorter than rhinophores. Ground color translucent rose pink. **Cerata opaque to translucent whitish, white to yellowish "cap", white tip, narrow purple subapical band**; gray digestive gland duct visible via translucence. Rhinophores ground color to whitish on proximal two-thirds, then a broad purple band and a yellow apex. Oral tentacles and foot corners ground color basally (usually basal third for oral tentacles), bright purple centrally, rest yellowish. Tail purplish with a yellowish line or a spot near the tip. Found on algae, hydroids, and hard substrates near reefs. Probably feeds on hydroids. Spawn is a tangled string of white eggs. Depth: 6-66 feet. Northwest and Southeast Florida, Bahamas, Turks and Caicos, Puerto Rico, Virgin Islands, Saba, Guadeloupe, Bonaire, Aruba, Grand Cayman. Rare. Formerly in the genus *Flabellina*. The similar *Coryphellina marcusorum* has different cerata – they taper gradually to a point instead of being abruptly rounded, the purple bands are broader and diffuse, and they lack white pigment basally.

Riviera Beach, Florida

Cozumel

Riviera Beach, Florida

with a parasite; Guadeloupe

with eggs; Riviera Beach, Florida

***Coryphellina marcusorum* (Gosliner and Kuzirian, 1990)** **purplering flabellina**

Size: to 30 mm Flabellinidae

ID: Elongate, tapers to a long pointed tail (half the body length); front foot corners tentacular, short, recurved. Cerata narrow at base, widen upwards, then taper gradually to an elongate conical tip; a line of 7-9 separate clusters per side of back, clusters configured as lines and arches, frontmost one slightly elevated above mantle. Rhinophores elongate, smooth but papillate on middle two thirds of rear face (up to 100 papillae), as tall as cerata; oral tentacles smooth, thin, taper to a rounded tip, longer than rhinophores. Ground color translucent pinkish or orange. Internal organs clearly visible as a bright orange mass along entire body length. Cerata multi-colored: proximal 2/3 translucent with a broad brown to pale-orange duct of the digestive gland, followed by a narrow purple band, a narrow blue band, and an elongate white/yellowish conical cap. The ceratal color sequence also occurs on the rhinophores, oral tentacles, and foot corners, but the proximal portion is ground color to grayish. Tail purple with a median white line. Crawls by day on algae and hard substrates near reefs, less active at night. Probably feeds on hydroids. Spawn is a tangled string of pink eggs. Depth: 10-73 feet. Cerata of the similar *Coryphellina hamanni* differ:abruptly rounded; short sharp tip instead of gradually tapered; purple bands narrower, not diffuse, usually white basally. Gulf of California, western Central America, amphiatlantic. Florida, St. Eustatius, Saba, Dominica, Bonaire, Curacao, Venezuela, Costa Rica, Cozumel, Grand Cayman, Brazil. Uncommon to rare.

Pompano Beach, Florida

Bonaire

Flabellina dana **Millen & Hamann, 2006** — **Dana's flabellina**

Size: to 12 mm — Flabellinidae

ID: Body elongate, extremely slender, higher than wide; head oval; foot narrow, tapering to a **very long, thin, bluntly-pointed tail,** front corners long and recurved. Mantle margin reduced to a low ridge (notal brim) along each edge of the back. Cerata elongate, cylindrical, tips pointed, a row of 5 well-separated clusters along each side on broadened areas of the notal brim (indistinct between clusters), the number of cerata per cluster decreases from front to back. The **rhinophores are annulate**, ~10 rings, long, tips pointed; oral tentacles long, slender, taper to a point, slightly shorter than the rhinophores. Ground color translucent white to pinkish-white to gray. **Oral tentacles and head opaque white**, the pigment continuing back between the rhinophores onto the dorsum as a **broad median white stripe** that extends onto the tail, while expanding in the interspaces between ceratal clusters. The annulate portions of the rhinophores are opaque white to pinkish-white to yellowish; the sides of the body below the notal brim are opaque white. **Cerata bright red to reddish-orange with long white tips**, sometimes with a darker band just below the tips. Found on sand, rubble, and on *Sargassum* attached to a substrate. Like many of its congeners, it probably feeds on hydroids. Depth: 2-66 feet. Southeast Florida, Bahamas, St. Lucia, Bonaire, Curacao, Costa Rica, northeast Brazil. Rare. This is the only tropical West Atlantic *Flabellina* with annulate rhinophores.

↑↓ Riviera Beach, Florida ↓↑

Flabellina dushia **(Ev. Marcus & Er. Marcus, 1963)** — **dushia flabellina**

Size: 20 mm — Flabellinidae

ID: Body elongate, slender, higher than wide; head small, rounded, mouth terminal; foot narrower than body, tapers to a **long thin trailing tail** up to one-third total length, front bilobed with median notch, front corners with short recurved tentacles; reduced mantle with a well-developed continuous edge (notal brim) that expands as it passes around ceratal clusters; cerata elongate, cylindrical, smooth, tips narrowly pointed, grouped in **up to 7 clusters** of 1-5 obliquely vertical rows along each side of back, number of rows (and cerata) per cluster decreases rearwards; **rhinophores smooth, slender, conical**, tips rounded, as long as or longer than cerata; oral tentacles like rhinophores but usually slightly longer; pericardial hump between first two ceratal clusters; genital pore below first ceratal cluster on right side; anus between first two clusters on right side, below rear of pericardial hump. Ground color translucent grayish-white. **Opaque white pigment on back, sides, and head** except near bases of ceratal clusters. Cerata translucent, with a wide **digestive gland duct that is red to orange-yellow to brownish-orange to dark brown; tips translucent whitish**, usually **bordered below by a dark red/orange ring** (dark brown if duct is brown). Proximal third of oral tentacles and rhinophores ground color, rest opaque white. Foot dull white. Found on sand, rubble, *Thalassia* seagrass, and floating *Sargassum* in areas with hydroids, on which it feeds. Spawn is a translucent gelatinous coil (2 whorls) containing white eggs. Depth: 3-65 feet. Amphiatlantic. Bermuda, Southeast Florida, Bahamas, Saba, Guadeloupe, Martinique, St. Vincent, Grenada, Curacao, Venezuela, Roatan, Jamaica, Grand Cayman, Brazil. Uncommon.

St. Vincent

Bonaire

Saba

Brazil

Pompano Beach, Florida

with eggs; Riviera Beach, Florida

Flabellina engeli **Ev. Marcus & Er. Marcus, 1968** — **Engel's flabellina**

Flabellinidae

Size: to 25 mm

ID: Body long, narrow, tapers to a thin pointed tail; front foot corners are short lateral tentacles. Cerata long, slightly inflated looking, tips short/pointed; a row of **5-8 well-separated clusters per side**, a cluster's cerata sharing a peduncle on a transverse swelling; pericardial hump between clusters 2 and 3. Rhinophores as tall as cerata, a long stout club on a short stalk, club with 20-30 **crowded lamellae**. Oral tentacles very long, tapering. Ground color translucent pinkish-gray to bluish-gray. Back and flanks usually pale orange, hump purplish; **irregular opaque white/cream patches and streaks** span gaps between ceratal clusters and partly wrap around their bases, merging at rear to continue as a **white median stripe to tip of tail**; white areas sometimes bordered by yellow patches/spots; **a median orange blotch** in front of rhinophores flanked by white/yellow blotches between bases of rhinophores and adjacent oral tentacles; white/yellow patch between bases of oral tentacles. Cerata pinkish proximally, rest pale orange to yellowish with a diffuse **dark orange to brown narrow band** about one third way down from tip. **Rhinophores usually dark up to mid-club and then white, but often with a broad white band at mid-length**. Oral tentacles with orange tints near base, rest whitish. Foot with submarginal row of white-to-blue spots, front tentacles bluish-white. Found on sand, rocks, and rubble, but mostly on the hydroids it feeds on. Deposits spawn on a hard substrate as a spiral coil of gelatinous string with round, crowded, purplish-white capsules containing one egg each. Hatching takes ~4 days. Depth: to ~50 feet. East Florida, Antilles, Brazil, Venezuela to Yucatan.

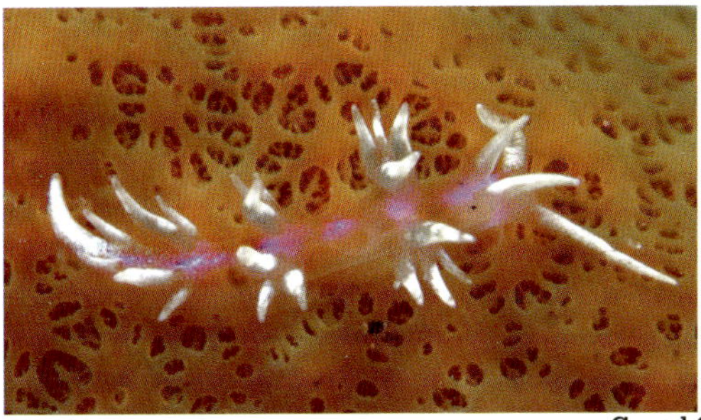

Grand Cayman

***Flabellina* sp. 1**

Size: ~10 mm Flabellinidae

ID: Body elongate, slender; foot as wide as body, tapers to a pointed tail. Cerata slender, conical, pointed, a row of 4 clusters per side, number of cerata per cluster (1-4) and cluster height both diminishing from front to rear. Rhinophores and oral tentacles smooth, conical, pointed; oral tentacles longer. Transparent. Dorsum with a median line of purple spots. Cerata with a grayish to orange digestive gland, distal half white. Rhinophores and upper surfaces of oral tentacles and tail white. Found among algae and hydroids in shallow water. Depth: 10-20 feet.

***Flabellina* sp. 2**

Size: ~10 mm Flabellinidae

ID: Body elongate, tapers to the rear. Cerata are long, slender, cylindrical, their tips pointed, a row of ~5 well-spaced clusters per side. The rhinophores and oral tentacles are similar – slender, conical, smooth, as long as the cerata. Ground color translucent gray. On the back, between each transverse pair of ceratal clusters, there is an irregular median white streak having an elongate central orange spot; there are additional orange spots along the midline and the sides; there is a transverse broken orange band between clusters 1 and 2 and another behind the rhinophores. The proximal three-quarters of the cerata is translucent gray with a greenish digestive gland duct, the distal quarter bluish, and the tip unpigmented. Rhinophores are white with two diffuse orange bands: one basal, one at mid-length. Its small black eyes are behind the rhinophores in unpigmented patches. Found in shallows.

Islamorada, Florida

Islamorada, Florida

***Flabellina* sp. 3**

Size: ~20 mm Flabellinidae

ID: Body elongate, slender. Cerata long, slender, pointed, in several well-separated clusters along each side. Rhinophores and oral tentacles smooth, conical, pointed, longer than cerata. Ground color translucent gray. Head with broad median white stripe from front margin to slightly behind rhinophores. An undulating and sometimes bifurcating longitudinal white stripe from base of each oral tentacle to rear of body, passing around or through bases of ceratal clusters. Cerata orange, covered with white dots, tips white. Rhinophores grayish near base, then white. Oral tentacles grayish/bluish near base, then white. Among algae and hydroids to ~10 feet. Similar to the white-speckled variant of *Coryphella verta* but the white striping is quite different.

Riviera Beach, Florida

Panacea, Florida

***Flabellina* sp. 4**

Size: ~15 mm Flabellinidae

ID: Body elongate, slender, tapers to a sharply pointed tail. Cerata long, slender, cylindrical, tipped with long and pointed cnidosacs; in a row of six well-separated sparse clusters along each side. The rhinophores and oral tentacles are smooth, conical, pointed, and considerably longer than the cerata. Ground color translucent gray. Head with a broad median white stripe from its front margin to slightly behind the rhinophores; a longitudinal white stripe from the base of each oral tentacle to the rear of the body, passing around or through the bases of the ceratal clusters. Cerata are clear with a broad reddish-orange internal axis, some white dots, a subapical bright red band, and a whitish tip. The rhinophores and oral tentacles are ground color proximally and whitish distally. Found among algae and with hydroids in shallow water. Depth: ~10 feet. The similar *Flabellina* sp. 3 has unbanded cerata, but might be the same species.

***Flabellina* sp. 5**

Size: ~5 mm Flabellinidae

ID: Body elongate and slender. Cerata long, slender, tips sharply pointed, in several well-separated clusters along each side; the median pericardial hump is between the front two ceratal clusters. The rhinophores are smooth, slender, conical, shorter than the longest cerata. The oral tentacles are short and slightly recurved. Ground color translucent gray, entire dorsal surface opaque white except for the extreme front of the head, which is unpigmented. Rhinophores are white except for a basal pale orange band that continues downwards as a lateral blotch. Oral tentacles are white. Each ceras is yellowish-orange with a narrow red ring immediately below a white cnidosac. Found on a hydroid at a depth of ~10 feet. Possibly a variant of *Flabellina* sp. 4.

Riviera Beach, Florida

Riviera Beach, Florida

***Flabellina* sp. 6**

Size: ~20 mm Flabellinidae

ID: Differs from *Flabellina dushia* (q.v.) by having a much more pronounced pericardial hump between the first and second ceratal clusters, and by having more inflated cerata. The ground color is translucent grayish-white. There is opaque white pigment on the back, sides, head, and tail. Cerata are red, sometimes with white stippling; their tips are translucent whitish. The proximal third of the oral tentacles and rhinophores is ground color, the rest opaque white. Found on hydroids. Depth: ~10 feet.

Riviera Beach, Florida

***Flabellina* sp. 7**

Size: ~10 mm Flabellinidae

ID: Body elongate, slender, tapers to a short rounded tail. Cerata long, slender, apical cnidosacs pointed, in five well-separated sparse clusters along each side. Rhinophores and oral tentacles are smooth, conical, gradually tapering to a point; rhinophores are as long as the cerata, the oral tentacles much longer. Front foot corners tentacular and recurved. The eyes are small, behind the rhinophoral bases. Ground color translucent gray. The head and body appear translucent orange-yellowish overall, due largely to clearly visible internal organs and eggs. A reddish blush behind the rhinophores includes the pericardium. There is a large red triangular blotch in front of each rhinophoral base. Cerata are dark red with clear to whitish tips. Rhinophores and oral tentacles are ground color proximally, white distally. Eyes black. Found crawling on sand and rubble. Depth: ~10 feet.

***Flabellina* sp. 8**

Size: ~5 mm Flabellinidae

ID: Body elongate, slender. Cerata long, slender, tips rounded to bluntly pointed, in several well-separated clusters along each side. The rhinophores are smooth, slender, conical, as long as the longest cerata. The oral tentacles are short, slightly recurved. Ground color is translucent gray. Front of head and mid-dorsum with opaque white pigment; a broad median opaque white stripe extends back from the rhinophores, an orange-red vertical bar extends down the head from the base of each rhinophore. Each ceras is yellowish-orange, its apex red with white spotting. Rhinophores are pale orange basally and distally, translucent gray between. Found on a hydroid at ~10 feet.

Riviera Beach, Florida

Saba

***Flabellina* sp. 9**

Size: ~10 mm Flabellinidae

ID: Body elongate, slender, tapers to a long sharply pointed tail. Cerata short, slender, smooth, tips with pointed cnidosacs, in five well-separated sparse clusters per side. Rhinophores and oral tentacles are smooth, conical, taper to a point, are similar in length and much longer than the cerata. Front foot corners tentacular and recurved. Eyes small, behind the rhinophoral bases. Ground color translucent gray. The head, body, and tail are opaque white. Rhinophores and oral tentacles are ground color basally, white elsewhere. Cerata are whitish to pale yellow with a narrow orange band at the base of the white cnidosac. Eyes black. Found crawling on sand and rubble. Depth: ~10 feet.

Bonaire

***Flabellina* sp. 10**

Size: ~10 mm Flabellinidae

ID: It is similar to *Flabellina* sp. 9 but the cerata are more slender. Ground color is translucent gray. The head, body, and the tail are opaque white. The rhinophores and the oral tentacles are ground color proximally, white distally. The yellow cerata have a broad carmine band at the base of the white cnidosac. Found crawling on a hard substrate covered with attached algae. Depth: ~10 feet.

Family GLAUCIDAE Bergh, 1889

A family of pelagic aeolid nudibranchs found in tropical to warm temperate seas, consisting of five species in two genera. The genus *Glaucilla* is a complex of four cryptic Indo-Pacific species while *Glaucus* has a single cosmopolitan species, *Glaucus atlanticus,* which has been recorded in the tropical West Atlantic.

Body elongate, slender, widest just behind the head and tapers to a pointed tail; head short; oral tentacles short, simple, well separated; short rhinophores low on the sides of the head; cerata in 3-4 clusters along each side of the body, progressively diminish in size posteriorly, each cluster at the end of a laterally extending peduncle, the frontmost peduncle longest and very thick, each cluster an arch-like array of horizontally projecting conical "fingers" in one or more flat planes; anus cleioproctic.

As they drift on the surface of open water at the mercy of winds and currents, they prey on other surface floaters such as the colonial hydrozoans *Physalia*, *Porpita*, and *Velella*, from which they ingest nematocysts and sequester them in the tips of their cerata for defensive purposes. The large and highly toxic nematocysts of Physalia are particularly strong deterrents, even capable of causing painful stings and respiratory distress in humans that inadvertently touch a glaucid, especially when they occasionally wash up en masse onto shores and beaches. Being silvery dorsally and dark ventrally they float upside down to take advantage of their countershading – camouflaged both from above (feeding birds) and from below (feeding fish). They control their buoyancy by periodically swallowing air and maintaining a gas bubble in their stomach. A rudimentary form of swimming is accomplished by combining up and down body flexions with ceratal rowing movements. The body length range is 20-40 mm.

Riviera Beach, Florida

Bermuda

Brazil

drifting to shore; Gulf Shores, Alabama

Glaucus atlanticus Forster, 1777 — blue glaucus

Size: to 40 mm — Glaucidae

ID: Elongate, slender, widest just behind the head, tapers to a long pointed tail; head short, not distinctly separate from its body. Oral tentacles very short, smooth, well separated. Foot broad, well developed, front corners rounded. Rhinophores are low on the sides of the head, short, and smooth. There are **up to 3 groups of pointed conical cerata per side, each group projects sideways with its cerata forming an oblique open fan, the most dorsal are longest, the rest shortening as they get more ventral**; cerata in the frontmost group share a short thick stalk, the others attach directly to the side of the body. **Ground color silver**. Rhinophores and oral tentacles electric blue. **Body and cerata blue or blue-and-white ventrally, silvery-gray dorsally. Foot has a median silver stripe with a blue stripe on each side**, the blue stripes merging at the tail tip and joining transversely at the head. As suggested by its countershading, this pelagic species spends its life floating upside-down at or near the surface in open water, thereby blending into the background whether viewed from above (by birds) or below (by fish). It maintains buoyancy by swallowing and releasing air, as necessary. It drifts but can also swim slowly by flapping its cerata, which helps it feed on *Physalia physalis* (Portuguese Man O'War), as well as on other siphonophores such as *Velella velella* and *Porpita porpita*. Because it harvests stinging cells from its prey for defensive purposes, it can produce a painful sting when touched. It can easily autotomize its cerata. Spawn consists of floating linear threads of oval white egg capsules within a transparent mucous tube. Veligers emerge and swim away in ~3 days. Wind, waves, and currents occasionally drive these pelagic animals onto beaches in large numbers. Circumglobal in temperate and tropical waters. Reported from throughout the tropical West Atlantic. Rarely seen by scuba divers.

mass mortality; Brazil

Family HANCOCKIIDAE MacFarland, 1923

A very small family of nudibranchs with only a few species in one genus (*Hancockia*) distributed widely but sparsely in temperate to tropical seas. Rarely found.

Body and foot elongate and narrow; along the mantle margin on each side of the body is an indistinct ridge bearing a row of cerata; each ceras is a cluster of long tapering filaments on a shared base, in a configuration ranging from a circle to a fan-like shape; rows of clusters are laterally symmetric or variously offset; there is no well-defined oral veil but on each side of the mouth there is a flattened lobe (oral tentacle) having several finger-like projections with rounded tips; rhinophores have an elongate stalk, a distal club, vertical lamellae, apical knob, and project from a papillate elevated sheath with a flared rim; anal and genital pores are below the anterior cerata on the right side of the body; the radula has rachidian and lateral teeth, the rachidian with a large median cusp flanked by denticles, the laterals irregularly triangular and usually smooth.

They prey on hydroids, from which they extract nutrients for sustenance as well as nematocysts for defensive purposes. The latter are sequestered in multiple muscular cnidosac-like structures at the tip of each ceras and, in some species, also in pustules within the rhinophoral sheaths. When alarmed they discharge the nematocysts to ward off the intruder. Animal lengths are in the range 10-20 mm.

St. Vincent

Hancockia ryrca **Er. Marcus, 1957** **white-spotted hancockia**

Size: to 16 mm Hancockiidae

ID: Body elongate, narrow, fusiform; mantle edge an inconspicuous ridge; foot not wider than the body; tail short, obtuse, tip slightly indented. There is a broad, flat, multi-fingered (5-7) oral tentacle on each side of the mouth. There are 4-5 bushy-looking transversely paired cerata along each side of the back; each ceras has 5-8 coplanar lobes (like an open hand), with the lobe edges having pointed cusps. The rhinophores are retractile, have a vertically lamellate club at the top followed by a thin cylindrical knob-tipped vertical projection; the sheaths are elevated and funnel-like with a jagged rim. Ground color translucent gray. The body and all appendages are densely spotted with white; the back and cerata have conspicuously visible yellow-orange to dark brown internal organs. The rhinophoral lamellae are gray, the apical projection white; the sheath rim white. Feeds on hydroids. It creates a defensive weapon by recycling stinging cells from its prey into the cnidosacs near the tips of its cerata. Depth: to ~20 feet. Bahamas, Nevis, St. Vincent, Barbados, Curacao, Venezuela, Brazil. Rare.

Family JANOLIDAE Pruvot-Fol, 1933

Since first being defined in 1933, the Janolidae have undergone a complicated taxonomic history. Initially it was a standalone family but, at various times, it was moved into and out of the families Proctonotidae and Zephyrinidae, until its most recent extraction (2019) from the Proctonotidae back to standalone status. Currently it consists of about thirty species in the four genera *Antiopella*, *Bonisa*, *Galeojanolus*, and *Janolus*. They are found circumglobally in temperate to tropical seas, with only a few members of the most speciose genus, *Janolus*, recorded in the tropical West Atlantic.

Body elongate, depressed; head broad; oral tentacles short and cylindrical; the broad foot tapers to a thin tail; the sides of the body and the front margin of the head have many smooth and slender to bulbous and tuberculate cerata, the cerata are either with digestive gland branches (*Janolus*) or without; rhinophores tall, lamellate, nonretractile, a caruncle between their bases; anus on the posterior dorsal midline; jaws large, wing-like, with denticles (*Janolus*) or without; radula broad, the rachidian tooth and laterals smooth or denticulate (both smooth in *Janolus*).

They feed on bryozoans. Some species can autotomize their cerata when alarmed. Lengths range from 6-80 mm.

Roatan

Janolus comis **Er. Marcus, 1955**

Size: to 9 mm Janolidae

ID: Body narrow, tapers to long pointed tail; front of head flattened, rounded. Cerata present on both head and body, the body ones may form an irregular line of clusters per side of back, each cluster a transverse row; cerata long, papillate, pedunculate basally then conical to bulbous, tapering to a pointed tip with a short cylindrical knob. Rhinophores close-set, papillate on the sides and rear; basal joint occupied by a longitudinal crest ("caruncle") composed of 3-7 transverse wavy folds of fleshy tissue. Ground color translucent gray/cream/brownish with a peppering of brown and opaque white dots. Papillae are cream to white. Varying amounts of dark blotching on back and sides (adults darker than juveniles), sometimes stippled with black dots, usually a prominent dark brown/purplish blotch above the pericardium. Cerata with a visible pale brown branch of digestive gland reaching mid-length, its surface with some brown dots and frosted with white spots. Rhinophores ground color. On sand, rocks, algae, arborescent bryozoans. Feeds on bryozoans. Can move quickly because of its broad foot; when disturbed it elevates its cerata. Spawn is a wavy white string deposited as a flat spiral coil on a substrate. Depth: 0-80 feet. South Florida, Cuba, Curacao, Costa Rica, Roatan, Brazil.

Riviera Beach, Florida

Janolus costacubensis **Ortea & Espinosa, 2000**

Size: to 6 mm Janolidae

ID: Body elongate, narrow, tapers to rear; foot broad, front not indented, tapers to a thin pointed tail; head broad, flat, front rounded. Oral tentacles conical, short. Cerata on back and head often have random orientations that produce a **characteristic "disorganized" look;** cerata long, papillate, pedunculate, tallest near front of back; rhinophores close-set, **lower half papillate, upper half with several thick lamellae that do not extend onto the front face of the rhinophore**, a small knob at the apex. Between the rhinophoral bases there is a caruncle composed of 6-7 transverse wavy folds of fleshy tissue. Ground color cream, semi-opaque. **Dorsum with a median line of white spots** interrupted by a **purplish spot on the pericardium**. Back and sides with some white spots of various sizes. Cerata with reddish-brown dots, whitish papillae, cream to orange digestive gland branches. Rhinophores ground color with reddish-brown spots, tips white. Reddish-brown dots along foot edges. Tail and oral tentacles are white spotted; **oral tentacles with a distal reddish-brown band.** Found on sand, rocks, algae. Depth: 0-80 feet. Southeast Florida, Bahamas, Costa Rica, Cuba.

Riviera Beach, Florida

Janolus* cf. *comis **Er. Marcus, 1955**

Size: to ~9 mm

Janolidae

ID: Differs from *Janolus comis* by its rhinophores having a small apical mucron atop a larger inflated "bulb". This prominent feature has not been mentioned in any of the scientific literature for *J. comis*, including in the book *Caribbean Sea Slugs* (Valdés et al, 2006) which shows a photo of an animal that does have that feature. This may or may not be a variant of *J. comis*. The overall appearance is whitish to yellow to pale orange. All papillae are white. The dorsum is variously brownish to reddish brown (viscera) with pale spots and mottling. The cerata have red streaks and spots, white tips, and a brown blotch at the base. Rhinophores are brownish with a white apical bulb and a mucron. Eyes black. It has been found crawling on various hard substrates and on arborescent bryozoans. Spawn is a wavy white string deposited as a flat spiral coil on a hard substrate. Depth: to ~10 feet.

Riviera Beach, Florida

Janolus flavoanulatus **Pola & Gosliner, 2019** **yellow-band janolus**

Size: to 50 mm Janolidae

ID: Body elongate, broadest at the front, elevated near the middle (cardiac region), tapers to the rear where the foot protrudes considerably. Cerata elongate, variably sized, inflated centrally, taper to a sharply pointed short tip; usually prominently tuberculate on the distal half; they are crowded, irregularly distributed in 8-12 longitudinal rows of 4-5 cerata per side, including at the front of the head; most of them have the digestive gland entering their base and splitting upward into branches that terminate near mid-height (frontmost cerata may lack branches). Rhinophores are elongate, slender, have 20-25 transverse to oblique lamellae (some incomplete) and a **large convoluted fleshy caruncle between the rhinophores**. A small eye at the rear outer edge of each rhinophoral base. Oral tentacles short, finger-like. Ground color translucent white. Mantle has a white dorsal midline flanked by purple pigment and a few scattered white spots and patches, especially behind the caruncle and around the rhinophores. Cerata translucent gray on proximal two-thirds, followed in sequence by a broad **white-to-yellow band, a narrower dark maroon band, a pale purple-to-blue cap, and a translucent yellowish-white tip**. Ceratal tubercles yellowish-to-white; digestive gland branches reddish-brown, black-dotted. **Rhinophores purple with white tips**, caruncle white to yellowish-white. Usually found on rubble and slopes on arborescent bryozoans, on which it feeds. Spawn is a wavy orange string configured as a spiral coil with about 3 whorls. Depth: in Florida: ~10 feet; elsewhere: 60-100 feet. Endemic to the Indo-Pacific. Recently discovered in Southeast Florida. Externally it appears identical to the Indo-Pacific animal but positive identification would require an examination of its anatomy and/or genetics.

Janolus mucloc **(Er. Marcus, 1958)**

Size: to 26 mm Janolidae

Brazil

ID: The body is elongate, its width and height are about one-fourth its length, the sides of the dorsum vertical. The foot has projecting front corners; it tapers to a short pointed tail. The cerata are along the entire perimeter of the body in 3-5 longitudinal rows per side, with the row heights decreasing ventrally; they are elongate, fusiform, and smooth. The rhinophores are close-set and have: a smooth shaft that supports a club of up to 12 oblique lamellae, a short knob at the apex, the space between their bases occupied by a caruncle with folds and frills. The oral tentacles are short and conical; the small eyes are far behind the rhinophores. The anal papilla is located one-fourth body length from the rear, to the right of median line, surrounded by a mass of glands. The ground color ranges from translucent whitish to transparent. The dorsum has a white median stripe that extends from in front of the rhinophores to the tip of the tail. Cerata are translucent with a subapical orange band, three short vertical white stripes running down from the band, a white tip, and visible brown digestive gland ducts. The rhinophores are ground color, each with a vertical white streak along the proximal inner face of its stalk, the streaks almost connecting with the front of the dorsal stripe; apical knob white; caruncle grayish. Found with algae. Depth: tidal zone. Brazil. In 2020 it was also reported from the Cape Verde Islands. Rare.

Destin, Florida

***Janolus* sp. 1**

Size: ~25 mm Janolidae

ID: Body elongate, covered with cerata. The cerata are crowded, highly inflated looking, apices rounded; their sizes decrease laterally and ventrally from mid-back. Small, slender, conical rhinophores; proximal third smooth, the rest with ~18 very oblique lamellae; a prominent ruffled caruncle between their bases. The pinkish dorsum has a bright white median stripe beginning at the front of the head. Cerata are translucent whitish, with a banded apex: a light brown "cap", below it a pale blue band, then a cream band with a reddish outer edge. Thin dark brown digestive gland branches are clearly visible in the lower half of each ceras. Rhinophores ground color near the base, blackish centrally, white apically. Caruncle pale pink. Found on sand, on a bryozoan, and on a hard substrate. Depth: ~80 feet. Destin (Florida) and Beaufort (North Carolina).

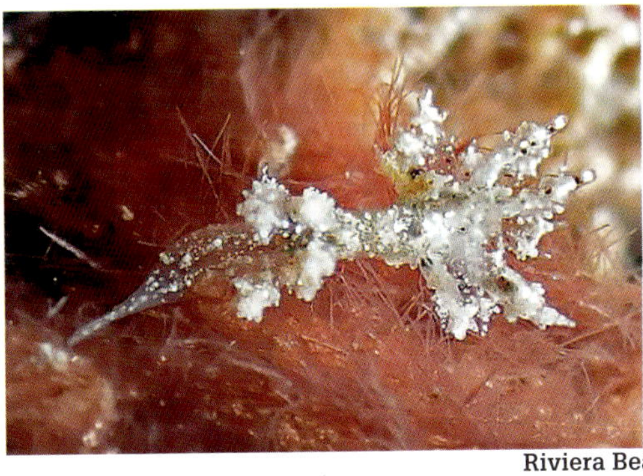

Riviera Beach, Florida

***Janolus* sp. 2**

Size: ~6 mm Janolidae

ID: Body elongate, slender; foot broader than body, tapers to a long sharply pointed tail. Two well-separated clusters of cerata – one at the front, one at the rear; **front cluster:** ~10 cerata along a horseshoe that begins behind the rhinophores and arcs around the front of the head; very tall cerata at rear, smaller ones at front; **rear cluster:** 2 cerata per side. All cerata are very tuberculate, their tips are pointed; the interspace between the two clusters has scattered minute tubercles. The rhinophores have a tuberculate subapical "club" and a narrow rounded apical projection. Ground color translucent yellowish-gray. Body with white tubercles and some dark dots; cerata white; rhinophores white with a few black spots on the clubs and a black band at the base of the apex. Foot translucent and white-spotted. Found on a branching bryozoan. When stationary it appeared as a "fuzzball". Its forward motion was "rhythmically jerky", with the cerata moving backward and forward in unison. Depth: ~10 feet. This slug is very similar to the Indo-Pacific *Janolus mirabilis* Baba & Abe, 1970, but there are notable differences. That species was described as having a black inter-rhinophoral caruncle and distinctly perfoliate rhinophores that are dark brown proximally. These features are not evident in the accompanying photos. The description also makes no mention of a black subapical band on each rhinophore. However, animals that look virtually identical to those pictured here can be found on the internet labeled as *J. mirabilis*. If this animal is, indeed, that species, it would be the first record of it in the tropical West Atlantic.

Family LOMANOTIDAE Bergh, 1890

A one-genus family consisting of a few species of *Lomanotus* that are distributed circumglobally as widely scattered populations in temperate to tropical seas.

Body elongate, slender, soft; each lateral mantle edge typically forms a raised undulating ridge that bears cerata in a single row; cerata short, pointed, conical to flattened or concave, they may or may not contain digestive gland branching; without gills or accessory gill-like appendages; oral tentacles present as 1-3 pairs; rhinophores club-like with a few to many transverse lamellae and an apical mucron; sheaths are elevated, rims have a few shallow to deep papillate cusps; front foot corners short to long to absent, rounded to tentacular, sometimes grooved; anus almost mid-lateral on right side behind pericardium; jaw elements usually denticulate along the masticatory border; radula without a rachidian tooth, the innermost laterals typically short and denticulate.

They feed on various hydroids, usually very cryptically. Some have been recorded as swimming. Lomanotidae can reach a length of almost 100 mm, but those in the tropical West Atlantic are usually 6-40 mm long and found shallower than 20 feet.

Riviera Beach, Florida

Roatan

Riviera Beach, Florida

San Andres, Colombia

Lomanotus phiops **Er. Marcus, 1957**

Size: to 8 mm Lomanotidae

ID: Body elongate, tapers to a pointed tail; the oral veil is rounded at the front with a short and pointed tentacle on each side; front of foot broad, smooth, crescent-shaped, the projecting corners short and recurved. Rhinophores with distal lamellate club (6 lamellae) and a small cylindrical apical mucron; **sheath elevated, its flared rim with 4-5 vertical lobes**. Mantle edge with a **wavy ridge that extends back from each rhinophore**, the undulations along one edge paired with those on the other and diminishing in height toward the rear; each wave with 4 peaks/valleys in a plane inclined toward the dorsum and with up to 20 short conical cerata inserted along each wave. The upper cerata point up-in, the lower ones point down-out, the upper middle ones largest; each ceras with knobby branches of the digestive gland. Ground color translucent gray to pale brown to blackish with many white dots; flanks are usually darker. The foot is transparent but its edges are sometimes brown spotted. The visible branches of the digestive gland in the rhinophores and cerata are usually light brown. Rhinophores are white-tipped, sheath lobes have white and brown patches. Cerata are tipped with clear to whitish cnidosacs. It is well camouflaged on its hydroid prey and has been found on anchored and floating algae. Depth: to at least 15 feet. Southeast Florida, Bahamas, Colombia, Honduras, Cozumel, Brazil. Rare.

Cozumel

Roatan

Grand Cayman

Riviera Beach, Florida

juvenile; Riviera Beach, Florida

***Lomanotus* cf. *phiops* Er. Marcus, 1957**

Size: ~8 mm

Lomanotidae

ID: External morphology is consistent with that of *Lomanotus phiops* Er. Marcus, 1957. Ground color white. Small reddish to dark brown spots on the cerata, sides, and rhinophores. The digestive gland branches in juveniles are suffused with pale yellowish-orange. Found on hydroids. Depth: to at least 15 feet. Southeast Florida, Grand Cayman. Rare. This is probably an undescribed color variant of *L. phiops*.

Southeast Florida

↑with spawn

***Lomanotus vermiformis* Eliot, 1908**

hydroid lomanotus

Size: to 40 mm

Lomanotidae

ID: Body very long, very narrow, tapers to a bluntly rounded tail; foot thin, sides project slightly. Oral veil rounded, prehensile, with a pair of pointed oral tentacles. Each mantle edge has a line of up to 30 cerata: short, broad, and spoon-like (concave up), undulating folds lacking branches of the digestive gland; each ceras is tipped with a dimple. The rhinophores have an apical club with up to 12 oblique lamellae followed by a narrow, bluntly-rounded vertical projection; each rhinophore is retractile into an elevated thick-walled sheath with 8-10 low papillae on its flared rim. Eyespots are behind the sheaths. Ground color gray to pale brown, sometimes with yellow lines. Along the length of the dorsal midline and each side there is a broad dark stripe composed of irregular, sometimes branching, close-set, dark brown lines superimposed on a pale brown background; there are some dark brown spots beneath the lateral stripes. Body and appendages with a tangled network of iridescent white threads, especially concentrated on or near the brown stripes. Cerata are translucent gray with white tips. Rhinophores are dark brown; the sheaths translucent gray with a white rim. Using a sticky mucous it adheres tightly to the stinging hydroid *Lytocarpus philippinus*, thought to be its exclusive prey. Found in dense aggregations on the tops and sides of boulders, within crevices, and under overhangs, where the hydroid provides it with excellent camouflage among its white feathery branches and dark brown stalks. When disturbed it can swim with lateral flexions of its body. Spawn is a transparent convoluted string containing ~2000 white egg capsules (1 egg each), deposited along the hydroid's stem. Depth: to ~15 feet. Circumtropical. Southeast Florida, Cuba, Bahamas, Panama, Grand Cayman.

Family Madrellidae Preston, 1911

A two-genus family consisting mainly of a few *Madrella* species that have been infrequently recorded, primarily in the Indian and Western Pacific Oceans but also in the Mediterranean Sea, Red Sea, Portugal, and a single record in the tropical West Atlantic.

Body elongate, tapers to the rear; oral veil prominent; no oral tentacles; rhinophores a stalked papillate club, nonretractile; mantle broad, margin ridged; cerata along the edges of the body and front of the head, usually long, thin, distally bulbous, apices curved, penetrated by digestive gland branches; anus lateral; jaws short, triangular; radula triseriate, the rachidian tooth usually broad, the laterals with denticles on a separate wing.

Usually found on their prey – encrusting bryozoans. Some species have glands at the bases of their cerata that can release a bright yellow to orange secretion (the color possibly diet-derived) in response to a threat. They can also autotomize some of the cerata as a diversionary tactic. Maximum lengths are in the range 15-45 mm.

***Madrella* sp.**

Size: ~25 mm Madrellidae

ID: Body elongate, tapers to the rear; the tail short and triangular, its tip rounded; entire perimeter of body with numerous long papillae (cerata), mid-dorsum without papillae; rhinophores with a thick stalk and apical club. Ground color dark red with many scattered white dots. The body's papillate area is brownish, the papillae dark red with black at the base. White dots coalesce to form a white cross between the rhinophores. Species of this genus are usually found on or near their bryozoan prey, which is the same color as the animal. Depth: ~20 feet. *Madrella* species have been recorded from the Indo-Pacific, Sweden, and Georgia (USA). This may be the first record from the Caribbean Sea.

Roatan

Family Myrrhinidae Bergh, 1905

Based on recent (2019) morphological and phylogenetic analyses a few genera of aeolid nudibranchs were removed from the large family Facelinidae and grouped as the resurrected family Myrrhinidae. The only genera that currently have representation in the tropical West Atlantic are *Dondice*, *Godiva*, and *Nanuca*. Their features are described below.

Body elongate, slender, moderately deep; foot narrow, tapers to a long pointed tail, front foot corners are tentacular; oral tentacles are long, smooth, tapering, often curled; rhinophores are long, slender, smooth or distally annulate, usually shorter than the oral tentacles; cerata are long, thin, tubular to fusiform with pointed tips, in several well-separated clusters of arches and rows along the sides of the back, the clusters often on slightly raised cushions; the anus and gonopore are typically on the right side between the first and second ceratal clusters; radula uniseriate, cusp flanked by a few denticles; jaws short.

The body is usually translucent with visible internal organs, including the jaws. Most species feed on hydroids in depths from intertidal to 100 feet. The length range is 10-50 mm.

Dondice arianeae

García-Méndez, Padula & A. Valdés, 2022

Ariane's dondice

Size: to 14 mm Myrrhinidae

ID: Body very elongate, slender, tapers to rear; front foot corners short, narrow, recurved, rear of foot projects as a long slender tapering tail. Rhinophoral distal club with 8-9 **annuli**. Cerata in a row of 6 well-separated clusters per side of back; the number per cluster decreases rearward from 28 to 2; cluster heights decrease rearwards and increase dorsally. Translucent overall. Body with **3 broken longitudinal white-cream stripes**: a median slightly pink one from behind the eyes to the rear of the back, and another one on each side running between ceratal clusters from the base of each oral tentacle to the end of the foot. **Orange line from head to rear along each side of body**. Mantle and tail with scattered pinkish-white cream spots (some blue pigmented), especially at bases of ceratal clusters. Head with a frontal pinkish-white "butterfly" patch, from which a median orange line passes between the rhinophores to behind the eyes; a streak of bright blue spots along the head margin from oral tentacle to rhinophore. Cerata are distally white-cream spotted; sometimes with a subapical pinkish ring; brownish digestive gland branches, white-cream apex. Oral tentacles mostly white-cream-spotted; proximally translucent with some bright blue spots. Rhinophores are white-cream basally, translucent up to the club, then with white-cream spots. Front foot corners have cream-white-spots (some bright blue); foot margin with bright blue spots; bright blue tail, maybe also white-cream-spotted. It usually feeds on hydroids in seagrass meadows, algae, sand patches, mangroves. Sometimes also on upside-down jellies. Depth: 3-7 feet. Florida Keys, Bahamas, maybe Jamaica. The slenderest *Dondice* in the West Atlantic.

Florida Keys

continued on next page

Key Largo, Florida

Riviera Beach, Florida

Riviera Beach, Florida

spawn; Florida Keys

Guadeloupe

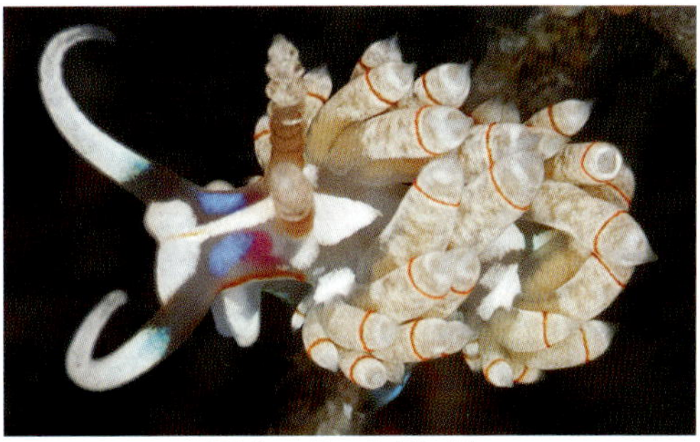

Riviera Beach, Florida

Dondice freddiemercuryi

García-Méndez, Padula & Á. Valdés, 2022

Size: to 19 mm Myrrhinidae

ID: Body elongate, slender, tapers to rear; front foot corners short, recurved; tail long, thin, pointed. Oral tentacles long, smooth, pointed. Rhinophores are long, distal club has 9-10 **annuli**. Cerata long, fusiform, a row of 6 clusters on each side of the back; cerata per cluster decreases rearward (16→2), cluster height decreases rearwards, height within a cluster increases dorsally. Eyes small, behind the rhinophores. Ground color is translucent gray. From head to tail, **a row of ~5 opaque pinkish-white diamond-shaped patches**, their median tips touching, sizes decreasing toward tail, with some bright blue edges. **A pinkish-white line on each side of back runs through ceratal clusters while connecting with lateral tips of diamonds**. Front of head with 2 adjacent opaque pinkish-white patches crossed by a longitudinal orange stripe reaching midpoint between rhinophores; streak or patch of bright blue along head margin from oral tentacle base part-way to adjacent rhinophore; thin longitudinal orange line below ceratal clusters from base of oral tentacle to rear of body. Cerata **opaque pinkish-white on most of the distal two-thirds**, darkening toward tip; **reddish-orange subapical ring**; digestive gland branches brown, cnidosac tip translucent whitish. Oral tentacles translucent proximally with some bluish pigment, becoming pinkish-white toward the tips. Rhinophoral stalks translucent, club pinkish-white. Front foot corners white-bluish distally; **foot margin with a band of bright blue spots or a blue line** to the end of the tail. Eyes black. Found on algae and sandy patches where it feeds on hydroids. Depth: 3-33 feet. Florida, Bahamas, Guadeloupe, Martinique, Cayman Islands.

Riviera Beach, Florida

Jupiter, Florida

Riviera Beach, Florida

Key Largo, Florida

with spawn; Riviera Beach, Florida

Dondice juansanchezi García-Méndez, Padula & Á. Valdés, 2022 **Juan's dondice**

Size: to 10 mm Myrrhinidae

ID: Body elongate, slender, tapers to rear; front foot corners short, wide at base, recurved; rear of foot projects as long thin pointed tail; oral tentacles long, smooth, tapering; rhinophores long, distal club with 14 **annuli**; cerata long, fusiform, in a row of 6 clusters per side; cerata per cluster and cluster height both decrease rearward, height within a cluster increases dorsally. Ground color translucent with many variably sized pinkish-white opaque spots. A **bright orange wave-like line extends from behind each rhinophore to the end of the body**, waves flank dorsal midline and travel in sync, curving in and almost touching in gaps between the ceratal clusters. Front of head with dense cluster of opaque pinkish-white spots crossed by a median orange stripe passing between the rhinophores to rear of head, followed by a light blue broken line extending to tip of tail. Cerata, oral tentacles, rhinophores, and front foot corners with **pinkish-white spots and dots; cerata ones very crowded and sparkly. Cerata with light brown digestive branches; translucent white cnidosacs;** usually with 2 pale pinkish-orange broad bands – one near the tip, one closer to mid-length; sometimes very indistinct. Oral tentacles with **a broken blue stripe from base to near mid-length**. Front foot corners have light blue tips; **foot margin with a bright blue band**; tail translucent with a bright blue median stripe. Feeds on hydroids in sandy patches. Depth: 3-17 feet. Southeast Florida. The above color details are based on the original description. Some of the pictured animals differ slightly, e.g. the orange lateral line below the cerata and the blue spots along the foot margin were not mentioned in that description. Those discrepancies can be attributed to intraspecies color variabilities that were not present in the examined specimens.

Jupiter, Florida

Riviera Beach, Florida

Riviera Beach, Florida

Panacea, Florida

Riviera Beach, Florida

with spawn; Jupiter, Florida

Dondice jupiteriensis **García-Méndez, Padula & Á. Valdés, 2022** **Jupiter dondice**

Size: to 40 mm Myrrhinidae

ID: Body elongate, slender, tapers to rear; front foot corners short, wide, recurved, rear of foot projects as a slender tapering tail; oral tentacles long, smooth, tapering distally, tips often curled. Rhinophores long, with a distal club, 10-14 annuli. Cerata fusiform, elongate, a row of 7 clusters along each side of the back; cerata on double arches in clusters 1-2, single arches in clusters 3-4, transverse rows in clusters 5-7, number of cerata per cluster decreases rearward (20 → 2), so does the cluster height, the height within a cluster increases dorsally. Eyes small, behind the rhinophores. The body, foot, and appendages are translucent. Body with three broken light blue stripes – a median one from behind the eyes to the rear of the dorsum and one on each side running between the ceratal clusters to the end of the foot. Head with a pinkish-orange front margin, a median orange line from front to behind eyes, two bright blue streaks along the head margins from oral tentacle bases to the adjacent rhinophores. Cerata translucent pinkish-orange, digestive gland branches chestnut brown, apex light blue with a translucent white cnidosac, **a subapical broad yellowish-orange band with a central reddish-orange ring**. Oral tentacles bright blue near middle, pinkish-orange toward tip. Rhinophores pinkish-orange distally. Foot has bright blue pigment on distal half of front corners and foot margin from front corner to rear. Tail bright blue. Eyes black. Inhabits shallow sandy patches near hydroids. In Tampa Bay (Florida) it is known to inhabit the colonial hydroid *Eudendrium carneum*. Regarded as a predator that feeds on the hydroid's polyps and probably stores undischarged nematocysts in its cerata for defensive purposes. Depth: 3-20 feet. Southeast and Northwest Florida, Bahamas, North Carolina.

Dominica

Bermuda

Grand Cayman

Bermuda

Roatan

with spawn; Grand Cayman

Dondice occidentalis **(Engel, 1925)** **fringeback dondice**

Size: to 50 mm Myrrhinidae

ID: Body elongate, narrow, tapering. Front foot corners tentacle-like, recurved, short; tail long, thin, pointed. Cerata long, slender to slightly inflated, cnidosac apices pointed, a row of 6-7 well-separated clusters per side on low cushions. Rhinophores long, tapering, **5-9 annulae distally**. Oral tentacles longer than rhinophores, slender, pointed, smooth, often curled. Ground color translucent gray; body may be suffused with yellow, yellowish-brown, or pinkish-brown, sometimes white-dotted with some dorsal black flecks. **Color patterns extremely variable**. Usually: a **median white or blue stripe** (sometimes broken) widens and narrows from rear of body to rhinophores, meeting or encompassing a **yellow/orange/red line from between the rhinophores to a white patch between the oral tentacles**; a blue patch/stripe in front of rhinophore may extend onto base of adjacent oral tentacle; **a thin red lateral stripe** from oral tentacle base to rear of body, interrupted by ceratal clusters. Ceratal proximal half to two-thirds with a yellowish-brown to brown to orange digestive gland; **distal bands** of yellow, red, orange, orange/brown, blue, white; usually several bands but sometimes only one thin orange or white subapical ring; cerata may be densely white-dotted. **Rhinophores gray to brown with white annuli**. Oral tentacles translucent gray/bluish basally, white elsewhere. Tail blue. Usually on algae and hydroids in habitats ranging from shallow mangrove-lined lagoons to reef rubble bottoms to coral reefs. Feeds mainly on hydroids (*Eudendrium* spp.) but sometimes on bryozoans (*Amathia* sp.). Stores undischarged hydroid stinging cells in its cnidosacs for defensive purposes. When alarmed it discards cerata as a wiggling distraction while it escapes. Spawn is a spiral coil (4-5 whorls) of a gelatinous tube attached at regular intervals to a substrate, with free loops between points of attachment; tube contains crowded capsules each with one white/pink/orange egg (color varies with diet). Depth: 0-100 feet. Bermuda, South Carolina to Florida, Bahamas, Brazil, Antilles to Mexico. The broad range of external appearances may encompass undescribed species. Molecular analyses across this range would be useful.

Roatan

Martinique

Panama

with spawn; Riviera Beach, Florida

with *Cassiopea xamachana;* Riviera Beach, Florida

Dondice parguerensis **Brandon & Cutress, 1985** **cassiopea dondice**

Size: to 48 mm Myrrhinidae

ID: Body elongate, slender, widest at the pericardial hump, tapers to the rear. Foot front corners long, broad, recurved; tail slender, pointed, about one-fourth body length. Oral tentacles about one-third body length, slender, pointed, smooth, tips often curled. Cerata long, tubular, tips pointed; a row of 9 clusters per side, each cluster on a slightly raised cushion; number of cerata per cluster decreases rearward (from 28 to 1), the height within a cluster increases dorsally, front clusters have the largest cerata, the front 2 clusters being widely separated by the pericardial hump. Rhinophores are slightly shorter than the longest cerata, tapering, their **distal third with 10-12 annuli** and smooth below. **Ground color translucent brown**. A **dorsal, white, broken median stripe extends from the front of the head to the rear of the body; some white patches along the sides of the back**, most concentrated at bases of ceratal clusters. Ceratal coloration is variable: brownish with a diffuse whitish band near the top, above it a brownish band and a white apical cap; or brown proximally, white distally, with a narrow subapical brown band; or with iridescent yellowish speckling and a subapical brownish band. When visible, the ceratal digestive branches are thin axial orange-brown tubes just after feeding. Rhinophores and oral tentacles are opaque white distally, the tentacles sometimes with bluish tips. Front foot corners have a white front edge; tail translucent whitish. Usually in calm shallows of mangrove stands, living exclusively and parasitically among tentacles of the upside-down jellyfishes *Cassiopea frondosa* and *Cassiopea xamachana*. The orange-brown color of its digestive branches are due to ingestion of the jellyfish's zooxanthellae. It stores stinging cells and releases them to repel predatory fishes. Spawn is a gelatinous tangled string of white spherical eggs deposited among a jellyfish's tentacles, the mass typically containing ~34,000 eggs. Hatching occurs in 2-6 days. Depth: 1-15 feet. Puerto Rico, Guadeloupe, Martinique, Venezuela, Panama, Roatan, Southern Florida.

Jupiter, Florida

***Dondice* sp. 1**

Size: to 10 mm Myrrhinidae

ID: Body elongate, slender; oral tentacles long, smooth, tapering distally; rhinophores shorter, distal **annulate club** (~10 annuli); cerata long, fusiform. Ground color translucent gray. Median orange and white narrow stripe extends from front of head to a white-spotted orange region behind the rhinophores; several larger scattered white spots on the rest of the body and foot, the foot edge is blue; an orange lateral stripe from the base of each oral tentacle to the rear of the body. Cerata translucent with a broad brown/orange digestive gland, white tips, and a few small white spots. Rhinophores are whitish distally. Oral tentacles bright blue, shading to whitish-blue distally. Found on a hydroid. Depth: to ~15 feet. Perhaps a variant of *Dondice juansanchezi*.

Jupiter, Florida

***Dondice* sp. 2**

Size: ~15 mm Myrrhinidae

ID: For external morphology see *Dondice occidentalis*. The bright reddish-orange cerata, the orange blush basally on the oral tentacles and rhinophores, and the absence of blue pigment on the front foot corners conflict with the most recent descriptions of *D. occidentalis*. Nevertheless, this could be a color variant. On rubble near hydroids. Depth: less than 15 feet.

***Dondice* sp. 3**

Size: ~15 mm Myrrhinidae

ID: For its external morphology see *Dondice occidentalis*. The back is reddish-brown with **three longitudinal blue stripes** – one median and an irregular one along the inner edge of each row of ceratal clusters; small blue spots concentrated at the bases of the clusters; a lateral red stripe extends from the base of each oral tentacle to the rear of the body below the cerata. A median orange stripe from the front of the head to just behind the rhinophores. Front foot corners are bluish-white; tail and foot margins are bright blue. Cerata reddish-brown proximally, pinkish-white distally, a narrow subapical reddish-orange band/ blotch, and a white tip. Oral tentacles reddish-orange with a blue dorsal stripe proximally, shading to whitish distally. Rhinophores dark basally, the rest pale orange. On a sand/ rubble bottom. Depth: less than 15 feet. Differs from *D. occidentalis* primarily by the presence of three blue dorsal stripes.

Jupiter, Florida

Roatan

Dondice sp. 4

Size: ~15 mm Myrrhinidae

ID: For its external morphology see *Dondice freddiemercuryi*. This animal's coloration is similar to that species' except for its pale orange ground color and a pinkish-blue median dorsal stripe instead of a string of white "diamonds". Possibly a variant of that species. Found on an algae leaf. Depth: ~40 feet.

Dondice sp. 5

Size: ~15 mm Myrrhinidae

ID: The external morphology of this animal is much like that of *Dondice occidentalis* (Engel, 1925) but its coloration seems to be a composite of two known variants of *D. occidentalis* as described by Garcia-Mendez et al (2022). In particular, the oral tentacles, rhinophores, and cerata are like those of their Panamanian specimen and the dorsal pattern is like that of their Jamaican specimen. This could be yet another variant of that broadly variable species. No ecological information has been made available for the pictured animal. Depth: ~15 feet.

Bocas del Toro, Panama

Martinique

Riviera Beach, Florida

Key Largo, Florida

truncated dorsal stripe; Riviera Beach, Florida

Godiva rubrolineata **Edmunds, 1964**

redline godiva

Size: to 21 mm Myrrhinidae

ID: Body elongate, narrow; front foot corners tentacle-like, recurved, half the length of the oral tentacles, tapers to a long pointed tail. **Rhinophores and oral tentacles smooth**, tapering, about one-fourth animal's total length. **Cerata thin, taper to a sharp point, half the length of the rhinophores**; 5-6 well-separated clusters per side. Ground color translucent grayish-white. White blotches on head and back; a white median stripe from oral tentacles to rhinophores; **median orange stripe** from front of head almost to tail with breaks at positions of ceratal clusters, stripe sometimes ends at first ceratal cluster; **orange cheek blotch** usually continues as another orange stripe (similarly broken) along each flank to start of tail. Cerata translucent, white-tipped, digestive gland **gray to reddish-brown to blackish**, surface may be white-dotted. Tops of tail, foot corners, distal half (or more) of each rhinophore and oral tentacle are white. On sand and mud among mangrove roots and crawling on hard substrates near algae. Presumably feeds on hydroids. Depth: 0-70 feet. Southeast Florida, Cuba, Jamaica, Puerto Rico, Guadeloupe, Martinique, Curacao, Venezuela, Brazil. Rare.

Using an integrative approach combining mitochondrial and nuclear gene sequences, morphological and biogeographical information, and several species delimitation methods, Henryco et al (2025) determined that what used to be the single species *Nanuca sebastiani* is actually a complex of at least 6 unique cryptic species: *N. alboviridis* (U.S. Virgin Islands), *N. cahuitensis* (Panama, Costa Rica), *N. leali* (Southeast Florida); *N. sebastiani*, *N. perolata*, *N. smaugi* all from Brazil. Moreover, they also point out that additional studies based on more comprehensive sampling for this genus could broaden their distributions and/or undercover the existence of additional species. In the following pages, the *Nanuca* images are grouped according to how closely the pictured animals' external appearances agree with the descriptions provided in this paper. Significant deviations are considered to be indications of undescribed species, but could also be examples of heretofore unknown intraspecies variability.

Riviera Beach, Florida

Nanuca leali **Henryco, Meirelles, García-Méndez, Camacho-García, Á. Valdés, Schrödl & Padula, 2025**

Size: to 10 mm Myrrhinidae

ID: Body elongate, slender, tapers to a long thin tail. Oral tentacles long, fingerlike. Rhinophores robust, long, annulate: four large well-separated cuplike rings with radii diminishing toward the apex. Cerata cucumber- shaped, tip constricted and bluntly pointed, 3-4 clusters per side. Ground color beige. Oral tentacles white-spotted. Rhinophores with a few white spots; distally diffuse white. Cerata beige to orange with longitudinal white stripes; darker basally. Dorsum with median patches between the ceratal clusters, each patch typically a white to yellow cross with a central orange spot and one in each arm, superimposed on a slightly larger white to bluish background cross. A white to bluish lateral stripe just below the ceratal clusters contains a line of orange spots/ocelli. Margin of foot translucent, white-spotted; tail with median line of orange dots. Usually on mixed algae with attached hydroids (its prey) in depths from the intertidal to 20 feet. The animals in the topmost two photos above are the closest match to the original description and its accompanying photos. The other photos above seem to be of the same species but could also be of additional cryptic species. Collection and analyses of more specimens is required to fill the knowledge gaps.

Grand Cayman

Nanuca cahuitaensis **Henryco, Meirelles, García-Méndez, Camacho-García, Á. Valdés, Schrödl & Padula, 2025**

Size: to 12 mm Myrrhinidae

ID: Body elongate, slender, tapers to a thin tail, head elongate. Oral tentacles digitiform up to a quarter of body length. Rhinophores long, annulate with 4 rings. Cerata cucumber-shaped, tip constricted and bluntly pointed, 3-4 clusters per side; each cluster a horseshoe; each ceras with almost no stalk, dorsal cerata longest. Mantle gray to translucent green; oral tentacles and rhinophores translucent gray, both with beige or white distal parts. Orange median line usually from near mouth to behind rhinophores. Dorsum with many white spots and a few blue-to-white patches with orange spots that form a cross-shaped pattern. Cerata yellowish/beige with faded longitudinal white stripes. Tail orange-dotted. Depth: to at least 11 feet. Reported from Sargassum. Caribbean coasts of Costa Rica and Panama, Venezuela, possibly Martinique.

Isla de Aves, Venezuela

Bahía de Río de Janeiro, Brazil

Nanuca sebastiani **Er. Marcus, 1957**

orange-spotted nanuca

Size: to at least 20 mm Myrrhinidae

ID: Body elongate, slender. Oral tentacles long, fingerlike, about a quarter of the body length. Rhinophores long, annulate with 3-5 rings distally, ring radii diminishing toward apex. Dorsal cerata in two rows of 3-5 clusters per side, each cluster a horseshoe; each ceras with almost no stalk at base. There are two color variants. **Brown morph**: body light brown; oral tentacles and rhinophores brown, with scattered cream-white dots (not on their bases); cerata beige with longitudinal white stripes; yellow/orange elongate patch on the head extends back as a broken stripe; 2-3 patches on the mantle, somewhat cross-like, consisting of five small round spots; a row of small orange spots along each upper flank from the head to the start of the tail; some irregular white dots along the foot margin. **Orange morph**: the body is reddish-brown; the oral tentacles/rhinophores are pale grayish distally but unspotted; an orange head patch extends to the first ceratal cluster; the back has three groups of broad elongate orange patches and lateral spots, the patches connected by an orange line along the flank; cerata are dark red with white longitudinal stripes and clear to opaque white tips; top of the tail with a broken median orange line. The pictured animal closely resembles the orange morph of this species, but it is not a perfect match relative to the above description. Typically found with algae and hydroids in depths of 4-15 feet.

The following set of photos represents individuals of *Nanuca* that cannot be reliably identified by external features alone based upon the species complex descriptions in the paper by Henryco et al (2025). Some are probably one of the six species described in the paper but there might be additional undescribed members of the complex. They could all have been displayed together under the heading *Nanuca* spp., but labeling them individually by number may simplify the references to these photos in future publications that expand the knowledge-base for this genus.

↓Martinique

Nanuca **sp. 1**

Florida↑

Nanuca **sp. 2**

↓Guadeloupe

Nanuca **sp. 3**

Puerto Rico↑

Nanuca **sp. 4**

↓Little Cayman

Grand Cayman↑

Nanuca **sp. 5**

Guadeloupe

Nanuca sp. 6

↓Brazil

Guadeloupe↑

Nanuca sp. 7

Nanuca sp. 8

↓Cozumel↑

Nanuca sp. 9

Nanuca sp. 10

Nanuca sp. 11 Guadeloupe

Family Phylliroidae Menke, 1830

The Phylliroidae are three species of pelagic nudibranchs occurring circumglobally in tropical surface waters. Two of the species are in the genus *Phylliroe* and one is in *Cephalopyge*.

Body elongate, laterally compressed, transparent, side profile oval and somewhat fish-like with the rear of the body like a slightly forked to truncate to rounded tail; rhinophores are behind the head, long, smooth, contractile, usually erect or curved slightly forward and sometimes directed laterally; the head slopes forward from the rhinophores to a terminal mouth (*Phylliroe*) or extends forward as a cephalic disk above the mouth (*Cephalopyge*); the foot is either an anterior vestigial pedal gland somewhat behind the rhinophores, or (in *Cephalopyge*) a less reduced structure that can contract or expand considerably; gills are absent, may be replaced by small, oblong, elevated "beads" along the dorsal and ventral margins and on the upper flanks; the anus is at the middle of the right side (*Phylliroe*) or dorsal just behind the rhinophores (*Cephalopyge*); the gonad consists of several ball-like structures.

They are good swimmers that keep their flat bodies vertical while propelling themselves forward using left-right body flexions, achieving speeds up to 150 mm/s. Prey consists of planktonic coelenterates (jellyfishes) which they latch onto with their foot remnant and manipulate with their rhinophores. Species of *Phylliroe* are bioluminescent. Body length range is 25-55 mm.

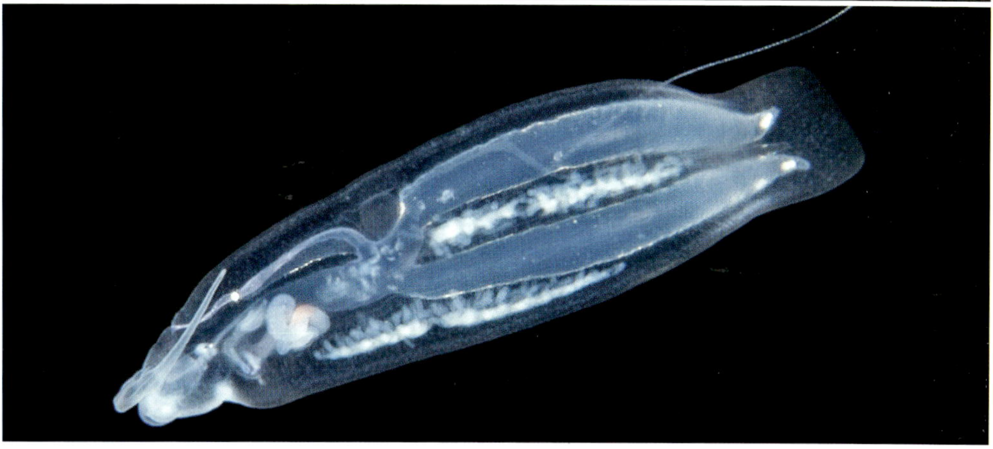

all photos from Southeast Florida

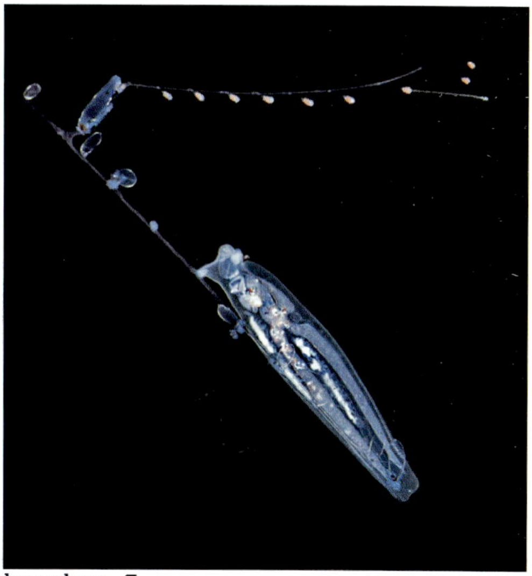

↖ eating a siphonophore ↗

***Cephalopyge trematoides* (Chun, 1889)**

Size: to 25 mm

Phylliroidae

ID: A pelagic nudibranch. Body elongate, laterally compressed, its side profile a long flattened oval (length ~7 times height); rear of the body (tail) slightly truncate; cephalic disc on top of the head above the mouth; rear of the cephalic disc straddled by two rhinophores that are long, smooth, usually erect or pointed slightly forward, almost completely contractile into pockets; the foot is a ventral structure somewhat behind the rhinophores that can contract or expand considerably; a ventral groove from the mouth to the foot; the dorsal anus is just behind the rhinophores. Body transparent and colorless. The body wall is densely covered with epithelial gland cells, especially dense on the dorsal and ventral margins where they form bands of opaque spots. Viscera are visible through the body wall, most being nearly colorless and with varying opacities but the intestine is pinkish and the ampulla of the reproductive system is reddish-brown. It is usually found at or slightly below the sea surface. Rapid undulations of its body enable it to attain swimming speeds up to 12 cm/s. It is a carnivore that feeds on small animals living near the surface of the sea (the "pleuston"). Depth: 0-3 feet. Circumglobal in warm to (occasionally) temperate marine waters. One of the very few planktonic nudibranchs.

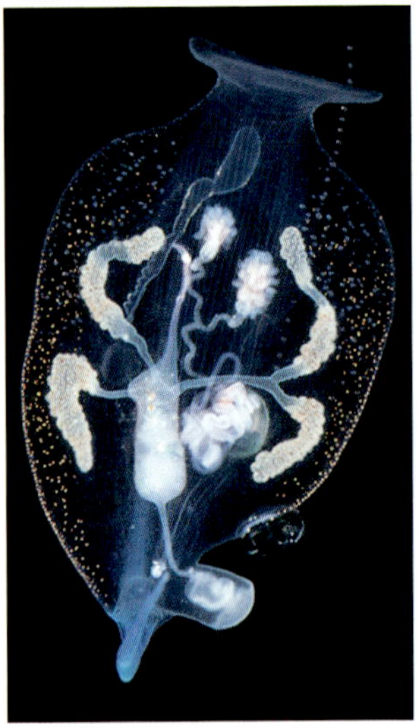

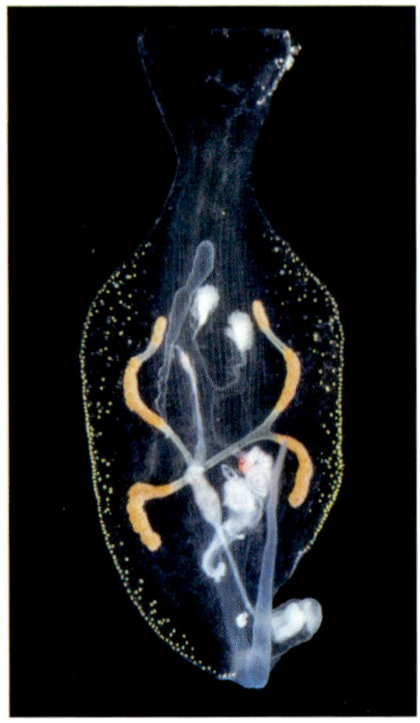

↑ spawning ↑

Phylliroe bucephala **Lamarck, 1816**

Size: to 80 mm — Phylliroidae

ID: One of very few planktonic nudibranchs. Body gelatinous, fragile, elongate, very compressed laterally, **side profile somewhat fish-like**, dorsal and ventral edges "sharp"; **its tail is a fin**, about one-sixth body length, separated from the body by a peduncle, rear edge truncate to **slightly emarginate**; head angled downward; mouth at front of head, wide, tip swollen; proboscis retractile; there are **two long conical tentacle-like rhinophores behind its head**; gills are small, oblong, elevated "beads" along the dorsal and ventral margins and on the upper flanks; **two ovotestis lobes**. **Body transparent**; gill beads are yellowish-white; **viscera are visible as a white/pink/gold mass surrounded by a cloud of orange dots**, the **digestive gland is X-shaped**. It inhabits upper layers of warm water, where it swims upright by undulating its body from side to side. Apparently photophobic, staying deep during daylight hours and migrating to the surface for night feeding. Dive lights make it swim deeper. It is highly **bioluminescent** – glandular cells on body surface emit flashes of light apparently at will, but especially when it is disturbed. The juvenile lives exclusively within the umbrella bell cavity of the medusa *Zanclea costata*, where it attaches to the bell by its foot gland, feeds on the medusa, and grows, eventually discarding the bell's remnants. The adult's prey is more diverse, including hydrozoans (e.g. *Z. costata*, *Aequorea* spp., siphonophores) and tunicates (e.g. *Oikopleura albicans*). Reproduction is hermaphroditic, with reciprocal fertilization of a mating pair – they join head-to-tail, intertwine their long penises, and swim in a spiral pattern for ~6 minutes. Depth: 0 to ~200 feet. Circumglobal in warm to temperate marine waters (18-25 °C). Scientific information for *Phylliroe* species is quite sparse and sometimes ambiguous. All of the photos are from southeast Florida.

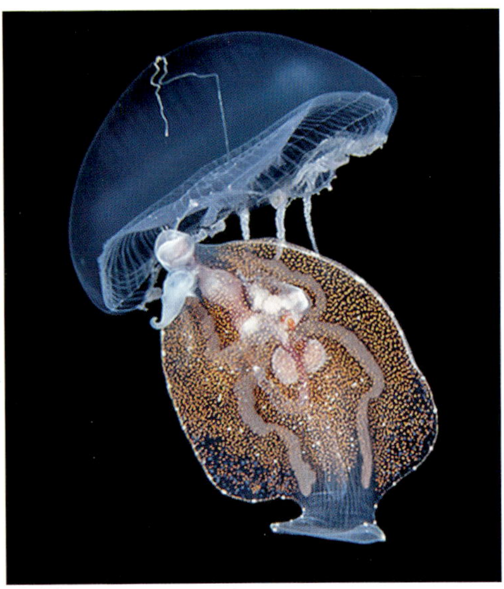

Southeast Florida

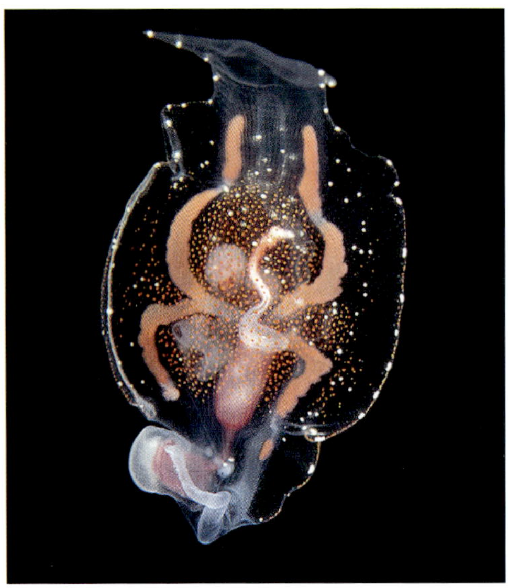

Phylliroe **sp.**

Size: to 80 mm — Phylliroidae

ID: This is tentatively considered a variant of *P. bucepala* because its extremely dense orange spotting warrants further study. Specimens are currently undergoing examinations, both morphological and molecular, to clarify this animal's taxonomic placement. It **could** be an undescribed species.

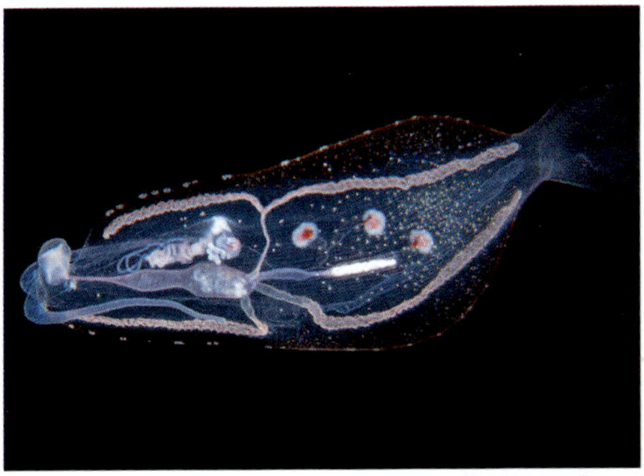

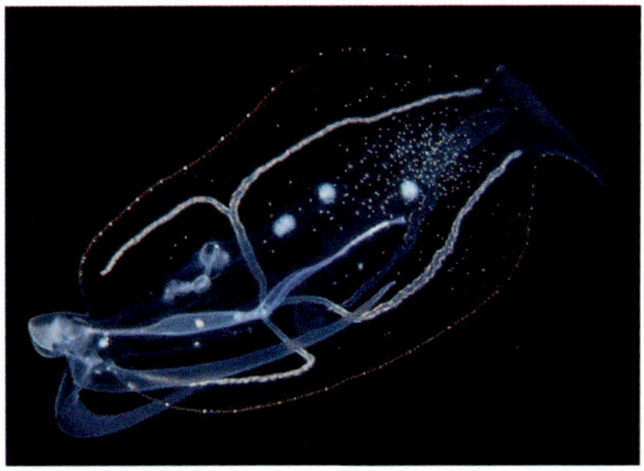

all images from Southeast Florida

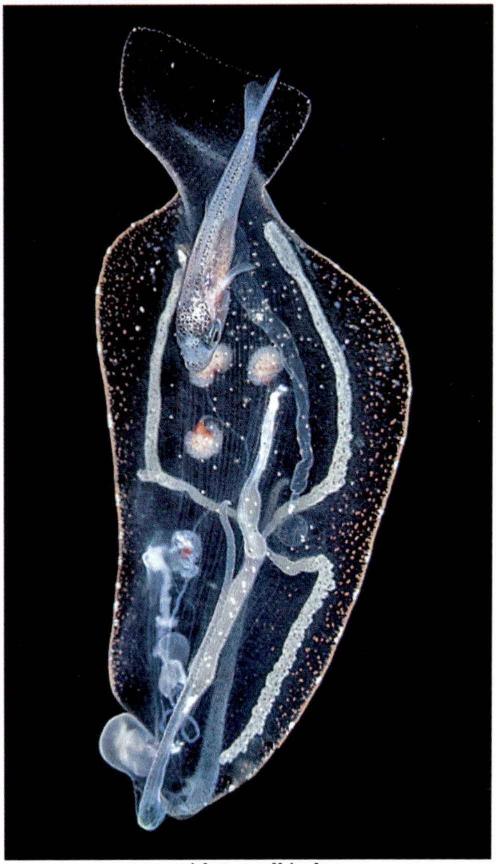

with a small jack

Phylliroe lichtensteinii **(Eschscholtz, 1825)** **pelagic nudibranch**

Size: to ~55 mm Phylliroidae

ID: A pelagic nudibranch. Body gelatinous, fragile, elongate, very compressed laterally, **side profile somewhat fish-like**, dorsal and ventral edges "sharp"; **its tail is a fin**, about one-sixth body length, separated from the body by a peduncle; the rear edge is truncate to **slightly rounded**; head angled downward; mouth at front of head, wide, tip swollen; proboscis retractile; two tentacle-like **long conical rhinophores behind the head**; gills are small, oblong, elevated "beads" along the dorsal and ventral margins and on the upper flanks; **usually three ovotestis lobes**; foot vestigial, reduced to an inconspicuous small gland. **Body transparent**, gill beads yellowish-white, **viscera visible** as a white to pink mass surrounded by a cloud of orange dots; **digestive gland X-shaped**. Inhabits upper layers of warm water where it swims upright by moving its tail from side to side. Prefers to stay deep during daylight hours, migrating to the surface at night for feeding. Prey includes hydrozoans (e.g. *Zanclea costata*, *Aequorea* spp., siphonophores) and tunicates (e.g. *Oikopleura albicans*). It is bioluminescent, emitting flashes of light when disturbed. Depth: to ~200 feet. Circumglobal in warm to temperate marine waters (18-25 °C). Scientific information for *Phylliroe* species is sparse and sometimes ambiguous. Currently, *P. lichtensteinii* is visually distinguished from *P. bucephala* by the number of ovotestis lobes (gonads) – 2 for *bucephala*, 3-5 for *lichtensteinii*. However, more studies are needed to validate this criterion. The pictured individuals have more than 2 gonads.

Family Scyllaeidae Alder & Hancock, 1855

There are about ten species of very cryptic nudibranchs in the genera *Crosslandia*, *Notobryon*, and *Scyllaea*, distributed worldwide in temperate to tropical seas. Gill concealed within the laterally-opening mantle cavity; the radula is usually absent but sometimes vestigial. Shell internal, reduced to a small translucent plate; it is either membranous or wholly/partially calcified.

Body elongate, somewhat thick and elevated, the rear with a dorso-median keel; each dorso-lateral margin has either one or two large flat lobes ("cerata") bracketing the midbody, either joined basally or not, but paired transversely; the inner ceratal faces have transparent, tubular, branched gill tufts; the oral veil is bilobed to indistinct; the rhinophores are relatively small, with a lamellate club and an apical knob; their sheaths are large and flat, like the cerata but slightly smaller; the anus is on right side, either lateral or laterodorsal; jaws cover the pharynx; the radula has several lateral teeth, and may or may not have a rachidian tooth.

These slugs are able to swim by combining lateral body flexions with the flapping of their broad flattened rhinophoral sheaths and cerata. They feed on hydroids that grow on the surfaces of brown algae and seagrasses. Some species spend their lives on fixed bottom substrates, but others drift with algae. Lengths are typically to 50 mm, but they can reach over 100 mm.

Riviera Beach, Florida

***Crosslandia* sp.**

Size: ~3 mm — Scyllaeidae

ID: Body elongate, narrow, deep. Rhinophores are lamellate, protrude slightly from their sheaths; sheath high, rim smooth and flared, rear margin smoothly continuous with the mantle edge. A pair of large lobes along the mantle edge on each side of the body; lobes broad, transversely flattened, edges jagged, the bases of each pair merge into a common long base. The ground color is greenish-orange. Dark brown pigment on the rear edge of a rhinophoral sheath continues posteriorly along the mantle edges and lobes. Some scattered dark brown spots. White spots along rim of rhinophoral sheath. It was found on a hydroid at a depth of ~10 feet. This might be a juvenile of *Crosslandia daedali* Poorman & Mulliner, 1981, which can reach a length of 25 mm. That species is known from the Eastern Pacific and has been (tentatively) recorded from the Caribbean coast of Costa Rica.

Dominica

U.S. Virgin Islands

U.S. Virgin Islands

St. Vincent

***Scyllaea pelagica* Linnaeus, 1758** — **sargassum nudibranch**

Size: typically to 50 mm, but can reach 102 mm — Scyllaeidae

ID: Elongate, flattened, stout; tapers to a **short blunt tail**. Mantle edges are usually folded up, with **two upright lobes per side**, transversely paired; lobes high, flat, rectangular/rounded with **jagged edges**. Delicately branched tufts **("gills") on lobe inner faces** and the rear of the back. There are **two high flattened rhinophoral sheaths at the front**, the rim widely flared, a crest on the rear face. Rhinophores finely lamellate, thin, protrude slightly from the sheaths. **Rear of body with a flat upright "keel-like" median crest**; sides with **small conical tubercles**. Ground color translucent **yellowish to green to brownish**. Surfaces with many scattered reddish-brown to blackish dots and spots, crowded with tiny white beads. **Yellowish individuals usually with prominent concentrations of reddish-brown pigment along mantle edges and partly up lobes**; scattered white markings, especially on and around the side tubercles; sometimes a row of **bright blue spots** medianly on the back and along each side. Spends life on *Sargassum*, usually on floating and drifting patches but also on clumps attached to a substrate. Blends well while feeding on hydroids that grow on its branches. Able to swim by repeatedly bending its body from side-to-side and flapping its dorso-lateral lobes and rhinophoral sheaths. Its spawn is a convoluted string of yellow eggs wrapped around *Sargassum* branches. Depth: 0-50 feet. Circumtemperate/tropical. Massachusetts to the Caribbean and the Gulf of Mexico. Seasonally common to locally abundant in Florida and the Bahamas.

Cozumel

Grand Cayman

at 65 feet; Grand Cayman

St. Vincent

Roatan

St. Vincent

Notobryon caribbaeum **Caballer & Ortea, 2014**

Size: to 45 mm

Scyllaeidae

ID: Body elongate, slender, laterally compressed, with a scattering of small tubercles. Oral veil bilobed, edges with some crenulations and small papillae. Two lobes per side of back, well separated, front pair larger; each lobe high, broad, and flattened, its margins crenulated with papillae; their bases are long and oriented longitudinally; each lobe's inner face has filamentous tripinnate gills. Rhinophores are narrow, retractile; have ~10 lamellae, pointed tips, and protrude slightly from a high robust sheath with a flared papillate rim; the sheath's rear face has a low vertical papillate ridge that continues to the lobes and tail. The tail is moderately long, has a blunt tip, and its sides curl up to stand vertically or overlap to form a tube. The ground color is translucent orange. There are white spots scattered on all surfaces, but most concentrated along the margins of the oral veil and dorsal lobes, on the ridges, some in larger patches on the back and sides. There are reddish-brown small spots and accents along the ridges as well as a few purplish and blue spots. The papillae and rhinophores are white-tipped; the gill filaments transparent and white-spotted. Internal organs are visible as orange spheres. It crawls over sand and rubble to algae patches (e.g. *Halophila stipulacea*) where it feeds on hydroids. Depth: to at least 65 feet. Southeast Florida, Guadeloupe, St. Lucia, St. Vincent, Roatan, Cozumel, Grand Cayman. Rare.

Family Tethydidae Rafinesque, 1815

A family of about 20 species in two genera, *Melibe* and *Tethys*, of which Melibe is by far the most speciose. These nudibranchs have a well-documented presence in the tropical Indo-Pacific, Mediterranean, temperate eastern Pacific, and southern Africa, but are quite rare in the Atlantic. In the tropical West Atlantic there are a few scattered records for two species of *Tethys*, most due to dredging from very deep water, and for only one recently described species of *Melibe*.

Body elongate; mouth region with small to large expandable and extendable oral hood fringed with rows of sensory tentacles; rhinophores with apical clubs having several lamellae; sheaths large, cylindrical to flattened; ceratal shape varies from oval to cylindrical to broad and flat with a somewhat cylindrical basal peduncle, surfaces may be tuberculate or papillate, sometimes with large apical digitations, usually fluid-filled but without digestive gland branches, organized as one long dorso-lateral row per side of body; tufted gills usually on inner ceratal faces, sometimes on outer ones and dorsum, or in pairs at ceratal bases; no radula, replaced by true jaws or extra-oral thickenings; stomach often with a girdle of 10-20 chitinous toothlike plates, presumably for crushing ingested prey.

They feed on small crustaceans that they trap in their oral hood. When disturbed, some larger species can swim away by elevating their cerata and repeatedly flexing their bodies from side to side. Since missing cerata can be regenerated, they can also discard some cerata to temporarily distract and confuse a predator. Body length range is broad: 10-300 mm.

Guadeloupe

Grand Cayman

Grand Cayman

Guadeloupe

Melibe arianeae Espinoza, DuPont & Á. Valdés, 2013

Ariane's melibe

Size: to 13 mm Tethydidae

ID: Body elongate, front somewhat compressed laterally, tapers to long conical tail; body and appendages covered with minute tubercles; back with a few variably sized blunt papillae; **anterior oral hood circular with 2 rows of simple elongate papillae**. A row of 7 cerata on each side of dorsal midline, alternating transversely instead of paired. Cerata inflated, oval, tuberculate, apices blunt or doubly/triply forked, sizes variable (some very large). Rhinophores at rear edge of oral hood, 3-4 lamellae, no tubercles; rhinophoral sheaths high, cylindrical, slightly inflated looking, rim flared with 2-3 papillae at rear. Body and appendages nearly transparent. Minute tubercles white; **all surfaces orange-flecked and white-blotched**; dorsal papillae white-tipped; internal organs visible as orange-brownish mass. Cerata transparent, branches of digestive gland visible as whitish to brownish axis. A bottom-dweller that feeds on small benthic crustaceans, including fast-moving and free-swimming ones. Unlike a typical nudibranch, it does not have radular teeth for catching and ingesting prey. Instead, its oral veil has evolved into an extendable hood with sensory papillae, which it sweeps from side to side as it crawls along the bottom. When the sensors record a "hit" the hood contracts rapidly around the crustacean to trap it and gradually convey it to the mouth. Prey remains alive in the gut until killed by digestive chemicals. Depth: to at least 10 feet. Southeast Florida, Virgin Gorda, St. Eustatius, Grand Cayman, Honduras. Rare. The identifications are tentative, based on *M. arianae* being the only member of this genus known to inhabit the tropical West Atlantic Ocean. Some of the pictured individuals could be closely-related cryptic species. Positive identifications require specimen analyses.

mouth; Guadeloupe

Family TRINCHESIIDAE F. Nordsieck, 1972

A family currently with 8 genera that were recently removed from the much larger traditional family Tergipedidae as a result of ongoing taxonomic revisions based on the integration of morphological and molecular phylogenetic data. The species are distributed circumglobally in temperate, subtropical, and tropical waters, with at least three of the genera – *Catriona*, *Selva*, *Zelentia* – being represented in the tropical West Atlantic.

Body elongate; mantle edge absent; a row of ceratal clusters along each side of back, each cluster an oblique row with several cerata, usually with more than one cluster in front of the pericardium (*Catriona* usually has >3); oral tentacles and rhinophores usually smooth; anus acleioproctic; the radula uniserial, its central teeth with a predominantly low cusp and long lateral denticles; pharynx narrow; male and female genital openings are close together below the right anterior cerata; penis with a hollow stylet.

Usually found in shallow rocky areas with hydroids, anemones, and algal growth but some are as deep as 80 feet. Most feed on hydroids but a few prey on hard corals. Their length range is 5-25 mm.

Roatan

Guadeloupe

Tenellia maua **(Er. Marcus & Ev. Marcus, 1960) white-band cuthona**

Size: usually 5-8 mm, max 12 mm Trinchesiidae

ID: Body elongate, width and height about one-third length; head narrow; foot narrower than body, front corners rounded, tapers to a short pointed tail. Cerata one-third body length, spindle-shaped, greatest swelling just past midlength, tapering to a blunt apex, up to 11 transversely diagonal rows per side of body, number of cerata per row decreasing from 5-6 at front to 1 at rear. Rhinophores conical, smooth, tips bluntly rounded, almost as long as cerata. Oral tentacles half as long as rhinophores, point sideways. Small eye at rear of each rhinophoral base; anus on right side, just above innermost ceras of the fourth row of cerata. Ground color translucent grayish-white. Scattered white blotches on the back and head, the back sometimes with a faint orange tinge and/or an indistinct reddish-orange patch behind the rhinophores. Cerata are dull translucent white with a broad white band at mid-swelling, a smaller white band just below the clear tip, and have a basal reddish-orange cast (the digestive gland). The oral tentacles and the top outer halves of the rhinophores are white-stippled; the rhinophores have a reddish-orange streak on the basal third of the rear face. It inhabits shallow areas with algae, sponges, rocks, and hydroids. Feeds on hydroids (*Pennaria* spp.) attached to *Sargassum*. Spawn is a one-whorl coil or a kidney-shaped tube containing ~70 eggs that hatch in ~2 weeks. Depth: to 26 feet. Amphiatlantic. Southeast Florida, Bahamas, Cuba, Guadeloupe, Martinique, Barbados, Aruba, Bonaire, Curacao, Honduras, Mexico, Jamaica.

with spawn; Guadeloupe

Riviera Beach, Florida

Selva rubra **Edmunds, 1964**

Size: to ~5 mm Trinchesiidae

ID: Body elongate; tail short. Rhinophores and oral tentacles long, slender, smooth; tentacles shorter. Long cerata in a row of ~5 clusters per side of back; 2-3 cerata per cluster (1 at rear); basally narrow, bulbous near/above middle, narrowing again to base of apical cnidosac. Ground color silvery-gray. Dorsum, head, top of tail white-mottled; flanks with a broken white band. **Cerata reddish-brown basally**; white with brownish mottling in bulbous portion; cnidosac whitish, orange to reddish band at base. Rhinophores and oral tentacles with a wide **crimson/orange band at midlength**, whitish elsewhere; tentacle bands less prominent. A **crimson stripe** from the first ceratal cluster curves inward and forward and continues up back of rhinophore to top of band. In areas with mangrove roots covered with hydroids and anemones; also in shallows with algae growth. Depth: to ~10 feet. Jamaica, Southeast Florida. Rare. The pictured animal matches Edmunds' description reasonably well, but lacks an orange streak from the proximal quarter of an oral tentacle, up the forehead, and onto the lower front half of the rhinophore. The identification, therefore, remains tentative.

Family Tritoniidae Lamarck, 1809

A family of cladobranchs with about a hundred species in a dozen genera distributed circumglobally in various seas and depths – frigid to tropical, shallow to deep. External morphologies vary considerably and sizes range from very small to very large.

Body broadly to narrowly elongate, vertical profile usually somewhat elevated; oral veil bilobed or not, with or without long processes; rhinophores smooth-stalked with an apical ring of vertical papillae or branched plumes; rhinophoral sheaths usually high, closed or partly flared, rim with or without lobes/appendages; cerata-like dorso-lateral appendages (secondary gills) usually distinct, branched, few to many, very large in some species (*Tritoniopsis*); mantle margin indistinct to distinct, sometimes remaining as a narrow fold between the dorso-lateral appendages, front corners absent to indistinct to present; stomach plates usually absent; rachidian tooth unicuspid or tricuspid, pointed lateral teeth few to many (11-200 per half-row); front of buccal bulb bordered by two strong jaw plates, exit from stomach sometimes with a band of toothlike "stomach plates".

They feed cryptically on soft corals or gorgonians, prey on which the shapes and colors of their secondary gills blend in very well with the coral polyps. Body length range is 11-300 mm.

Brazil

Tritonicula khaleesi **(F. V. Silva, de Azevedo & Matthews-Cascon, 2014)** **Khaleesi's tritonia**

Size: to 12 mm Tritoniidae

ID: Body elongate, slender, tapers abruptly at rear. Foot narrower than body, rounded at front, tapers to acutely pointed tail. Mantle margin reduced; edge usually with a line of 7 cerata per side (the gills), larger and smaller ones somewhat alternating; each ceras with a thick smooth stalk that usually splits distally into two branches, each with 2-3 smaller branches. Rhinophores smooth, cylindrical, retractable, tip of apical club with 5 bipinnate plumes; sheaths smooth, wide, high; rim flared, smooth, a long fleshy extension on its outer edge. Oral veil slightly flared laterally, rounded anteriorly, with 4 finger-like appendages, the inner 2 being longest ("oral tentacles"). Ground color translucent grayish. Dorsum with longitudinal white median stripe extending from in front of (or between) the rhinophores onto the tail, broadening/branching sideways at 4 locations – directly behind rhinophores and at 2nd, 4th, and 6th cerata. Stripe's front 2 branches extend up rear faces of rhinophoral sheaths. Thinner white lines join ceratal bases on each side. Cerata, oral tentacles, sheath rims, and rhinophoral tips are white. Digestive glands visible as an orange mass surrounded by white ovotestis. Currently known only from the tidal zone. Depth: to ~4 feet. So far, reported only from northeast Brazil. The species name derives from the character Daenerys in the television series "A Game of Thrones", who has long white strands of hair (reminiscent of the slug's white stripes) and acquires the title "Khaleesi" (queen).

Marionia limceana

F. V. Silva, de Meirelles & Matthews-Cascon, 2013

Size: to 31 mm Tritoniidae

ID: Body robust, elongate, widest near head, tapers to a short pointed tail; frontal veil bilobed, each lobe edged with 8 variously branched papillae, the papillae of one lobe paired symmetrically with those of the other lobe. Mantle with elongate row of small polygons along each side of mantle from veil to tail, also a patch proximally on each velar lobe. Rhinophores retractable, stalk narrow, apical club with 7 bipinnate projecting plumes; sheath very high, chalice-shaped, concealing all but the plumes. A row of up to 14 multi-pinnate gills along each mantle edge, symmetrically paired, sizes varying. Ground color whitish. Mantle with broad white median stripe from between rhinophores to near tail, branching laterally in several places to interrupt the rows of red-edged polygons. Both adults and juveniles may have small silver mantle spots (glands related to feeding) mainly over the centrally-positioned pericardium but sometimes covering the entire epidermis, excluding the foot but including the rhinophores. Rhinophores whitish, sheaths translucent white. Gills appear pale orange due to penetrating branches of the visible front and rear digestive glands. Discovered feeding only on the recently described encrusting octocoral *Stragulum bicolor*. When alarmed it will try to escape by "swimming" using strong body contractions. Depth: So far, known only from the intertidal zone.

northeast Brazil

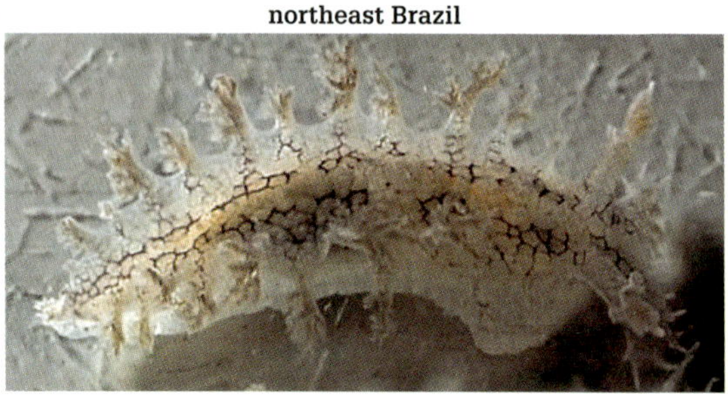

Marionia tedi Ev. Marcus, 1983

Size: 30-50 mm Tritoniidae

ID: Body elongate, length four times width, the width is slightly greater at mid-length, height slightly less than width, posterior somewhat upturned; frontal veil more or less bilobed, fringed with up to 15 papillae each having 1-3 tips. Mantle nearly smooth but with some tubercles that are probably glands (they also occur on the body's flanks). Rhinophores simple, sheath rims jagged. A row of ~10 multipinnate gills along each side of mantle; transversely paired. Ground color grayish-brown with whitish granulations. Rhinophoral sheaths, gills, and tubercles are opaque white. Has been found on rocky bottoms in deep water. Depth: 200-1200 feet. South Florida, Louisiana, Isla Margarita, Yucatan Strait.

Mississippi Canyon, Louisiana

Riviera Beach, Florida

Grand Cayman

Grand Cayman

Roatan

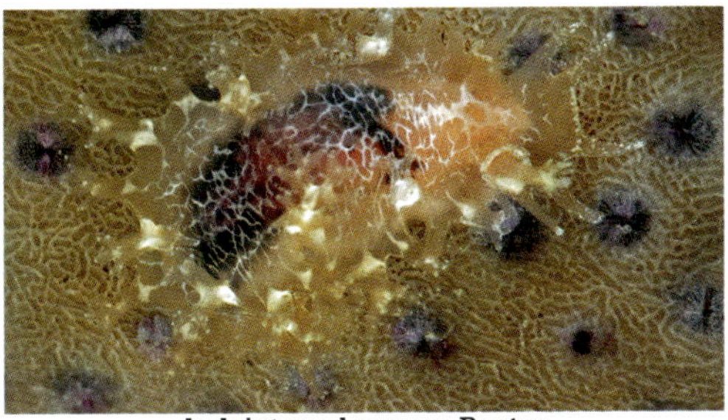

dark internal organs; Roatan

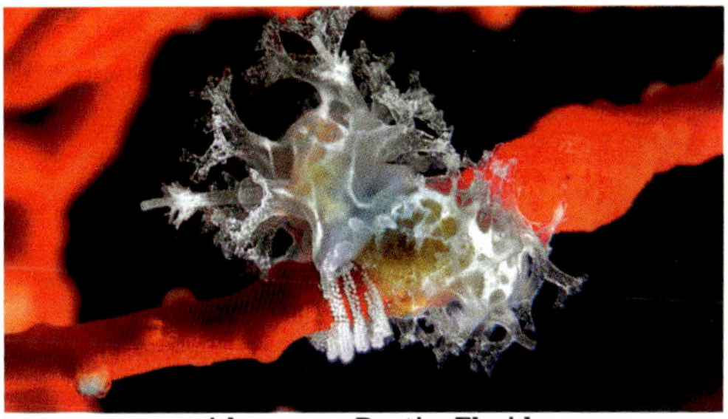

with spawn; Destin, Florida

Tritonicula bayeri (Ev. Marcus & Er. Marcus, 1967) **crisscross tritonia**

Size: to 11 mm Tritoniidae

ID: Body smooth, elongate, narrow; foot narrower than body, rounded at front, tapers to short pointed tail; oral veil flared laterally slightly, rounded anteriorly, **veil's front edge with four appendages:** a short rolled tentacle (ventrally grooved) at each corner, 2 finger-like papillae between them.; Mantle margin reduced; edge with a line of **up to 15 cerata per side**, larger and smaller ones somewhat alternating. Each ceras with a **thick smooth stalk supporting a few sturdy branches;** longer cerata more branched; **branching sparse not "bushy"**. Rhinophores smooth, cylindrical, apical club with a projecting **slender tip encircled by several delicate upright papillae**; sheaths smooth, wide, reach mid-length, **rim flared and smooth with a long papilla-like appendage on its outer edge**. Ground color translucent gray. Back with distinctive **opaque white net-like reticulations** formed by crisscrossing lines of various thicknesses. Internal organs visible as pinkish, dark brown, or blackish masses. Ceratal and rhinophoral apices white-tipped, ceratal stalks often white. Rhinophoral sheath rims white. Velar and sheath rim appendages white-dotted. **Tail with median white line**. Inhabits living reefs, coral rubble, and sandy areas near soft corals to feed on, e.g. *Pseudopterogorgia* sp., *Briareum asbestinum*, *Leptogorgia virgulata*. Spawn is a string of white eggs deposited as a spiral coil of 5-6 whorls. Depth: 5-250 feet. North Carolina to Georgia, South and East Florida, Lesser Antilles, Panama to Yucatan Peninsula and Grand Cayman. Common.

Dominica

Guadeloupe

Roatan

aggregation, Bermuda

Roatan

with spawn; Riviera Beach, Florida

Tritonicula hamnerorum **(Gosliner & Ghiselin, 1987)** **Hamner's tritonia**

Size: to 15 mm Tritoniidae

ID: Body elongate, narrow, smooth; tail long; mantle reduced, its edge lined with ~9 outward-pointing short branched cerata; rhinophores smooth, long, with a club-like tip consisting of a clump of vertical papillae; rhinophoral sheath high, with a lobed rim, the largest lobe being a triangular fleshy extension on its outer side; margin of semi-circular oral veil with 4-6 long tentacles. Ground color pale gray to pale orange to pinkish purple, tending to match the color of its sea fan prey. About 20 thin, white, irregular, longitudinal lines run along the length of the back. Branches of cerata white, stalks sometimes white. Rhinophores and oral appendages white-tipped. Rims of rhinophoral sheaths sometimes white. Feeds on the sea fans *Gorgonia ventilina* and *Gorgonia flabellum*, from which it secretes noxious chemicals for defensive purposes. Huge aggregations (up to 1700) have been reported on a single sea fan. Deposits a convoluted white ribbon of eggs pressed flat against the sea fan. Depth: 6-50 feet. Bermuda, Florida, Bahamas, Cuba, St. Barthélemy, Guadeloupe, Dominica, Martinique, Roatan, Mahahual (Mexico), Grand Cayman. Uncommon.

Roatan

Bermuda

Guadeloupe

Roatan

Bermuda

St. Vincent

Tritonicula wellsi **(Er. Marcus, 1961)** **Wells' tritonia**

Size: to 15 mm Tritoniidae

ID: Body smooth, elongate, tapers to long pointed tail; head with slightly indented "neck"; oral veil flared laterally, rounded anteriorly, **veil's front edge with 6-10 appendages** – a rolled tentacle at each corner bracketing shorter finger-like papillae. Mantle margin reduced; a line of **up to 18 cerata per side** (the gills), usually pointing outwards; sizes from large to very small. Each ceras has a **thick smooth stalk supporting a few short branches.** Rhinophores smooth, cylindrical, apical club with a projecting **slender tip encircled by several delicate upright papillae**; sheaths smooth, wide, reach mid-length, rim flared and smooth. Ground color translucent gray. Visible internal organs and branches of digestive system give dorsum a **pale brown to orange-brown cast**; its prominence depends on density of back's **opaque white spots**. Dark striations of digestive tract are visible in cerata and rhinophoral sheaths. **All appendages & tail heavily white-spotted**; tips of rhinophoral papillae & ceratal branches are white. Inhabits living reefs where it feeds on the sea whip *Leptogorgia virgulata* and is protected by camouflage – its branched cerata resemble the extended polyps of its soft coral host. Spawn is a string of white eggs deposited on a soft coral as a spiral coil of ~6 whorls, sometimes as a series of short segments. Depth: 10-120 feet. North Carolina, Bermuda, South and East Florida, Bahamas, Virgin Islands, St. Martin/St. Maarten, Guadeloupe, Martinique, St. Vincent, Panama, Costa Rica, Roatan, Belize, Grand Cayman, Sao Paulo (Brazil).

Bonaire

Grand Cayman

Bonaire

Guadeloupe

Lauderdale-by-the-Sea, Florida

spawn; Grand Cayman

Tritoniopsis frydis **Er. Marcus & Ev. Marcus, 1970** **tufted tritoniopsis**

Size: to 35 mm Tritoniidae

ID: Body elongate; tail short, pointed; cephalic veil corners with soft rolled tentacles, **front truncate with up to 13 short papillae**, longer and shorter ones alternating irregularly. The mantle is **covered with short papillae**, its edge lined with **up to 20 cerata per side**. Cerata alternate in size, **each ceras is a treelike gill**. Rhinophores are thin, cylindrical, smooth, shorter than the cerata; there is a **ringlet of upright threadlike papillae attached at midlength**; sheaths reach midlength and are wide, papillose, with a flared and lobed rim. The ground color is translucent white to brownish to yellowish to orange with a complex network of very fine white lines that are predominantly longitudinal. The back often has indistinct irregular pale patches. The dorsal papillae, cerata, and rhinophores have white tips; the sheath rims are white-spotted. Found typically in reef areas with a mix of coral, sand, sponges, debris, and often on living or dead sea fans, sometimes in small groups. Feeds on gorgonians of the genus *Plexaurella*. Its color seems uncorrelated with its food source – differently colored individuals are often together. Spawn is a clear gelatinous string with rows of white or yellow egg capsules, deposited on a soft coral as a spiral coil of 4-5 whorls; the string is sometimes a series of short sausage-like segments. Depth: 6-100 feet. Bermuda, Southeast Florida, Bahamas, Lesser Antilles, Venezuela to Cozumel and Grand Cayman.

Family Unidentiidae Millen & Hermosillo, 2012

Very few species in two genera, *Pacifica* and *Unidentia*, well known from relatively shallow coastal waters (10-70 feet) of the northeast Pacific Ocean and the Indo-Pacific.

Body elongate, slender; foot narrower to slightly wider than body, front bilobed, corners tentacular; tail narrow, flat, and pointed; mantle with raised anterior flange, lateral edges absent to indistinct; cerata in a longitudinal row of 4-12 clusters on each side of dorsum, each cluster an oblique row (angled down and back) of 1-12 cerata on a slightly raised cushion; cerata long, slender, cylindrical to fusiform, with long cnidosacs, digestive gland diverticulum fills a significant portion of each ceras; rhinophores smooth; oral tentacles smooth, slender, and tapering, longer than or as long as rhinophores; eyes behind bases of rhinophores; anus on right side, acleioproctic to pleuroproctic; radula uniseriate, rachidian with narrow cusp and distinct denticles. Feed on hydroids in stony subtidal habitats. Body lengths are 6-20 mm.

Unidentia angelvaldesi **Millen & Hermosillo, 2012**

Angel's aeolid

Size: to 20 mm Unidentiidae

ID: Cerata cylindrical, a row of ~11 clusters per side; less than 12 cerata per cluster; oral tentacles slightly flattened ventrally; rhinophores long, less tapering than oral tentacles. Ground color **translucent orange/red**. A median dorsal **purple stripe** and one per side under the cerata merge behind them into a single broad stripe ending on the white-tipped tail. Cerata ground color, cores darker orange or red, tips white, sometimes a thin purple band at base of white tip. **Rhinophores and oral tentacles banded** – ground color on proximal third, then whitish, purple, and a white tip; band widths, shades, and solidity varying somewhat across individuals. Lives and feeds on the orange hydroid *Corydendrium parasiticum*, where it deposits eggs in 5-6 slightly flattened orange coils. Ground color is food-dependent; a pale form that lives on a white hydroid exists in the Indo-Pacific. Depth: 10-70 feet. A common tropical Indo-Pacific and Eastern Pacific species recently found off Southeast Florida.

Lake Worth Lagoon, Florida

Family Pleurobranchaeidae Pilsbry, 1896

It consists of about twenty species of sidegill slugs in three genera, most of them in the genus *Pleurobranchaea*. The family is broadly distributed in temperate to tropical waters and is, so far, sparsely represented in the tropical West Atlantic by four species of *Pleurobranchaea*. Two of the species were recorded as preserved specimens dredged from deep waters, another from two preserved specimens found in 1955 off Rio de Janeiro at 30 feet, and only one species, *P. inconspicua*, that continues to be sporadically photographed alive and in situ by scuba divers. The morphological variability for this family is defined largely by the genus *Pleurobranchaea*.

Body elongate oval, front rounded, tapers to a rounded tail, top of tail usually with a caudal spur; mouth forms a proboscis; oral veil broad, trapezoidal, front corners produced as cephalic tentacles (sometimes notched), front margin irregular with a row of sensory papillae or small tubercles; rhinophores smooth, well separated, rolled, pointed to blunt, sometimes fused with the mantle; the mantle is rugose to smooth, fused anteriorly with the oral veil and head between the rhinophores, usually does not extend laterally beyond margin of foot; gill unipinnate or bipinnate with 17-25 pinnules, in the middle of the right side between the mantle and the foot, may or may not be covered by the mantle; no internal shell in adults; radula lacks rachidian teeth, laterals are bicuspid with 40-100 per side; jaw with 1-14 denticles.

They inhabit sedimentary bottoms in depths from intertidal shallows to >600 feet where they hunt opportunistically for invertebrates using their large oral veils to wrap and hold prey. Acidic secretions produced by skin glands are used for deterring predators. Body length range is 60-150 mm.

Pleurobranchaea inconspicua **Bergh, 1897**

inconspicuous pleurobranch

Size: to 60 mm Pleurobranchaeidae

ID: Body elongate but broadly oval, steeply convex. Mantle small, not covering foot, centrally smooth but bumpy with rounded papillae near edges. Oral veil broad, thin, front edge ragged with a few tiny rounded papillae, corner expansions (oral tentacles) deeply grooved for up to half their outer edges. Rhinophores rolled, well separated. **Foot projects as a narrow flange around body**; tail prominent, its tip pointed or rounded, with a **small rounded dorsal flap (spur). Gill exposed on right rear of body**, up to two-thirds body length, unipinnate, 20-26 pinnules, 7-10 free from the body wall. **Mantle edge curls over gill posterior to form a siphon**. Ground color translucent white to gray. Dorsal surfaces with a pattern of **fine dark brown reticulations** and **opaque white spots & blotches** (glands). Rhinophores are pale brown to grayish, darkly smudged near top, tips white. **Tail with a medial dark or yellowish stripe** from rear of mantle to **blackish spur at the tip of the tail**. Internal organs visible as a diffuse orange mass. Usually on rocky, sandy, muddy bottoms. *Pleurobranchaea* species are opportunistic carnivores with a broad range of prey – hydroids, polychaetes, isopods, copepods, ascidians, sponges, even their own spawn. Since cnidarians appear to be their preferred prey, it's likely they are the main diet of this species. Spawn is deposited on a hard substrate as a hollow cylindrical spiral with closely spaced loops of ~3 gelatinous strings crowded with white-to-pinkish capsules, each with 2-5 eggs. Depth: 0-20 feet, but has been found as deep as 500 feet. North Carolina to Florida, Dry Tortugas, Texas, Lesser Antilles, Venezuela to Yucatan, Suriname to Argentina.

Guadeloupe

continued on next page

Riviera Beach, Florida

Riviera Beach, Florida

juvenile; Guadeloupe

exposed gills; Guadeloupe

gills and spawn; Riviera Beach, Florida

Family Pleurobranchidae Gray, 1827

About seventy species of epibenthic sidegill slugs in seven genera broadly distributed globally in tropical to frigid waters.

Body elongate oval, sides elevated with flattish top, bulky looking; foot large; mantle relatively smooth to tuberculate (no wrinkles), with or without median anterior notch, front margin free, completely covers body; oral veil slightly protruding, wide, trapezoidal to almost triangular, front corners bluntly pointed; rhinophores antero-median above oral veil, rolled with longitudinal slit, bases fused, form erect or forward tilted "V"; gill along right side usually concealed by mantle flap, a usually smooth (sometimes tuberculate) longitudinal rachis with a row of appressed bipinnate tufts, rachis attached to body wall by a membrane for most of its length; internal shell present to absent, ovate and plate-like, a small remnant to a relatively large (half body length) shield above the viscera; anus and genital aperture near gill membrane; jaw elements smooth or denticulate; radula without rachidian teeth, laterals hamate and very numerous.

They are sedentary carnivores that typically inhabit the surfaces of intertidal and shallow subtidal substrates (sometimes under rocks) where they feed by day on sponges, tunicates, and other usually soft-bodied sessile invertebrates. Some can swim by undulations of the foot and mantle edge. Most have epidermal mantle glands and a median buccal gland that are able to secrete and release acidic substances for defensive purposes. One species has the ability to autotomize (and regenerate) portions of its mantle to confuse a predator. Body length range is 15-200 mm.

Guadeloupe

Martinique

Destin, Florida

Berthella agassizii (MacFarland, 1909)

Agassiz's berthella

Size: to 25 mm Pleurobranchidae

ID: Body convex, gelatinous; oral veil broad, trapezoidal, front straight, front corners pointed. Foot narrow, shorter than body, the front is bilobed, the rear slightly pointed. Mantle fleshy, surface smooth in juveniles, reticulate-looking in adults due to closely-set low bumps, covers foot entirely. Rhinophores rolled, fused from base to halfway up, directed forward. Gill along right rear of body, half to third body length, usually concealed by mantle flap. Shell internal, oblong, up to two-thirds body length, left and forward of mid-dorsum. Ground color translucent pale pink to reddish-pink; grayish-white in juveniles. Back has a scattering of small white spots. Rhinophores rosy, oral veil and gill whitish, internal organs visible as a dark area below the shell. Found among algae and under rocks in shallow water. Feeds on sponges. Spawn – usually attached to the underside of a rock; a translucent gelatinous ribbon of almost circular cross section, ~2 mm diameter; a compressed spiral coil of ~2 whorls containing spherical capsules with one white egg each. Shortly before hatching the spawn changes color from white to violet. The hatching of around 20,000 veligers occurs after ~7 days of development. Depth: to 20 feet. Bermuda, Northwest Florida, Bahamas, Guadeloupe, Martinique, Curacao, Venezuela to Belize, Cayman Islands, Brazil.

↓Bonaire

Martinique↑

Berthella sp. 1

Size: to 40 mm Pleurobranchidae

ID: Body somewhat convex; oral veil projects ahead of mantle; rhinophores rolled, fused near base. Ground color translucent cream. Mantle appears reddish-orange due to crowded reddish dots centered on small tubercles except for a broad area along the mantle margin. A mid-dorsal round blackish spot within a broad gray patch which, in turn, may be encircled by a few somewhat stellate white blotches. Rhinophores, oral veil, gill, and foot ground color. Found on bottoms with sand, rocks, mixed algae and coral, rubble, and detritus. Depth: 30-66 feet. Bonaire, Martinique. Rare. Externally very similar to the Eastern Atlantic *Berthella africana* (Pruvot-Fol, 1953) and the Indo-Pacific *Berthella caledonica* (Risbec, 1928).

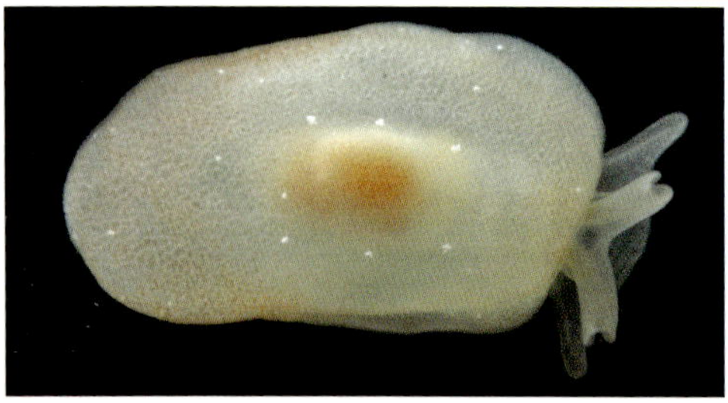

Colombia

***Berthella* sp. 2**

Size: ~20 mm Pleurobranchidae

ID: Body oval; mantle finely reticulate; oral veil broad, trapezoidal, front edge slightly rounded and corners fingerlike; rhinophores rolled, fused at base; foot edges protrude slightly beyond mantle edges; the shell is mid-dorsal, length about one-fifth body length. Ground color translucent white with some faint orange shading and a few small white spots. Shell region orange (internal organs), flanked by a longitudinal row of three larger white spots per side. Rhinophores and oral veil translucent white. Under a rock. Depth: 2 feet. Colombia. The mid-dorsal pattern of white dots seems to be more than a random arrangement. Otherwise, this could be an unusually pale version of *Berthella agassizii* (q.v.).

Islamorada, Florida Keys

***Berthella* sp. 3**

Size: ~6 mm Pleurobranchidae

ID: Body oval, somewhat convex; mantle finely granular, many scattered small tubercles; oral veil broad, trapezoidal, front edge straight but with prominent median notch, front corners pointed; rhinophores rolled, probably fused near base; foot edges protrude slightly beyond mantle edges. Ground color translucent white. Mid-dorsum with prominent opaque white blotch crossed by some blackish scrawls; dorsal tubercles opaque white; rhinophores and oral veil with some opaque white spots; internal organs visible as a brownish mass. Crawling on a hard substrate. Depth: ~2 feet. Florida Keys.

Pleurehdera vialactea **(Ghanimi, Schrödl, Goddard, Ballesteros, Gosliner & Á. Valdés, 2020)**

Size: to 8 mm Pleurobranchidae

ID: Body elongate, oval; mantle broad, smooth, covers foot. Oral veil broad, trapezoidal. Rhinophores between oral veil and mantle, rolled, **fused up to mid-length**. Gill on right side under mantle, **a half body length**, bipinnate, 6-10 pinnae per side of gill rachis, 2-4 free of body wall. Shell internal, oval, convex, covers entire body. **Ground color translucent white to milky white with many opaque white spots**. Mantle with mid-dorsal **white blotch – T- to Y-shaped, or irregularly-shaped**. Oral veil and rhinophores translucent white with some white spots; shell translucent white. Found in various habitats – mangrove, sand, rock, rubble, detritus. Depth: 3-50 feet. Guadeloupe, Martinique, Roatan. Recent molecular and morphological analyses (Ghanimi et al, 2020) of worldwide specimens that had hitherto been considered *Berthella stellata* (Risso, 1826) has revealed a complex of at least 8 distinct species. Two of them are Caribbean: *P. nebula* and *P. vialactea*.

Guadeloupe

Roatan

Missouri Key, Florida

Roatan

Big Pine Key, Florida

spawn, 14 mm across; Big Pine Key, Florida

Pleurehdera nebula **(Ghanimi, Schrödl, Goddard, Ballesteros, Gosliner & Á. Valdés, 2020)**

Size: to at least 10 mm Pleurobranchidae

ID: Body elongate, oval; mantle broad, smooth, covers foot. Oral veil broad, trapezoidal; rhinophores between oral veil and mantle, rolled, **fused up to two-thirds length**. Gill on right side under mantle, **less than half body length**, bipinnate, up to 10 pinnae per side of gill rachis, 5-6 not attached to body wall. Shell internal, oval, convex, covers entire body. **Ground color translucent white, off-white grayish-white, with many opaque white spots**. Some of the mantle's white spots usually cluster into a **mid-dorsal T- or Y-shaped blotch**. Oral veil translucent white with few to many opaque white spots near front and side margins. Rhinophores translucent white, apex with opaque white spot. Shell translucent light honey-brown. Florida animals were found under a stone in very shallow water in a patch of blue *Oscarella* sponges they were feeding on. Depth: ~3 feet. Florida (Fort Pierce, Keys), Puerto Rico, Guadeloupe, Martinique, Jamaica, Caribbean coast of Panama, Roatan, and Mexico.

Roatan

Pleurobranchus **sp.**

Size: to 70 mm Pleurobranchidae

ID: Body elliptical, moderately elevated; mantle slightly broader than foot, indented anteriorly, densely covered with variably sized low rounded tubercles, smallest near mantle margins; oral veil broad, trapezoidal, corners extended; rhinophores rolled, smooth, bases close together, situated at rear of mantle indentation. Ground color purplish. Tubercles grayish with diffuse purplish tips, some to many with **bright white ring around base and small black spot at tip**. Oral veil and rhinophores grayish; dorsal surface of foot grayish with prominently **white front margin**, ventral surface white. Found under rubble in shallow water. Depth: 10 feet. Roatan.

Riviera Beach, Florida

St. Vincent

Riviera Beach, Florida

Guadeloupe

orange *crossei* form; St. Vincent

whitish *crossei* form; Bahamas

Pleurobranchus areolatus **Mörch, 1863** **warty pleurobranch**

Size: to 150 mm Pleurobranchidae

ID: Body elliptical to elongate, moderately elevated. **Mantle deeply indented at front**; margin often undulate; **crowded with soft-tipped polygonal tubercles** of various shapes and sizes: (i) low, rounded, broad; similar in size or largest centrally/anteriorly; smaller near margins (ii) uniformly small, pointed, some conical and with projecting filaments. **Rhinophores rolled, transversely "wrinkled" from base to tip** (but smooth in juveniles), **bases either fused or very close together**, at rear of mantle indentation. Gill under mantle flap on right side; bipinnate, two-thirds body length; ~20 pairs of pinnae, half of them free from the body wall. Shell internal, small, oval, above the pericardium. There are color variants. **Adults with low broad tubercles**: ground color brownish-red to brownish-purple; whitish to yellowish central patches; some larger/higher mid-dorsal tubercles are much darker to blackish; mantle edges orange, margins with white radial blotches; rhinophores ground color to much darker; juveniles whitish to pinkish, with larger light-to-dark maroon tubercles having a prominently dark basal ring. **Adults with similarly-sized smaller tubercles, occasionally with apical filaments ("the *crossei* variant")**: ground color translucent gray; irregular patches of white to purplish-brown tubercles; darker tubercles of variable intensity, usually white-tipped with a dark basal ring; some animals mostly white; rhinophores orange, white-dusted; mantle margin orange with short white radial streaks; other individuals in shades of yellowish-orange, sometimes uniform in color. Apical filaments, when present, are white. The ***crossei*** **variant juveniles:** translucent gray, mantle densely white-dotted, margin bright orange, some brown to orange spots above the shell, the more developed animals with a scattering of large orange tubercles, rhinophores white. **Shell:** white to brownish with gold highlights. Inhabits shallows with rocky and rubble bottoms where it hides under rocks and debris and within crevices. Feeds on sea squirts. During spring in the Florida Keys they are often seen in pairs depositing spawn as white to purplish convoluted ribbons on tidal flats with *Porites* corals, *Halimeda* algae, and *Thalassia* turtle-grasses. Depth: usually <20 feet but it has been dredged from as deep as 230 feet. Amphiatlantic, possibly circumtropical. Bermuda, Florida, Bahamas, Lesser Antilles, Venezuela to Yucatan, and Brazil. Recent analyses (Goodheart et al, 2015) based on molecular and morphological evidence indicate that the following species are synonymous with *P. areolatus*: *P. atlanticus*, *P. evelinae*, *P. crossei*, and *P. reesi*. This gives *P. areolatus* a broad range of appearances, e.g., the "*crossei* variant". A re-analysis of Brazilian species of *Pleurobranchus* (Alvim and Pimenta, 2016) concluded that many Brazilian sightings of *P. areolatus* had actually been misidentifications of *Pleurobranchus reticulatus* Rang, 1832. Since their reliable distinguishing features are primarily anatomical, it is possible that some of the animals in the accompanying photos might be *P. reticulatus*.

continued on next page

juvenile, St. Vincent

juvenile, Roatan

Riviera Beach, Florida

St. Vincent

Roatan

↑ *crossei* form juveniles ↑

egg ribbons; Riviera Beach, Florida

Ilhas do Rio, Brazil

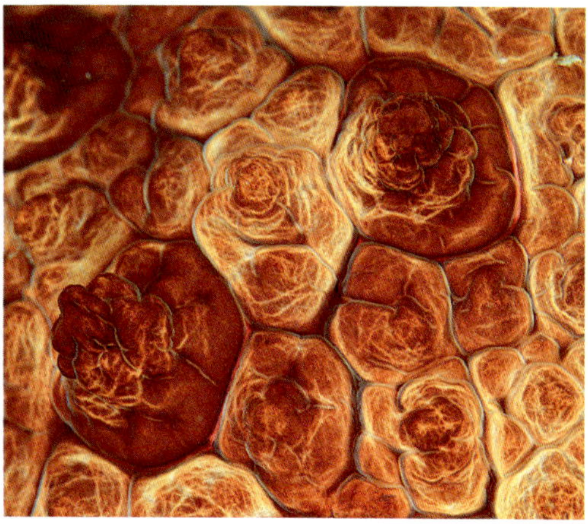

Pleurobranchus iouspi **Ev. Marcus, 1984**
Size: to 99 mm Pleurobranchidae
ID: Body elongate oval to quadrate, moderately depressed; foot narrower than mantle. The mantle is covered with tubercles that are irregularly-shaped and distributed; very large pointed ones centrally and smaller rounded ones near mantle edges. The slightly protruding oral veil is broad and thick with furrowed sides (the "oral tentacles"). Rhinophores short, stout, rolled, fused at their bases. Gill beneath the mantle on the right side; half a body length; up to 24 tripinnate leaves, half of which are fused to the body wall. Mantle, tubercles, rhinophores, and top of veil are densely spiculose. The internal shell in the left front of the body is fragile and small (< 10% of the body length). There are two main color patterns: (i) pale orange or bright yellow, sometimes with white markings; mantle margin orange or white; some pointed tubercles with a basal pink ring (ii) orange to red to dark red with white interspaces between the tubercles; the largest pointed tubercles are usually dark red and sometimes have a basal pink ring. The mantle margin may be white. The color patterns of the rhinophores, gill, and foot are the same as the mantle's. It has been found on sandy/muddy bottoms, crawling on hard substrates, and once in an underwater cave. It can contract its tubercles when alarmed. Stomach contents of dissected specimens included ascidians, bivalves, and plant debris. Depth: to at least 150 feet. Southern Brazil, Colombia (Santa Marta). Recent comparisons of nuclear DNA from a Brazilian *P. iouspi* and the Mediterranean *P. testudinarius* Cantraine, 1835 suggest that they are the same species, but the synonymy remains tentative pending more detailed molecular analyses.

Pleurobranchus **cf.** ***iouspi*** **Ev. Marcus, 1984**
Size: ~20 mm Pleurobranchidae
ID: The external morphology and coloration of this animal are very similar to that of the above-described *P. iouspi*, but there are several notable differences: the largest tubercles on its mantle have rounded apices instead of having central pointed tips; the interspaces between the tubercles and their inner faces are white; rhinophores have white tips; the broad mantle margin is opaque white. It was found crawling on a hard substrate in shallow water. Depth: ~15 feet. This could be a variant of *Pleurobranchus iouspi*, a juvenile, or perhaps an undescribed species. A physical examination could remove the uncertainty.

Riviera Beach, Florida

St. Vincent

viscera visible; St. Vincent

with egg ribbon; St. Vincent

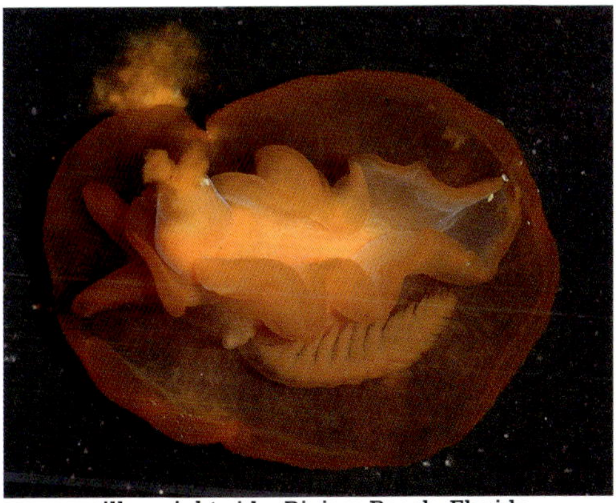

gill on right side; Riviera Beach, Florida

aggregation; Bonaire

Berthellina quadridens **(Mörch, 1863)** **apricot sidegill slug**

Pleurobranchidae

Size: to 80 mm

ID: Body oval to elongate, gelatinous, semi-translucent, inflated, seems smooth but is densely covered with very small retractable papillae that may give it a granular appearance. Oral veil trapezoidal, its wide front parallel to its narrow rear, front corners are short "oral tentacles". Rhinophores rolled, fused at the base, behind the veil, longer than the oral tentacles, smooth, contractile. Gill obscured by the mantle on the lower right flank. Shell above the heart, slightly ahead of center and left of the dorsal midline, 15-20 percent of the body length. **Adults**: semi-translucent yellow to orange to red. **Juveniles**: translucent gray; the mantle, rhinophores, and oral tentacles densely white-dotted; one or more small red to reddish-brown spots directly above the shell; the mantle's edge is orange. Viscera are visible as a darker patch. The shell is yellowish-white. Usually found on shallow coral flats under rocks and reef rubble. It probably feeds on corals (e.g. *Orbicella* spp.), anemones, and possibly on sponges. When at rest, it rhythmically raises and lowers its mantle to irrigate its gill. Spawn is usually found under a rock as a gelatinous grayish-white ribbon, short to very long. Depth: 0-165 feet. East and South Florida, Bahamas, Antilles, Venezuela to Mexico. It does not occur in Brazil, where it is replaced by its look-alikes *Berthellina circularis* (Mörch, 1863) and *Berthellina ignis* Alvim & Pimenta, 2015.

Family Costasiellidae K. B. Clark, 1984

A family of about twenty sacoglossans in two genera, *Costasiella* and *Panderevela*, distributed in subtropical to tropical Indo-Pacific and West Atlantic waters and the Red Sea. A few of the species occur in the Caribbean Sea.

Costasiella: Body elongate oval; rhinophores vary: simple and smooth with blunt apices, long, thin/thick, entirely flattened or just at the base, grooved, with or without a basal lobe; eyes median, close together between rhinophores; cerata fusiform, often with mammillate tips, may be penetrated by branches of digestive and albumen glands; pericardium small to large, low to elevated, round to elongate, with/without dorsal vessels; foot wider than body, front usually rounded, corners sometimes pointed or tentacular, tentacles thin to spatulate; tail pointed, often like a ceras; pharyngeal pouch usually present; oesophageal pouch present or absent; radula uniseriate, teeth smooth and blade-shaped; 2-4 genital openings, male gonopore below right rhinophore, female one just behind it. *Panderevela* is distinguished by having median eyes behind rather than between the rhinophores, not having foot tentacles or protruding front foot corners, and by having rhinophores with lateral extensions that extend onto the front of the head to form crests bordering a marked cleft.

They feed on green algae, e.g. *Avrainvillea* spp., with species usually having their own preferences. Typical habitats include substrates with soft muddy sediments in intertidal areas. Lengths are usually 5-10 mm.

St. Vincent

Riviera Beach, Florida

Riviera Beach, Florida

with spawn; Riviera Beach, Florida

Costasiella arenaria

K. R. Jensen, Krug, DuPont & Nishina, 2014

sand costasiella

Size: to 8 mm Costasiellidae

ID: Body oval; front edge of head prominently indented; pericardial hump elongate, elevated, narrows to rear. Cerata in 3-4 longitudinal rows on each side of the body, inflated, tips short and pointed. Rhinophores long, smooth, taper to a rounded tip, bases flattened. Foot with tentacles at front corners; tail short, pointed. Eyes small, elongate oval, centered between rhinophores, almost touching. Ground color translucent grayish with variably crowded white dots on head, body, and appendages. Head usually whitish between rhinophores and ahead, with a diffuse brown patch along each side; back darker brown. Ceratal color variable: usually shades of green with dense white stippling, sometimes brown with white/green/blue/red stippling, rarely transparent to translucent whitish with brownish/greenish stippling, tips usually white but sometimes brownish. Rhinophores ground color with brown front, rear vertical edges. Eyes black. There is still very little scientific information about this species. It has been found on sand and algae in shallow water. Molecular analyses of ceratal digestive gland contents identified the green algae *Bryopsis hypnoides* and a *Pseudocodiaceae* sp. as two of its food sources. It deposits eggs on an algal surface, the mass consisting of a coil of encapsulated white eggs distributed randomly within the gelatinous string. Depth: to ~15 feet. Riviera Beach (Florida), St. Vincent, Roatan.

Roatan

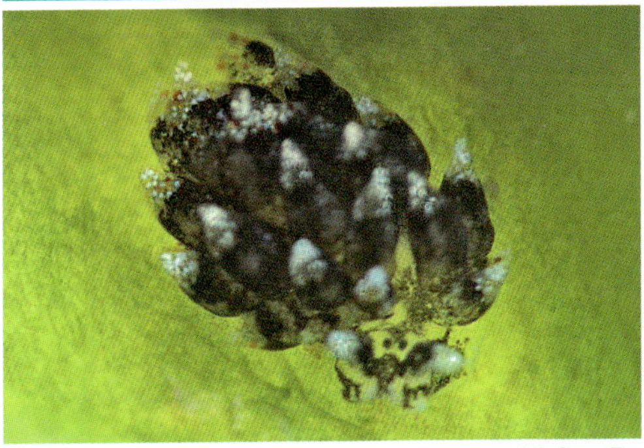

Riviera Beach, Florida

Costasiella nonatoi
Ev. Marcus & Er. Marcus, 1960
white-masked costasiella

Size: to 5 mm Costasiellidae
ID: Body short and narrow (width and height one-fourth length); head broadened at the front, bilobed; foot wider than the body anteriorly, front bilobed with slightly rounded corners. Cerata fusiform, height one-fourth body length; ~10 oblique rows with 5 cerata each on each side of the back, the topmost cerata largest; knob-like diverticula of the digestive gland are sometimes visible. Rhinophores short, broad, **front vertically grooved along whole length**, tips bluntly rounded. **Eyes straddle the dorsal midline between the rhinophoral bases**. A longitudinal furrow on each side separates the back from the foot. The small pericardial hump is behind the eyes. Black to brown with cream to white stippling and spotting. The foot sole and front border are white. Extending back from this border, along each side of the head, is a white streak that continues up the outer side of each rhinophore to its white tip. Cerata have some pale speckling and white tips. The black eyes are near the inner edges of "mask-like" contiguous large white circles. Very few individuals have been recorded so far – two from Brazil found on a sandy bottom with clay and unspecified algae, four from Puerto Rico on unspecified algae, one from Florida on the green algae *Avrainvillea*. Depth: to ~12 feet. Most online images purporting to be of *C. nonatoi* are either misidentifications of *Panderevela ipse* (see below), or are indeterminate.

Roatan

Fort Lauderdale, Florida

Panderevela* cf. *ipse **Ortea, Moro & Espinosa, 2015**

Size: to 5 mm Costasiellidae
ID: The following describes *Panderevela ipse*: Body short, narrow; tail long and narrow; front of head broadly bilobed, scoop-like. Cerata fusiform; 3-4 crowded rows per side, topmost cerata largest. Rhinophores are short, stout, bases anteriorly fused, tips bluntly rounded; anterior muscular process begins at midlength, broadens down/forward as a flap that either expands transversely outwards or folds transversely inwards onto the head; in either case the flap appears to originate from a pre-rhinophoral "crest" adjacent to a median cleft. **Eyes behind rhinophoral bases**, large, median, close together. The predominant color is **black to very dark grayish-brown**. The head and back with some **white patches**, the tail white. Cerata and distal third of clear rhinophores with **bright white dots**. Ceratal tips whitish. Black eyes in round whitish patches that nearly touch medianly; may be separated by a faint black line. Lives mainly on the alga *Avrainvillea nigricans* (sometimes on *Bryopsis* and *Pseudochlorodesmis* species) where it burrows into the thallus to feed on its tissues and deposit eggs. Has been seen burying itself in sediment next to *A. nigricans* with only the tips of its rhinophores and cerata protruding. Depth: 5-65 feet. The accompanying photos illustrate what **may** be *P. ipse* but they show anomalous features – orange subapical ceratal bands, orange dots on the head's front margin and near the rhinophoral tips, blue dots on the cerata and rhinophores, a short tail. These differences could be ascribed to insufficient sampling of this rare species. Moro & Ortea (2015) suggested that most (all?) post-1960 records of *Costasiella nonatoi* (see above) are in fact records of *P. ipse*.

Guadeloupe

Riviera Beach, Florida

Roatan

Riviera Beach, Florida

Key Largo, Florida

juveniles; Guadeloupe

Costasiella ocellifera **(Simroth, 1895)**

eyespot costasiella

Costasiellidae

Size: to 10 mm

ID: Body oval; head blunt; tail long, narrow, pointed. Cerata inflated, tips long, gradually taper to fine point. Rhinophores long, narrow, taper to fine point. Eyes between rhinophores, closely spaced. Ground color translucent clear with a dense distribution of black and white spots. Head has some orange/yellow patches. Small median blue spot (sometimes indistinct) edged with black, brown, or orange just behind rhinophores. Cerata greenish to whitish to almost black (**sometimes yellowish in small individuals**) due to visible branches of digestive gland; surface with small black and white spots to just below apex where there is an orange-yellow band (width < one-fourth ceras length), followed by a white-spotted band, and a small cluster of black dots at the tip. Rhinophores white-spotted, often white or yellow/orange near base, the rest variously spotted/banded with black, sometimes with white tips. Tail ground color with median black streak. Eyes black, in an unpigmented area. Usually found living on the surface of and slowly eating from the very thin filaments of the green algae *Avrainvillea nigricans*, less often on *A. longicaulis* and *A. mazei.* Sequesters chloroplasts in its cells for harnessing energy from photosynthesis, possibly for use during periods of food depletion. Deposits a translucent spiral ribbon (1.5-2.5 whorls) containing white eggs on its host alga. Depth: to 30 feet. Reported from Bermuda, Florida, Bahamas, Antilles, Costa Rica to Bay of Campeche, Brazil (Sao Paolo). The very similar *Costasiella patricki* (see next page) differs externally by being lighter green and having a narrower orange-yellow band on each ceras (less than one-eighth ceras length).

Jamaica

St. Vincent

Costasiella cf. *patricki*
Espinoza, DuPont, & Á. Valdés, 2014
Patrick's costasiella

Size: to ~10 mm Costasiellidae
ID: Morphology like that of *Costasiella ocellifera*. Differs from it only by often being lighter in color and having the orange-yellow band on each ceras narrower (less than one-eighth ceras length) to almost indistinguishable. Positive identification of such animals requires an internal examination and/or molecular analysis. Found on or near algae in shallow water. Depth: to ~20 feet. The original description of *C. patricki* was based on specimens from the Bahamas. Photos indicate that it may be more widespread, including Cuba, Jamaica, and St. Vincent. The pictured individual from Jamaica is the closest match to *C. patricki*. The other two photos are included here because of their very small ceratal orange bands and pale green color, but they could be variants of *C. ocellifera*. Such specimens should be physically examined to clarify their species identity.

Guadeloupe

Panderevela dacilae **Moro & Ortea, 2015**
Size: to 4 mm Costasiellidae
ID: Body short, narrow; tail short, triangular. Up to 30 fusiform cerata uniformly distributed in ~5 longitudinal and transverse rows. The pericardium is small, hemispherical, median between the frontmost cerata. The rhinophores are cylindrical to conical; each with a broad anterior flap that folds inward, the 2 meeting as a pre-rhinophoral veil. The eyes are large, medianly adjacent, behind the rhinophoral bases. The body, head, and cerata are dark green to almost black. The eyes are black, each one at the inner edge of a post-rhinophoral whitish patch, patches nearly touching. The rhinophoral proximal half is black, the rest white; frontal lobes with a bright blue border in larger animals. Pericardium and ceratal apices are white. **Sides of the head, rhinophores, and cerata (including the apices) with numerous brick red and bright blue dots**. Found among *Avrainvilla* algae fronds on which it probably feeds. Spawn is a spiral gelatinous cord of 2.5 whorls containing two rows of white eggs in oval capsules. Depth: to 80 feet. Rare. One photo from Bermuda, and five specimens from the East Atlantic (Canary Islands).

Spanish Point, Bermuda

Grand Cayman

***Panderevela* sp.**
Size: < 2 mm Costasiellidae
ID: See the description for *Panderevela* cf. *ipse*. This animal differs by having red-spotted cerata with blue tips. The different color pattern could be a heretofore unrecognized intraspecific variability for that species or a diagnostic feature for a distinct species. It was found near sand level on *Acetabularia crenulata* green algae. Depth: ~50 feet.

Family Cylindrobullidae Thiele, 1931

A family of small slugs having external shells, consisting only of a few species in the genus *Cylindrobulla*. They are distributed globally in temperate to tropical waters but, so far, only two of the species, *C. beauii* and *C. gigas*, are represented in the tropical West Atlantic. For many years there was uncertainty regarding this family's systematic taxonomic position, e.g. whether it should be grouped with the Cephalaspidea or the Sacoglossa, because of gaps in the knowledge of their anatomical features. Eventually, the conclusions based on the most reliable morphological, anatomical, digestive, and reproductive studies were confirmed by molecular phylogenetic analyses (2010) – the Cylindrobullidae are sacoglossans.

Body oval, small, no operculum or parapodia; cephalic shield present, furrowed medianly, the furrow either with a shallow cleft and no inter-lobe gap or with a deep cleft and gap; eyes usually visible; radula uniseriate, teeth broadly quadrate with large denticles or needle-like with tiny denticles; ascus sack may be present; esophageal pouch usually present. Shell external, cylindrical, fragile, coiled, apex conical to blunt, protoconch not visible in the adult, outer lip usually straight, the shell usually longer than foot, large enough for the entire animal to retract into.

Little is known about their ecology, but they seem to be herbivores that ingest cell-sap from species of *Halimeda* or *Caulerpa* (and possibly *Thalassia*) algae, either by burrowing into sediment to their roots or climbing onto the stalks. The uncertainty relates to the quadrate shape of their teeth, which is quite different from that of confirmed suctorial feeders, i.e. long and sharp, apparently optimized for a piercing and sucking mode of feeding.

Bocas del Toro, Panama

Guadeloupe

Cylindrobulla beauii P. Fischer, 1857

Beau's paperbubble

Size: to 8 mm Cylindrobullidae

ID: The shell is external, thin, fragile, cylindrical, and is rounded at the apex; it has thin growth lines parallel to the longitudinal axis; the aperture is wide, circular, and as long as the shell, it can accommodate the fully retracted animal. The cephalic shield of the crawling animal has a shallow median furrow and is almost as long as the shell. The foot extends back under the front of the shell when the animal is moving. There are no parapodia. The shell is translucent whitish to yellowish to white-mottled, with white growth lines. The animal's body is cream-white with opaque white flecks. It burrows into sediment or crawls along rocky substrates, usually near *Thalassia* spp. or *Halimeda* spp. algae. Its diet is unknown, but it is thought to be an herbivore that scrapes nutrients from the rhizoids of marine plants. Depth: to at least 40 feet. Bermuda, Florida, Bahamas, Grand Turk, Antilles, Venezuela, Belize, Quintana Roo, Brazil, Panama.

Bahamas

Costa Rica

Cylindrobulla rozadora Ortea, 2020

grazer paperbubble

Size: to 3.5 mm Cylindrobullidae

ID: Cephalic shield shorter than shell, bilobed, medianly furrowed, rounded anteriorly, truncate posteriorly; foot as wide as the cephalic shield but somewhat longer, front rounded, rear slit, sides of the rear third thickened and bent upwards; no parapodia; diffuse folds of the gill visible through the shell. Shell external, thin, fragile, cylindrical, spire sunken and acentric, protoconch totally immersed, the aperture long and straight, the base obliquely truncate and entirely open. When it is outside of its shell the live animal appears white from its dense distribution of white dots. When inside its shell the overall coloration is cinnamon, the front third appearing lighter because of yellowish-white granules associated with the hypobranchial gland. Found burrowing in sediment or on rocky substrates, usually near *Halimeda* green algae. Depth: to at least 10 feet. Known from one animal found at Manzanillo, Costa Rica.

Family Hermaeidae H. Adams & A. Adams, 1854

A family of about sixty sacoglossans in seven genera distributed circumglobally in temperate to tropical waters. All but one of the genera are represented in the tropical West Atlantic.

Body narrowly elongate, often transparent; rhinophores auriculate, usually with a short lateral projection ("step") subapically, sometimes flared at the step; eyes behind the rhinophores, usually within white to unpigmented patches; front of foot rounded or with two oral lobes; these lobes are rounded, slightly above and flanking the mouth, extending somewhat sideways; ceratal shapes vary: elongate, fusiform, globose, flat and leaf-like, short and slender, irregular; cerata often configured as a single row along each side with long and short ones alternating, sometimes penetrated by visible digestive gland branches; pericardial swelling behind rhinophores; male gonopore below and slightly behind the right rhinophore, the female pore just behind it, usually followed by the anus as a papilla or a flush opening ahead of the pericardium; radula uniseriate, teeth sabot-shaped or short and blunt with denticulate cusps.

They are usually found in relatively shallow subtidal areas with an abundance of green or red algae to feed on suctorially. Length range is about 2-35 mm.

Martinique

Sao Paulo, Brazil

Bermuda

Caliphylla mediterranea **A. Costa, 1867** — **spotted sacoglossan**

Size: to 20 mm — Hermaeidae

ID: Body flat, elongate oval; foot broad; tail moderately long, tapers to a sharp point. Four longitudinal rows of crowded cerata per side; heights diminish laterally, (longest cerata about half a body length); cerata in outermost row short, fusiform, those in the other 3 rows are oval and leaf-like; digestive gland highly branched in each ceras, most branches end near the ceratal margin with a bifid tip. Rhinophores smooth, bifid; the longer fork grooved along its outer side, proximal two-thirds penetrated by digestive branches. Oral veil with a median recess above the mouth, short pointed lobes at front corners. Ground color translucent pale gray; dorsum with an irregular network of fine brown lines. Digestive system branches in cerata and rhinophores vary from green (recently fed) to brownish (no food for 1-2 days); scattered yellow granules on the ceratal margins and stalks. Usually in sheltered and shaded areas near or on the filamentous green algae it feeds on (e.g. *Briopsis* spp.). It randomly waves its cerata as it crawls,. When disturbed it contracts while its anterior cerata cluster over its rhinophores, and heads for protection. Depth: 1-17 feet. Amphiatlantic. Bermuda, Florida, Cuba, Puerto Rico, Virgin Islands, Guadeloupe, Martinique, Curacao, Brazil (Fernando de Noronha, Sao Paulo). Rare.

Guadeloupe

Caliphylla **sp.**

Size: to 20 mm — Hermaeidae

ID: External morphology is like that of *Caliphylla mediterranea* but the cerata are relatively longer and more slender. Ground color translucent pale gray. Cerata and rhinophores with highly ramified, thin, light green branches of the digestive gland. The body and cerata have patches of white dots. Found feeding on the filamentous green alga *Briopsis plumosa* in sheltered rocky areas or among mangrove roots. This may be the *Caliphylla* sp. from Costa Rica and Tobago shown in *Caribbean Sea Slugs* (Valdés et al 2006).

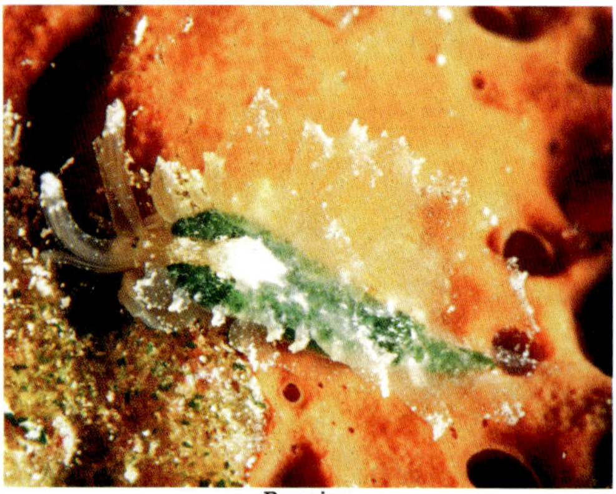

Bonaire

Cozumel

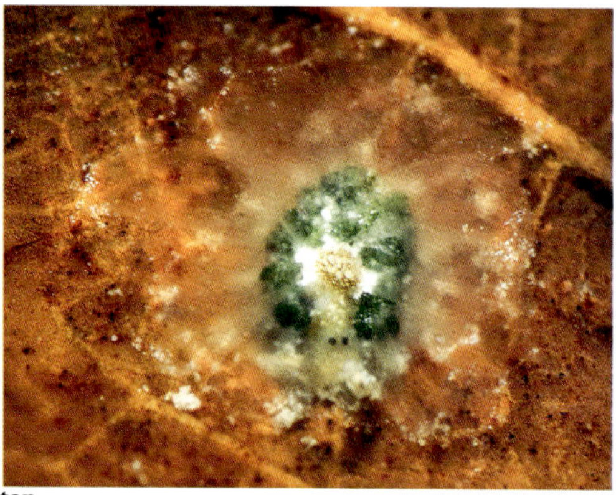

Roatan

Bocas del Toro, Panama

Guadeloupe

Cyerce antillensis **Engel, 1927** **Antilles glass slug**

Size: to 60 mm Hermaeidae

ID: Body narrowly elongate to ovate, flattened; tail pointed. Up to 10 large cerata per side of the back; more ventral smaller ones between; slightly inflated-looking, broadly rounded, narrow peduncular base; margins crenulate with rounded to pointed extensions (especially distally); **without branches of digestive gland**. Rhinophores long, bifurcated below middle, branches rolled. Oral tentacles cylindrical, short, smooth. Eyes large, close together between and behind rhinophoral bases; pericardial hump slightly behind head, irregular in shape; anal papilla at its right front. Mantle and foot translucent white, margins may be white-speckled. Body yellow to pale green to dark green, darkest when digesting algae. Head white to pale yellow; pericardium opaque white, sometimes brownish; anal papilla gray or white. Cerata translucent, white-specked; edges of distal extensions are white to yellow; variably crowded small **spherical clear/pale yellow/orange glandular inclusions** (close-set glands can create an overall yellow-orange cast); often a single **dark spot** below the tip or near the ceratal middle and a burgundy line at the inner ceratal base extending onto the body. Rhinophores and oral tentacles white to pale yellow with gray and burgundy patches and apical white dots. Usually under coral rubble/rocks or on algae in shallow coral/sand areas. Preys mainly on *Halimeda opuntia* but sometimes on *H. tuna*, *Udotea flabellum*, and *Penicillus dumetosus*. When alarmed it can readily discard some cerata and release a milky defensive fluid. Discarded cerata can wiggle for at least 24 hours to distract predators. Spawn is a string of crowded white egg capsules deposited on a hard substrate as a spiral coil of 4-5 loops. Depth: to ~20 feet. Bermuda, Florida, Bahamas, Antilles, Brazil, coasts of Venezuela to Bay of Campeche.

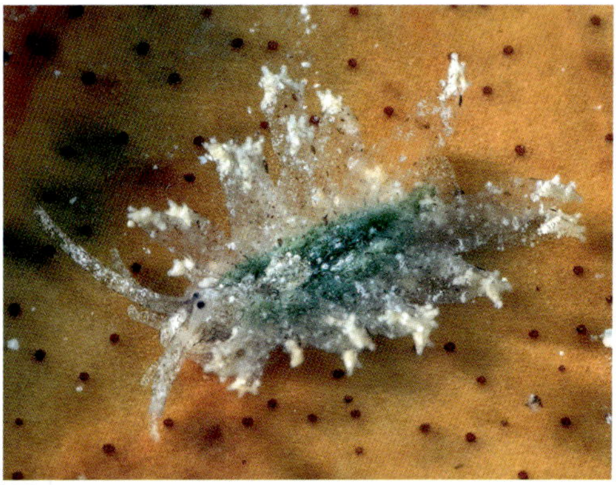

Roatan

Roatan

Riviera Beach, Florida

Cyerce nicholasi **K. Moreno, Rico, Middlebrooks, Medrano, Á. Valdés & P. J. Krug, 2023**

Size: to 5 mm Hermaeidae

ID: Body narrowly elongate to ovate, flattened; tail pointed. Cerata flat, narrow, elongate; many per side of back; each ceras with an unevenly crenulate distal margin due to rounded tubercle-like projections; no branches of the digestive gland. Rhinophores long, smooth, bifurcated below the middle, branches rolled. Pericardium slightly behind the head, oval to round, heavily papillose. Anal papilla at the right of the pericardium. Body pale yellow-green with tiny burgundy or brown specks and scrawls. Head white to pale green with burgundy and brown flecks and patches – **a burgundy/brown patch between the eyes extends diagonally as a thin line to the base of each rhinophore**; another patch is lateral, behind, and below each rhinophoral base. The pericardium is brown to burgundy but may appear totally white due to crowded white papillae; the anal papilla is bright white. Cerata are transparent, white-specked, with scattered white patches; sometimes with burgundy/brown specks and streaks; **a small burgundy/brown oval spot at each ceratal base and another near its apex**; its distal margin opaque white. Rhinophores are translucent with white flecks and patches, their tips whitish. Known to feed on *Penicillus dumetosus*, *Rhipocephalus phoenix*, and *Halimeda goreaui*. Depth: to ~40 feet. Southeast Florida, Bahamas, Dominica, Panama, Roatan, Cozumel. The original description makes no mention of spherical inclusions in the cerata (as in *C. antillensis*) but does note that the dark spot near the ceratal tip is "surrounded by a fine yellow ring". The presence of noticeable inclusions and total absence of rings in the accompanying images may be intraspecific variations.

Cyerce **cf.** ***nicholasi*** **K. Moreno, Rico, Middlebrooks, Medrano, Á. Valdés & P. J. Krug, 2023**

Size: ~5 mm Hermaeidae

ID: The morphology of the pictured animal appears to be virtually identical to that of *C. nicholasi*, which is described above. Its coloration is also largely in agreement with that species' pattern. However, based on the description mentioned above, this animal has a prominent feature that was not explicitly mentioned for *C. nicholasi*, and therefore distinguishes it: the presence of very crowded brown spherical inclusions (glands) throughout its cerata. It was found at a depth of about ~40 feet, but there is no additional information available for it. More extensive sampling of *Cyerce* specimens is required to determine if the presence of such prominent glandular inclusions is an additional feature that is present in *Cyerce nicholasi*.

Roatan

Jibacoa, Cuba

Pompano Beach, Florida

immature; Bonaire

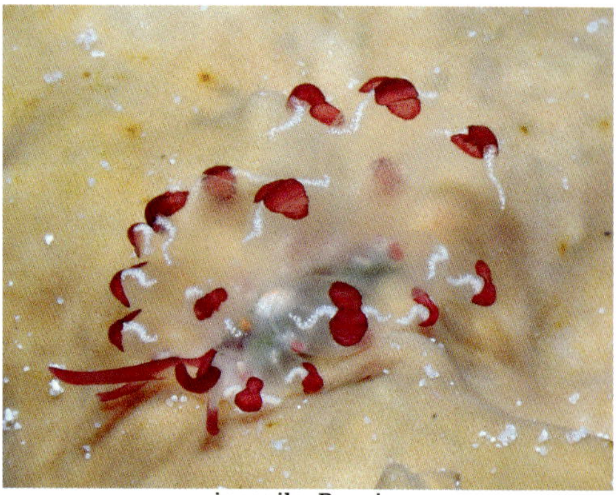
juvenile; Bonaire

juvenile; Bonaire

Roatan

Cyerce cristallina (Trinchese, 1881) **harlequin glass slug**

Size: to 50 mm Hermaeidae

ID: Body broad, oval, depressed, a slight pericardial swelling; foot broad, front corners rounded, narrows abruptly at the rear to form a pointed tail. Cerata are large, leaf-like, inflated-looking, variably sized, about 10-15 per side of body, erect when the animal is stationary. Rhinophores are long, smooth, rolled, bifid, one branch thicker/longer. Oral tentacles are short, smooth, rolled. Ground color translucent white. A median elongate-oval burgundy patch from the front of the head to behind the rhinophores; a translucent white oval patch superimposed on it behind each rhinophore, with a small black eye near its rear inner edge; a burgundy to brown ring or stripe or U-shaped patch on the pericardial swelling. The larger rhinophoral branch is burgundy with a white inner face proximally; the smaller branch is whitish, paler burgundy near the tip. Cerata are usually translucent white with opaque white or orange dots; **two burgundy patches** – proximal broad, apical narrower, sometimes converging; a white marginal "shoulder" on each side of the apical patch; in juveniles the apical patch is reduced to a burgundy margin. Sometimes the cerata have a dark burgundy apical blotch, white shoulders, and the rest suffused to varying extents with paler burgundy. The sides of the body may have a greenish band. The anal papilla is bright orange. It usually inhabits shallow bottoms with abundant algae to feed on; often under rocks on seagrass beds. When alarmed it can discard some cerata which then cause a distraction by wiggling. Cerata completely regenerate in about 7 days. It deposits white eggs in a flat ribbon configured as a spiral. Depth: 3-60 feet. Bermuda, Southeast Florida, Bahamas, Cuba, St. Martin/St. Maarten, Barbados, Bonaire, Curacao, Colombia, Belize, Cozumel, Jamaica.

Cyerce piercei **K. Moreno, Rico, Middlebrooks, Medrano, Á. Valdés & P. J. Krug, 2023**

Size: to 2.5 mm

Hermaeidae

ID: Body narrowly elongate to ovate, flattened; tail pointed; numerous flat and somewhat narrowly rectangular cerata along each side of the back; each ceras has an unevenly crenulate distal margin due to rounded tubercle-like projections, **without branches of the digestive gland**. Rhinophores are long, smooth, bifurcated proximally, both branches rolled; eyes close together between and behind their bases; oral tentacles cylindrical, short, smooth; the round pericardial hump is slightly behind the head, smooth or with some small papillae; anal papilla at hump's right front. Body mottled dark yellow, head white; **two dark burgundy patches originating from between eyes – one extending back to encircle pericardium and down sides of head, the other extending forward onto front of head as a triangle with sides skirting inner edges of rhinophoral bases**. Each side of the head also with a longitudinal dark purple line. The pericardium is light burgundy, sometimes with dark spots and/or white papillae. Mantle brown-flecked. Cerata are transparent; white-specked; a small burgundy or brown oval patch basally and a larger brown oval blotch near the ceratal apex; distal crenulate margin opaque white or brown speckled; each ceras with gray spherical glandular inclusions. The rhinophores and oral tentacles are translucent white with faint burgundy mottling. Anal papilla light to dark gray. Eyes black. It has been reported feeding on the algae *Penicillus lamourouxii*, *Penicillus capitatus*, *Halimeda incrassata*, *Udotea luna*, and *Udotea looensis*. Depth: 6-50 feet. West Florida, Bahamas.

Tampa, Florida

Tarpon Springs, Florida

***Cyerce* sp. 1**

Size: ~3 mm

Hermaeidae

ID: External morphology like that of *C. piercei* (see above) but appears to have spherical glandular inclusions not mentioned in the original description for that species. Ground color translucent gray. Dorsum yellow to pale orange with bright orange spherical glands; irregular carmine spots and patches along the inner edges of the ceratal bases. A small dark purple patch extends slightly forward from between the eyes and branches slightly toward the rhinophores. Cerata are transparent, abruptly transitioning to a broad ragged white-frosted distal margin; there is a **subapical to central bright white spot**, sometimes also a brown spot and a few small orange spots. The rhinophores and oral tentacles are clear and white-speckled. It was found on green algae at 49 feet. Specimen analyses are required to determine whether or not this is a variant of *Cyerce piercei*.

St. Vincent

***Cyerce* sp. 2**

orange-speckled glass slug

Size: to ~25 mm

Hermaeidae

ID: Its external morphology is similar to that of *C. cristallina* (previous page). The cerata are dark purplish-brown, with white edges and very crowded orange spots. It has been found on a shallow sandy bottom under a rock as well as on blades of seagrasses. Depth: to ~10 feet. So far, it is known only from Florida, Cuba, St. Vincent, and Cozumel. Possibly a color variant of *C. cristallina*.

Riviera Beach, Florida

***Cyerce* sp. 3**

Size: ~10 mm Hermaeidae

ID: Body elongate, broad, rounded in front; foot broad. The upper sides have many leaf-like cerata that are flattened, oval, their tops rounded or bluntly pointed and their edges studded with small pearl-like swellings (glands?); apparently no hepatic diverticula. Rhinophores are long, bases close together; bifid from near the base; one branch stout and tall, the other shorter and thinner, both branches rolled. The oral tentacles are thin and short. Ground color whitish-gray. Drab, white-specked everywhere. The back and cerata mottled with reddish-brown, yellowish, greenish; ceratal swellings white. Rhinophores and oral tentacles are clear with white spots/blotches and some orange spots. Found crawling on rubble. Depth: to ~10 feet. The similar *Mourgona germaineae* also has prominent glandular structures along the ceratal edges and reddish-brown pigment, but its cerata do have hepatic diverticula.

Roatan

***Cyerce* sp. 4**

Size: ~10 mm Hermaeidae

ID: Cerata are numerous, variably sized, long, narrow, rectangular, with pointed apices and a few projecting rounded tubercles. Rhinophores are long, bifurcated, one branch longer and stouter, both rolled. Oral tentacles are short and tapering. Eyes large, close together, just behind the rhinophores. Ground color translucent colorless. Viscera greenish/yellowish, the pericardium white, a brownish patch in front of the pericardium, **a brown triangle on the head with its base between the rhinophores and its vertex between the eyes**. A brown patch behind the eyes extends down the sides of the head and rearwards to encircle the pericardium. Cerata are transparent with **orange spherical inclusions**, surfaces white and brown speckled, distal margin white-frosted, and the rest of the margin brown to blackish. Rhinophores and oral tentacles are clear, white-speckled. Depth: ~15 feet. The dark triangle in front of the eyes and the pericardium's dark circumference suggest *C. piercei*, but the original description of that species' cerata does not mention the presence of either dark ceratal edges or spherical glandular inclusions. Nevertheless, it could be a variant.

***Cyerce* sp. 5**

Size: ~10 mm Hermaeidae

ID: The external morphology is similar to that of *Cyerce* sp. 4 but there are no spherical glandular inclusions apparent in its cerata, and there are several differences in the color pattern. For example: from base to tip each rhinophore is prominently brown with white spotting, except on its posterior face which is almost uniformly white; also, the pericardium has brown and white mottling. This animal could be a variant of *Cyerce* sp. 4, or perhaps of *C. piercei*. Depth: ~5-10 feet. No other information is available.

Cozumel

Lauderdale-by-the-Sea, Florida

Ceramium sp. camouflage; Riviera Beach, Florida

with visible ovotestes; Grand Cayman

Roatan

juvenile; Roatan

↑ Riviera Beach, Florida ↓

Hermaea cruciata A. A. Gould, 1870

cross-bearer hermaea

Size: to 10 mm Hermaeidae

ID: Body elongate, narrow; foot bilobed at front, corners pointed, tapers to a long thin tail; a prominent pericardial hump ahead of midbody. Cerata long and slender, in a simple row of up to 30 per side of back; ceratal apex somewhat conical: a broad base of usually 4 **transverse digestive gland branches forming a "cross" in side view** followed by a taper to a blunt tip, occasionally a tubercle at the end of each branch. **A single ceras on the dorsal midline in front of the pericardium and 1-3 behind it; dorsal branches of the digestive gland merge in the tail and in front of the pericardium**. Rhinophores almost as long as the longest cerata; unevenly rolled: a thin rear part of the open top projects up, the shorter front part extends forward, together resembling a claw. Ground color transparent to translucent cream, sometimes with a greenish cast. Brown/orange dorsolateral digestive tract branches originate at the tail's tip, run along the ceratal bases, and merge behind the head. Head and pericardium crowded with white dots. Translucent cerata have thin brown/orange digestive gland branches, a densely white-spotted "cross"-member, and a transparent tip with white spots. Rhinophores white-spotted, with brown/orange digestive branches posteriorly. Usually on/under rocks and debris and on wharf pilings in nutrient-rich waters. Feeds on filamentous red algae. It tolerates low salinity water. Rhinophores are held forward when the animal is moving. Spawn is a cylindrical, comma-shaped, gelatinous mass. Depth: to at least 25 feet. Southeast Florida, Bahamas, Trinidad & Tobago, Costa Rica, Roatan, Bay of Campeche, Cayman Islands.

Riviera Beach, Florida

Hermaea cubana **Caballer & Ortea, 2013** **Cuban hermaea**

Size: to 2 mm Hermaeidae

ID: Cryptic with *H. cruciata*, having a very similar external morphology, albeit at a smaller maximum size (2 mm vs. 10 mm). Some of its subtle visible differences have been identified as: relatively shorter rhinophores (shorter than longest ceras); no oral tentacles; eyes set in pigmented areas; anus **in** cardiac region instead of **ahead** of it; anus not stalked; **the 2 dorsal branches of the digestive gland converge only in the tail**. The specimen shown in the Caballer & Ortea paper also has some smaller ceratal digestive gland branches below the subapical cross. Body translucent with scattered **reddish or brownish pigment**. White patches on body and head. Rhinophores and cerata translucent with white patches; cerata also with a few small red spots/specks. Digestive gland reddish or brownish red. Front margin of foot with orange cast. Found on filamentous red algae attached to rocks. Depth: 3-26 feet. Northwest Cuba, Southeast Florida. Rare. Main features for identification purposes: **reddish marks on body; anterior lateral digestive gland branches not connected.**

northwest Cuba

Hermaea nautica **Caballer & Ortea, 2007**

Size: 2 mm Hermaeidae

ID: Body elongate, narrow; tail long, thin, tapers to a point; pericardium ovoid; 2 rows of ~5 cerata per side of back, long ones above short ones; **each ceras nearly smooth, elongately- globose, circular in cross-section, slightly "pinched" near apex**. Two prominent lateral digestive gland branches along ceratal bases begin (disconnected) in front of the pericardium and continue into the tail where they merge. Ceratal branch of gland is axial, thin, unbranched, sometimes bulged distally. Rhinophores as long as or shorter than longest cerata; unevenly rolled – thin rear part of scroll's top projects upwards to a point, shorter front part extends forward. Oral tentacles short, rounded. Eyes at outer sides of rhinophoral bases. Body, head, and appendages transparent with superficial patches of white granules. Ceratal concentrations of white granules primarily in two transverse bands – one near the base, one near the apex. A thin reddish stripe along each side of body from tail to front of head; a similar more dorsal stripe from midbody skirts the eyes and goes up the inner face of a rhinophore. All digestive gland branches are reddish. Superficial red dots on and slightly behind the transparent pericardial area. Found on rubble at 3-7 feet.

Grand Cayman

Hermaea **sp. 1**

Size: ~4 mm Hermaeidae

ID: This might be a variant of *Hermaea cruciata* (q.v.) but the cerata are quite inflated and the dorsal and ceratal digestive gland branches are prominently bright orange. In particular, the ceratal branches are uniformly orange from the base to the tip. The dorsal branches appear to end just behind the rhinophores, parallel to one another. The rhinophores are white-banded.

Riviera Beach, Florida

Roatan

Roatan

Lauderdale-by-the-Sea, Florida

Fort Lauderdale, Florida

***Hermaea* sp. 2**

Size: ~4 mm Hermaeidae

ID: Like *Hermaea cruciata* but: the two dorso-lateral branches of the digestive gland do not merge ahead of the cardiac area; a median ceras is not present at that location; the cerata are very inflated; the ceratal digestive gland branch can be entirely peppered with reddish to purplish specks. The ceratal surface pale specks are white to bluish.

Roatan

***Hermaea* sp. 3**

Size: ~4 mm Hermaeidae

ID: Orange coloration like that of *Hermaea* sp. 1 but the lateral branching inside the inflated cerata occurs at several levels.

Guadeloupe

***Hermaea* sp. 4**

Size: ~4 mm Hermaeidae

ID: The ceratal digestive gland stalk is thick; stalk and branches purple; rhinophores white with a transparent central band.

Guadeloupe

***Mourgona germaineae* Er. Marcus & Ev. Marcus, 1970**

Size: to 15 mm Hermaeidae

ID: Body elongate-oval. Cerata are flat, oval, crowded; the distal edges are lined with bubble-like glands, some also below either in clusters or scattered; the digestive gland penetrates them in the form of a narrow vertical axis almost reaching the apex, and having lobe-like side branches. The rhinophores are relatively long, bifurcated, with each branch inrolled. The oral tentacles are simple, shorter, and ventrally grooved. The ground color is translucent gray; head, rhinophores, and oral tentacles with white dots/patches; ceratal **digestive gland branches yellowish-green (upper cerata) to dark green (lower cerata). All surfaces peppered with small orange-brown dots**. It is cryptic among dense growths of green algae. Also reported from *Cladophora* or algae-covered rocks in the Bahamas, but its preferred host and food source in the Florida Keys is the alga *Cymopolia barbata*, from which it extracts cell sap for food and toxic chemicals for defensive purposes. When disturbed it curls up into a ball and may autotomize several cerata as a distraction and/or to have them secrete and release a sticky milky-white mucous. Depth: 2-40 feet. Florida, Bahamas, Puerto Rico, Guadeloupe, Martinique, St. Vincent, Costa Rica. Rare.

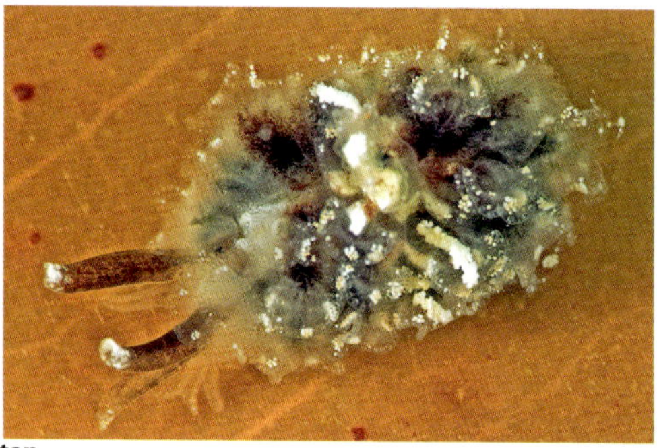

Roatan

Guadeloupe

***Mourgona murca* Er. Marcus & Ev. Marcus, 1970**

Size: to 11 mm Hermaeidae

ID: Cerata broadly oval, flat to somewhat globose; a row of 8-9 clusters per side; digestive gland branches fern-like; distal edge lined with up to 25 spheroidal glands, sometimes a few also scattered below the edge. Rhinophores are relatively long, distal two-thirds bifurcated. Oral tentacles are short, inrolled, and open ventrally. Ground color translucent grayish. Back yellowish-white/yellowish-green; flanks greenish; a diffuse reddish to purplish-brown median blotch behind the eyes, an elongate median white spot in front of them; usually a larger white to diffuse reddish blotch in the cardiac region. Cerata are translucent; many white and brown dots; digestive gland branches green to blackish-green, or brownish to yellowish if the animal is not well fed; the edge glands are clear to whitish, the inner ones opaque white. Rhinophores are translucent, the upper branch purplish-red with a white tip; the lower one whitish. Oral tentacles whitish, black-dotted dorsally. Reported from rocky bottoms with mixed algae; in Bahamas on *Cladophoropsis* or attached *Sargassum*. Depth: 5-60 feet. Bahamas, St. Vincent, Guadeloupe, Martinique, Curacao, Costa Rica, Roatan. Rare.

Mourgona siero **(Ortea, Moro & Espinosa, 2022)**
Size: to 10 mm Hermaeidae
ID: Body elongate-oval. Cerata leafy; peduncles thick somewhat cup-shaped; about 6 longitudinal rows per side; central axis of digestive gland reaches apex, up to 10 simple long branches per side (tips sometimes forked); many small oval/spheroidal glands along the distal edges of larger cerata. Rhinophores long, thick, top two-thirds split into two oppositely curved branches. Oral tentacles are much shorter, recurved, inrolled, tapered. Eyes large, behind the rhinophores. Ground color is whitish, dorsally more opaque and brighter. Cerata are dark green, distal margin and peduncle white-spotted, glands pearly. Rhinophores opaque white with some maroon spots proximally; oral tentacles whitish; eyes black. The only animal ever reported was found at a dock on loose rocks covered with algae and sedimentary detritus. Depth: ~6 feet. Rare.

Puerto Morales, Mexico

Guadeloupe

Mourgona sp.
Size: to 20 mm Hermaeidae
ID: Its external morphology matches that of *Mourgona germaineae* (previous page), so it might be a variant of that species. However, the **very prominent brownish-black coloration of the cardiac region** that extends forward to form an arc behind the rhinophores and spreads onto the oral tentacles is not a feature that has been specifically noted for it in the scientific literature. Although Marcus and Marcus (1970) stated "some reddish brown pigment on head and back" in their description of *M. germaineae*, it was not emphasized as being prominent. Found on and under rocks. Depth: ~7 feet.

photos from St. Vincent

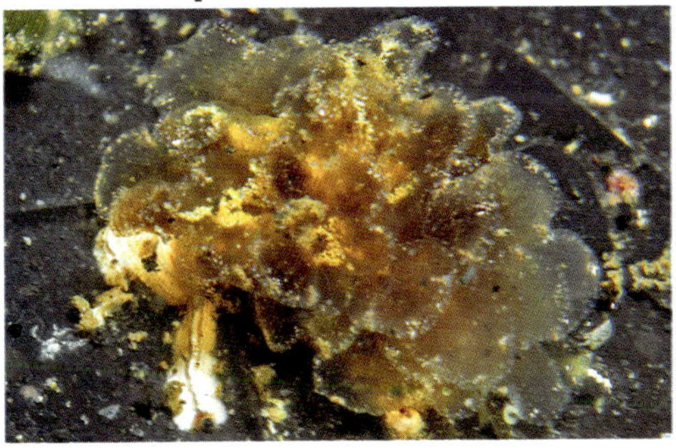

Polybranchia **cf.** ***schmekelae***
Medrano, Krug, Gosliner, Biju Kumar & Á. Valdés, 2018
Size: ~20 mm Hermaeidae
ID: Possibly a variant of *P. schmekelae* (see next page), but it differs by being prominently orange and has dense aggregations of gold granules near the edges of some cerata. It was found under rocks at a depth of ~10 feet.

Riviera Beach, Florida

Grand Cayman

Guadeloupe

Polybranchia schmekelae **Medrano, Krug, Gosliner, Biju Kumar & Á. Valdés, 2018**

Size: to 70 mm Hermaeidae
ID: Body oval, high, broader at front, tapers to rear; foot broad, tail short and triangular; pericardium large, smooth, at the rear of the front third of the back. **Cerata leaf-like**, length twice width, flat, pedunculate, several longitudinal series per side of back; each ceras penetrated by a **complex of narrow digestive gland branches** (often touching); **surfaces crowded with minute conical papillae; edges with papillae containing pearl-like glands**. Rhinophores are bifurcate, inrolled, with scattered small papillae. Oral tentacles are short and inrolled. Ground color translucent **pale-to-dark olive-green to yellowish-brown**. Back with scattered brown specks; pericardium paler and unspecked. Rhinophoral papillae white. Ceratal edge papillae/glands **prominently white**; surface papillae appear as **white dusting**; tops of peduncles white; dark digestive branches often prominent against pale background; smaller cerata usually darker, tips sometimes dark-spotted. By day it hides under rocks and deep within coral crevices in (often agitated) shallow water near high concentrations of green algae; emerges at night to feed preferentially on *Caulerpa racemosa*. Spawn is a gelatinous ribbon deposited as a white spiral coil of ~3 whorls containing ~25,000 capsules with 2-5 eggs each. When disturbed it can easily discard cerata. Depth: to at least 10 feet. Southeast Florida and the Keys, Antilles, Panama to Quintana Roo (Mexico). Rarely seen.

Polybranchia **sp.**

Size: ~10 mm Hermaeidae
ID: The rhinophores and distal edges of this animal's cerata are prominently studded with numerous tubercles that are white, small, elongate, pedunculate, and have bulbous tips, much like they are in several species of *Cyerce*. However, the presence of digestive gland diverticula within the cerata of this animal precludes its identification as a species of *Cyerce*. Perhaps it is a variant of *Polybranchia schmekelae* (see above). It was found among seagrasses and rocks at a depth of ~7 feet at one of Guadeloupe's sister islands.

Marie Galante, Guadeloupe

Family Juliidae E. A. Smith, 1885

A family of "bivalved gastropods" in the superorder Sacoglossa consisting of about forty species of small slugs in seven genera. They are distributed globally in temperate to tropical waters except in the Eastern Atlantic and Mediterranean where they have been recorded only as fossils. So far, only the two most speciose genera – *Berthelinia* and *Julia* – have been recorded in the tropical West Atlantic.

Body elongate, usually pale to dark green; mouth flanked by a pair of rounded oral lobes; rhinophores are long, usually rolled as hollow cylinders; eyes close together behind the rhinophores; the shell is external, oblong, and clam-like with two similar hinged valves; the valves are thin, translucent or thick, capable of completely enclosing the retracted animal; valve opening and closing is controlled by a muscle near the front of the viscera; the foot tapers to a point, its tip rarely extends beyond the rear of the shell; radula uniseriate, sides of teeth usually finely denticulate, the tips sometimes notched.

Although long considered as bivalves based on fossil records, the discovery of living specimens beginning in 1959 has revealed them to be something extraordinary – gastropods that developed bivalved shells, another example of "convergent evolution" in nature. They spend their lives in subtidal and lower intertidal areas with species of the green algae *Caulerpa* on which they feed suctorially and reproduce while highly camouflaged. In fact, both veligers and adults can produce an elastic mucous for tethering themselves to a plant to avoid being dislodged, even when entirely within the shell and the valves closed. When crawling, their necks elongate considerably.

Jamaica

left valve with protoconch; Jamaica

Bermuda

Edenttellina caribbea **(Edmunds, 1963)** **Caribbean bivalved snail**

Juliidae

Size: Body ~8 mm; shell ~5 mm

ID: Shell valves wedge-shaped, narrowed but rounded posteriorly; can enclose the entire animal but usually the head and long narrow neck protrude by about half a shell-length. Rhinophores are cylindrical and rolled; oral tentacles are just two slightly raised vertical ridges. Eyes are well behind the rhinophores. Valves are translucent green to grayish-green, with brownish/yellowish tinges and some whitish areas. Its clearly visible green mantle has irregular longitudinal reddish-brown to yellowish-brown lines and scattered white patches. Head and neck with white scrawls; the neck has a median dark line that bifurcates behind the rhinophores, a branch going to each inner base. Inhabits salt-water channels in mangrove swamps; lives and feeds on *Caulerpa verticillata* that grows on mangrove roots and mud/gravel banks. Spawn is a 1-4 mm long mass of 40-80 ovoid capsules deposited on an algal frond, each capsule with one yolky egg. It secretes and ejects an acidic white fluid for defensive purposes. Depth: to 10 feet. Bermuda, Florida, Bahamas, Jamaica to Guadeloupe, Panama to Yucatan, Brazil. Sometimes common but always rarely seen.

Family Limapontiidae Gray, 1847

A family of more than seventy sacoglossans in about a dozen genera, distributed globally in mostly temperate to tropical waters, but with a few species in frigid northern areas. The genera *Ercolania*, *Placida*, *Stiliger*, and *Olea*, which encompass about three-fourths of the species, are represented in the tropical West Atlantic.

Body oval to elongate, small; front of foot rounded to squared, corners sometimes tentacular; rhinophores (if they exist) are smooth, short to long, and spindle-like, usually in-rolled in *Placida*, sometimes with a basal lateral groove; eyes behind rhinophores, usually in clear patches; the reno-pericardial prominence is well behind the eyes, elongate or inconspicuous; cerata are variously shaped (slender, swollen, globular, fusiform), numerous to non-existent, penetrated by often visible digestive gland branches, and may also contain tributaries of the albumen gland; digestive gland branches may also extend into the head and rhinophores; anus at right front of the pericardium, as a papilla or just a surface pore; the male gonopore is below/behind the base of the right rhinophore, the female opening further back; radula uniseriate; non-denticulate teeth narrow, sharply pointed to blade-like to sabot-shaped.

Most species live and feed on green algae but some prey on the egg masses of other sea slugs. Some species (*Sacoproteus* spp.) mimic the appearance of their non-filamentous algal host. A few species are euryhaline. Usually shorter than 20 mm.

Guadeloupe

Roatan

Ercolania coerulea **Trinchese, 1892** **blue stiliger**
Size: ~10 mm Limapontiidae
ID: Body elongate; front of foot extended, broadly rounded; oral veil broad, front margin slightly concave. Cerata short, broadly fusiform, crowded, with a small apical protuberance; 3-5 longitudinal rows per side, the size decreasing ventrally. Rhinophores long, smooth, narrowly conical. Eyes small, well-separated, dorso-lateral behind the rhinophores. Ground color translucent yellowish. Head, foot, and appendages white-dotted, the dots concentrated at rhinophoral tips and coalesce into spots to form marginal white "arches" at the tops of cerata. Cerata transparent; digestive gland a thick central axis with finer side branches extending two-thirds the way up; color brown, green, yellow, orange, probably food-dependent. Highly branched fine green digestive ducts are visible in the head and almost to the tips of the transparent rhinophores. Eyes black. Usually on the green bubble alga *Dictyosphaeria cavernosa*. Feeds by coordinating its front foot lobes and oral lobes to grasp the alga's wide utricles and extracting their nutrients. Depth: 20-70 feet. Roatan, Guadeloupe. The pictured animals match most of the Mediterranean animals that have long been regarded as *Ercolania coerulea* in spite of lacking the feature that inspired its name – a large light blue spot on the front surface of each ceras ("coerulea" is Latin for "sky blue"). Recent molecular analyses have demonstrated that cryptic species exist within groups of what had been considered "intraspecies variants". This may also be the case for West Atlantic *Ercolania*.

Guadeloupe

Riviera Beach, Florida

↑Bermuda Riviera Beach, Florida↓

Riviera Beach, Florida

Ercolania fuscata **(A. A. Gould, 1870)**

dusky stiliger

Limapontiidae

Size: to 8 mm

ID: Body elongate, quasi-cylindrical, tapers to the rear; foot as broad as the body, tapers to a long pointed tail, has a front notch, may project beyond the head; head not distinct from body, front rounded, slightly concave medianly. Cerata in 2 longitudinal rows per side of back starting about one-third way down body; up to 16 per side, large ones above, small ones below, get smaller toward the rear; long, inflated-looking, pinched near base, inclined outwards. Rhinophores smooth, long, narrowly conical, tips bluntly rounded. Ground color **translucent gray to whitish to blackish-brown**. Back with **maroon or brown spots**, usually scattered but sometimes grouped into irregular transverse dusky bands. Back of head may be dusky; with a short stripe extending back from each rhinophore. Cerata translucent with dark specks and **white spots**; digestive gland branches sometimes visible as a pale orange/greenish mass; tips prominently white in brownish individuals. Rhinophores translucent gray, white-spotted. Eyes black, each in the middle of an unpigmented area. Usually on tangled colonies of filamentous algae of the genus *Cladophora*, either attached to mangrove roots or free floating; occasionally on *Chaetomorpha* spp. When they feed in groups they often cause algal mats to break loose from their substrate and drift, dispersing the individuals. When disturbed it thrashes its cerata and may discharge large white droplets from glands, or it may autotomize some cerata. Depth: to ~10 feet. Bermuda, Florida, Texas, Cuba, Puerto Rico, Curacao, Jamaica, Brazil.

Riviera Beach, Florida

Guadeloupe

Grand Cayman

Guadeloupe

Ercolania viridis **(A. Costa, 1866)** **variable stiliger**

Size: to 12 mm Limapontiidae

ID: Body elongate, narrow; head broader; foot as wide as the body, tapers to a pointed tail. Rhinophores long, simple, finger-like. Cerata in 1 or 2 rows per side of the body, outermost smallest; shape changeable from slightly inflated to bulbous, stalks short and thin, tips nippled. Back open; eyes behind the rhinophores, well-separated. Ground color translucent gray to black. Color patterns are variable. A pink, reddish-brown, or dark brown broad median stripe (may be absent) starts between the rhinophores, bifurcates at the start of the cerata, each branch heading to the tail as a narrow lateral stripe. Usually an irregular green digestive gland branch on each side of the head from the frontmost ceras, skirts the eye's inner edge, continues to the inner base of the rhinophore. The rest of the back whitish. Body, head, and foot densely white-dotted; rhinophores white-spotted. Cerata usually white-spotted with white tips, but the tips may be red, reddish-brown, purple; cerata sometimes tinged with red, bluish-green, blackish-green. Eyes black, in whitish patches. Lives and feeds on green algae in the genera *Chaetomorpha*, *Cladophora*, and *Cladophoropsis*, from which it can sequester chloroplasts that can continue to provide it with energy photosynthetically for up to 24 hours. Spawn is a broad cylindrical coil of 1-1.5 whorls containing white to yellowish eggs. Depth: 3-20 feet. Mediterranean and amphiatlantic. North Carolina to Florida, Bermuda, Bahamas, Puerto Rico, Virgin Islands, Martinique, Curacao, Costa Rica, Cozumel, Cayman Islands, Jamaica, Brazil.

Riviera Beach, Florida

Ercolania **sp. 1**

Size: ~10 mm Limapontiidae

ID: Possibly *E. viridis*, but: the digestive gland branches enter the rhinophores and almost reach their tips; cerata have a thick brown axis nearly reaching a **pinkish-red tip**, the distal half with thin green branches. Found on green algae at ~10 feet.

Riviera Beach, Florida

Ercolania sp. 2

Size: ~2 mm

Limapontiidae

ID: Like *Ercolania* sp. 1 but smaller and the cerata do not have pinkish-red tips. Possibly a juvenile of that species. Depth: ~10 feet.

Riviera Beach, Florida

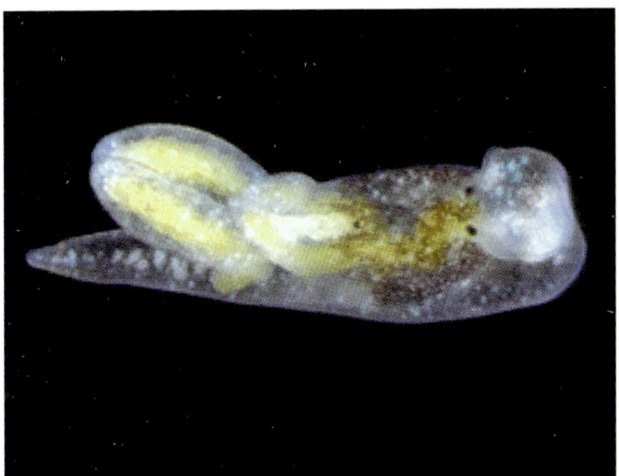
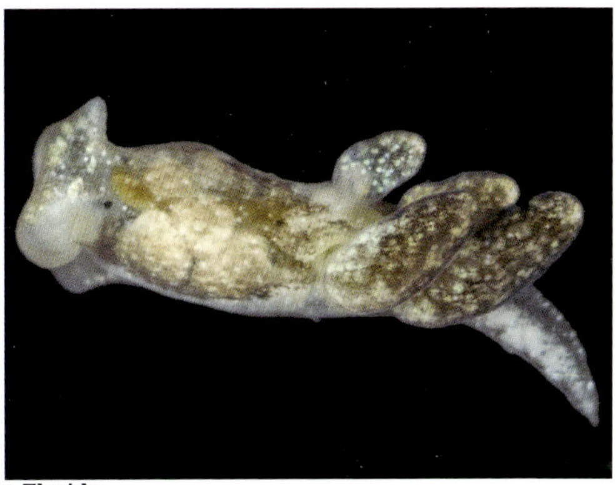

Cedar Key, Florida

Olea hensoni Galvão Filho, Paulay & Krug, 2019

Henson's olea

Size: to 5 mm

Limapontiidae

ID: Body elongate, smooth, tapers to a thick pointed tail; head broad, squarish, the broadest part of the animal, about one-fourth body length. Rhinophores are either very short or just small "bumps" at the rear of the head. An inconspicuous pericardial hump at midbody. A longitudinal row of **3 fusiform cerata per side between pericardium and start of tail**. Ground color translucent to cream. Back with diffuse yellowish and dark brown blotches; **small rounded white glands on body, head, and cerata**; larger bluish glands are scattered near the rhinophores and form a median row on the tail. **Digestive gland yellowish**. It was found on a sand bar feeding within the gelatinous egg masses of an unidentified headshield slug/bubble snail. Also seen feeding on algae. Spawn is a clear flat ribbon 2-3 whitish eggs across, wound into a loose to tight spiral coil of ~5 whorls. Depth: intertidal to ~10 feet. Florida. Rare.

Riviera Beach, Florida

***Placida* cf. *cremoniana* (Trinchese, 1892)** **West Indian burnt placida**

Size: to ~12 mm Limapontiidae

ID: Body elongate. Cerata numerous, cover most of the back, cylindrical, tips pointed, longest at midback. Rhinophores conical, rolled, longer than the cerata; oral lobes broadly rounded; foot rounded at front, tapers to a pointed tail; eyes small, behind the inner rear bases of the rhinophores. Ground color yellowish-orange. Cerata black on the upper two-thirds; rhinophores black with a long whitish posterior stripe; tail black dorsally; head and oral lobes blackish, except for a yellowish-orange patch behind each rhinophore that has a thin extension angling down and back; eyes black, prominently centered in the head patches. *P. cremoniana* has been found in shallow water on species of *Briopsis* and *Derbesia* algae. Spawn is a flat band deposited as a short spiral coil with white eggs. Depth: to ~10 feet. Florida, Venezuela, Grand Cayman. *Placida cremoniana* has long been considered a circumtropical species. However, recent molecular and morphological analyses of Pacific and Eastern Atlantic species (McCarthy et al, 2017) have revealed a complex of four species. Since West Indies specimens were not sampled for the analyses, the complex may consist of even more species. The pictured specimens **could** be *P. cremoniana*, but could also be another member of the complex.

Roatan

Guadeloupe

St. Petersburg, Florida

***Placida kingstoni* T. E. Thompson, 1977** **Kingston placida**

Size: to 15 mm Limapontiidae

ID: Body elongate, depressed, back exposed; **cerata very long, slender, smooth**, along the sides; rhinophores are long, slender, rolled on proximal half, cylindrical, tapering distally, often project forward. Translucent to transparent. **Green to brown digestive gland ducts** visible in body and appendages, giving body a mottled look; ceratal ducts long, highly branched. Cerata with some white speckling, sometimes with dark tips; rhinophores white-tipped. Pericardium behind head, **hump prominently opaque white**. Hides among and extracts sap from filamentous algae, primarily *Bryopsis plumosa* but occasionally *Penicillus lamourouxii* and *Cladophora* spp. Able to feed from a distance greater than its body length. As it moves, its cerata pulsate in a random jerky fashion. Although it ingests chloroplasts it's unable to store them for long-term photosynthetic energy production; completely digests them within 24 hours. Since chloroplasts lose their photosynthetic function after ~4 hours in the body, only continuous feeding can confer their benefits to this species. If it doesn't feed for a day its green coloration can fade or vanish. It can autotomize its cerata as a defensive mechanism. Depth: to 50 feet. Bermuda, Florida, Guadeloupe, Martinique, Venezuela, Costa Rica, Roatan, Jamaica, Yucatan.

Placida verticilata **Ortea, 1981** **whorled placida**

Size: to 10 mm Limapontiidae

ID: Body elongate, narrow, tapers to the rear; the foot tapers to a long pointed tail (about one-third body length); a pronounced pericardial hump; digestive gland ramifications present in the head and back, the tail has 2 branches that converge at the tip. Rhinophores are long, thin, rolled, grooved anteriorly, and crowded with digestive gland branches. Cerata are in a line of up to 8 transverse rows per side (up to 4 cerata per row); elongate, fusiform, tips pointed; along a ceras' entire length the digestive gland is a central axis encircled by groups (whorls) of up to 5 oblique stubby blunt-tipped extensions. The ground color is translucent whitish. **Extensive olive-green branching of the digestive gland in the appendages and most of the head and body**. Cerata with scattered white spots; white patches at/near apices; tips of internal whorls darker green. Flanks whitish. Lives, feeds, and reproduces cryptically on green algae, e.g. *Codium* spp. Stores ingested chloroplasts in its digestive system for several weeks as sources of energy. Spawn is a gelatinous spiral coil (2+ whorls) of a flat ribbon containing white eggs in round capsules. At rest, it bends its tail forward like a cat. Depth: 1-40 feet. Amphiatlantic. Bermuda, Florida, Bahamas, Cuba, Guadeloupe, Venezuela, Costa Rica, Jamaica. Rare.

Riviera Beach, Florida

Riviera Beach, Florida

Placida **sp. 1**

Size: ~10 mm Limapontiidae

ID: Body elongate, depressed. A row of slender, terete, smooth cerata along each side of back, each ceras with some small rounded surface tubercles. Rhinophores are long, slender, conical, tip rounded, also with a few small tubercles; prominent eyes behind rhinophoral bases. Pericardium small, flattish, median between frontmost cerata. Translucent to transparent. Yellowish brown to greenish brown digestive gland ducts extend along the inner ceratal bases and penetrate unbranched deeply into the cerata and the rhinophores. Cerata with distal white blotches, their **tubercles pale blue**. Rhinophores dark-tipped, their sparse tubercles white to pale blue. Pericardium pale pink. Eyes black. Found among green algae. Depth: ~10 feet..

Placida **sp. 2**

Limapontiidae

Size: ~10 mm

ID:. The body is elongate, the tail long, slender, and pointed. There is a row of ceratal clusters per side; the back is open and smooth; the cerata terete, slightly inflated, with rounded tips. Rhinophores are long and conical, their tips rounded, the eyes are behind the bases of the rhinophores. The pericardium is elongate, flattish, between the frontmost cerata. Ground color is pale tan. Dark brown digestive gland ducts extend along the sides of the back and penetrate deeply into the cerata and the proximal two-thirds of the rhinophores. The cerata have many white blotches that are especially concentrated near the apices, which have a prominent brown spot. The pericardium is white; the rhinophores white on the distal third. Found among green algae. Depth: ~10 feet.

Guadeloupe

Riviera Beach, Florida

Stiliger fuscovittatus **Lance, 1962**

brown-streak stiliger

Size: to 15 mm Limapontiidae

ID: Body elongate, broadest anteriorly, dorsally rounded, tapers to a bluntly rounded tail. Rhinophores are long, slender, smooth, and taper to a blunt tip; nonretractile. Cerata club-shaped, one row per side of back starting about one-fourth body length behind the rhinophores. Ground color is translucent white to grayish-white. Body with scattered **reddish-brown streaks, blotches, and dots**. An irregular **reddish-brown stripe from part-way up the rhinophore to its base**, and may continue past the eye. Cerata clear; axis brown, unbranched, ends below tip; white spots mostly on distal half. Inhabits estuaries and rocky shores, feeding on filamentous red algae such as *Polysiphonia* and *Callithamnion*. Tolerates temperature/salinity variations very well. Depth: to ~10 feet. It is a northeast Pacific species that has been recorded off Texas and Florida only as single individuals; probably introduced via shipping. Whether or not breeding populations exist in the tropical West Atlantic remains to be determined.

Family Oxynoidae Stoliczka, 1868 (1847)

A family of sacoglossan sea slugs consisting of about twenty species in three genera distributed globally in temperate to tropical waters. Only two of the genera – *Lobiger* and *Oxynoe* – occur in the tropical West Atlantic.

Body elongate, tapers to a very long thick tail; foot extensions form dorso-lateral "parapodia" either as a pair of narrow fleshy flaps (*Oxynoe*) or a pair of erect lobes (*Lobiger*), both papillate/tuberculate on their outer faces; the shell is an external dorsally central univalve with a broad aperture, fragile, too small to accommodate the entire animal during defensive retraction; the shell is oval and partially to fully covered by *Oxynoe* flaps, or elongate and surrounded by *Lobiger* lobes; the rhinophores are rolled, smooth to slightly tuberculate; *Lobiger* has a pair of broad flat oral lobes in front of the rhinophores; the mantle cavity has a very long opening on the right side, the gill at its opening has many elongate ciliated lamellae; the female gonopore is at the inside right front edge of the shell; anus posterior (*Lobiger*) or postero-lateral (*Oxynoe*); radula uniseriate, teeth blade-shaped and denticulate.

They feed primarily on *Caulerpa* species of green algae where their coloration gives them excellent camouflage. Other strategies for evading predators include the release of noxious chemical secretions, the autotomization of appendages (parapodial lobes, end of the tail) as a distraction, and even attempts to escape by swimming using their sturdy tails.

St.Vincent

Guadeloupe

Riviera Beach, Florida

Guadeloupe

Lobiger souverbii **P. Fischer, 1857** — **lined-shell lobiger**

Size: to 30 mm — Oxynoidae

ID: Body elongate, ovate, tapers to the rear; tail long, tapers to a blunt point; flanks and top of tail with **scattered small conical tubercles**; those on the tail form a "crest"; **external shell** dorsally central, very thin, translucent, cap-like, half a body length; **two parapodial lobes on each side of the shell**; the lobes are similar, long, broad, concave inwards when erect, **their edges bumpy and wavy-looking**, outer (ventral) surfaces tuberculate; **lobes often longitudinally "rolled in"**. **Rhinophores rolled**, slightly tuberculate, the inner basal edge of the ventral groove with a broad **downward-projecting flap**; rhinophores incline forward/up/out to significantly overhang the front of the head. Mantle cavity with a long opening on right side; its comb-like gill has ~60 parallel longitudinal ridges spanning the width of the cavity's roof; the top of the mantle edge has a groove lined with shell-forming glands. Ground color **light green. Whitish and pale brown pigment cells on mantle and lobe edges, lobe outer surfaces, and top of tail**, concentrated around/on tubercles. **Lobe inner surfaces with white a margin and an orange submargin**. All surfaces are crowded with a (usually) **faint mesh of randomly-oriented, very short, thin, dark brown line segments. Shell green to yellowish; with ~6 thin, irregular, longitudinal black lines** – some interrupted, short, and/or bifurcated; a few lines with **iridescent blue dots**. Rhinophores are white-tipped. It inhabits calm waters where it blends in with and feeds on species of *Caulerpa* green algae. Its long lobes can envelop the shell as well as intimidate predators when erect. When disturbed it may discharge a sticky milky-white fluid and swim away by flapping its lobes, and even cast them off. Spawn is a coiled string of yellow eggs. Depth: 3-140 feet. Circumtropical. Florida, Bahamas, Antilles, Venezuela to Bay of Campeche, Jamaica, Brazil. Uncommon.

Roatan

Riviera Beach, Florida

Riviera Beach, Florida

Bermuda

5 mm juvenile; Riviera Beach, Florida

with spawn; Grand Cayman

Oxynoe antillarum Mörch, 1863 **Antilles oxynoe**

Size: to 50 mm Oxynoidae

ID: The body is elongate and oval centrally; the head is prominent; parapodia short, fleshy, margins irregular, fully/ partially enclose its **dorsal external shell**; its exhalant siphon is visible through a small parapodial gap at the rear. The shell is about one-fourth the total length, ovoid, dorsally convex, smooth, fragile, has a wide ventral aperture, and protects the mantle cavity and viscera. The tail is long, fleshy, autotomizable, starts broad behind shell and tapers to a point. The parapodia and tail have papillae: scattered, conical, retractile, large and small, sometimes elongate, often larger and more concentrated along the parapodial margins and medianly along the top of the tail. Rhinophores smooth, elongate, rolled; a groove on the outer face extends back to the shell as a furrow along the head's edge. **Juveniles**: relatively shorter and wider rhinophores, more exposed shells. Ground color olive to dark green; sparse to dense yellow spotting/speckling (can be dark green to brown). Parapodial and foot margins, tail's dorsal midline, rhinophores, front of face, and oral lobes are yellowish with **irregular blue to purplish spots and streaks**; foot margin are uniformly-sized, evenly-spaced, in a row from the head to the tip of the tail. Parapodial papillae are whitish to yellowish and white-tipped. **Juveniles: yellowish to light green with bluish spots**. Shell translucent white. Inhabits relatively calm shallow water; sometimes under rocks but usually living and feeding on *Caulerpa* seaweed. Spawn is a transparent gelatinous white egg mass deposited on its host alga. Development is planktotrophic – the larvae start feeding immediately after hatching. When disturbed it secretes a milky white liquid and may discard its tail (it regenerates). Depth: 1-35 feet. Bermuda, Florida, Bahamas, Antilles, Venezuela to Yucatan, Brazil. Common. Adults of *O. antillarum* and *O. azuropunctata* Jensen, 1980 are nearly impossible to distinguish based on external appearance alone because of their high intraspecific color variability and a lack of consistent differences.

Family Plakobranchidae Gray, 1840

A family of sacoglossan sea slugs consisting of more than 140 species in four genera distributed globally in temperate to tropical waters. Three of the genera – *Bosellia*, *Elysia*, *Thuridilla* – occur in the tropical West Atlantic, the most speciose being *Elysia*.

Body usually long, narrow, higher than wide, sometimes flattened and smooth but it can be papillate; the head is differentiated; front of foot rounded, corners triangular to tentacular, sole with transverse groove ahead of visceral portion of body; rhinophores slender and auriculate, eye spots behind their bases; oral tentacles small to absent. Sides of body usually with flap-like extensions of the foot (parapodia) that extend along most of the body length and fold up to meet over the dorsal midline; parapodia can be narrow to broad; their edges can be delicate to thickened, smooth to papillate, straight to sinuous, sometimes forming dorsal gaps (siphonal openings); in *Bosellia* the parapodia appear to be permanently splayed wide open to give the feeding animal a flattened ovate to circular appearance; the pericardial cavity and renopericardial complex are behind the head, usually with a prominently visible associated network of branched dorsal vessels on the inner parapodial faces; branches of the digestive gland are also visible; the anus and female gonopore are in a lateral groove near the right front, the male gonopore at the outer side of the right rhinophore; radula uniseriate, teeth pointed, smooth or denticulate; front of radula rests in a "crop sack" where spent teeth accumulate.

Plakobranchidae feed primarily on green algae, with different slug species having their own algae preferences. Usually they simply puncture plant cells with their radula, suck out the cytoplasm, and digest the contents. However, some species practice "functional kleptoplasty" whereby the cell contents are only partially digested, with the algae's photosynthetic plastids (e.g. chloroplasts) being shunted into the cytosol of the epithelial cells lining the digestive gland tract. By remaining intact and functional the plastids remain photosynthetically active and continue to produce the oxygen and sugars that are thought to help the animal survive periods of starvation. Some of the plastids remain functional for weeks or months. The respiratory functions are provided by the dorsal vessels.

↓Roatan Bermuda↑

↓Roatan Bermuda↑

Guadeloupe

Bosellia mimetica **Trinchese, 1891** — **mimic bosellia**

Size: to 15 mm — Plakobranchidae

ID: Elongate oval when moving, almost circular when stationary, flat with a slight pericardial hump. Head with a rounded oral veil separated from the body by a distinct indentation and a pair of broad transverse folds. Parapodia are permanently opened/flattened. Rhinophores are very short and rolled; eyes are in an unpigmented oval blotch behind the rhinophoral bases. Ground color light to dark green to brownish, the color due primarily to contents of its widely branched digestive tract. The back is white-specked; often has yellowish-green and whitish mottling and sometimes has large irregular white patches at the front and going down its sides; body edge may be lined with white spots. Dark green individuals sometimes have crowded, small, irregular white spots. Rhinophores are ground color below, white-blotched grayish above; the apical rim white-spotted, sometimes reddish subapically. Eyes black. Usually on patch reefs in shallow turbulent water on or near algae, or in debris and under rocks. As it feeds on *Halimeda* spp. green algae its color and shape mimic the algae leaf (hence "*mimetica*"), rendering it almost invisible. Sucks nutrients and chloroplasts from cells; stores chloroplasts in its digestive tract to enable energy production by photosynthesis. As it feeds it creates white serpentine tracks 1 mm wide. Spawn is a tight yellow or white spiral of ~12 whorls deposited on the algae. Depth: typically to ~15 feet, but as deep as 330 feet. Amphiatlantic. Bermuda, Florida, Bahamas, Antilles, Costa Rica to Yucatan. Also Brazil (Bahia, Rio). Entirely white-backed animals don't match published descriptions but are probably this species. *B. curasoae* Er. Marcus & Ev. Marcus, 1970, which overlaps *B. mimetica* in appearance is regarded by some taxonomists as the same species. Molecular comparisons are needed.

Guadeloupe

Roatan

Guadeloupe

Cozumel

Grand Cayman

Elysia annedupontae **Ortea, Caballer, Moro & Espinosa, 2005** **Anne's elysia**

Size: to 20 mm Plakobranchidae

ID: Body elongate, slender, tapers to a short triangular tail (or none); the body and head with short conical papillae, often as a median row on the head; usually **one large white papilla between the eyes**. Rhinophores short, rolled, stout, blunt, many rounded papillae. Parapodia thick, partly open, their wavy edges producing siphonal openings; **edges vertically scalloped due to evenly-spaced erect 3-pointed papillae**; outer faces with variously-sized papillae. Renopericardial complex has a small globose pericardium followed by a straight to sinuous extension not reaching midback. Eyes large, well behind rhinophoral bases. Ground color whitish; **papillae white**; usually a scattering of **brown, orange, or red dots**, sometimes with dark stripes. Head and neck white to light brown; sometimes a white patch behind the rhinophores between 2 parallel diffuse brown stripes from front of pericardium to sides of rhinophores; or a blue patch behind/below each eye with a green bar beneath it. Dorsum with iridescent **blue, green, or reddish-orange speckles**, and larger **spots of dark green, brown, or black**. Parapodial outer surface pale to dark green, tan, or brown proximally, grading distally to white (juveniles) or brown (adults); usually some scattered red/orange flecks and **evenly-spaced thin bars of red dots crossing the margin**. Parapodial inner face greenish with blue and brown spots and white papillae. Rhinophores mostly white; a prominent orange-brown band at end of proximal third; sometimes a paler one in distal third. The renopericardial complex has: orange-brown dots; white pericardium; the radiating vessels of its clear renopericardial extension often with **iridescent blue-green speckles**. Feeds on *Penicillus* green algae. When alarmed it swims by flapping its parapodia. Sometimes mates in groups, the insemination occurring hypodermically by a thin flexible penis that can extend half a body length. Spawn is a spiral coil of ~5 whorls containing white egg capsules. Depth: to at least 15 feet. Bahamas, Florida, U.S. Virgin Islands, Guadeloupe, Dominica, Roatan, Cozumel, Grand Cayman, Jamaica. The similar *Elysia taino* Krug, Vendetti & A. Valdés, 2016 is currently listed in WoRMS as a synonym.

Destin, Florida

Riviera Beach, Florida

Riviera Beach, Florida

Bocas del Toro, Panama

Martinique

Guadeloupe

Elysia canguzua **Er. Marcus, 1955** **bluespot elysia**

Size: to 40 mm Plakobranchidae

ID: Body slender, tapering; tail long and pointed; flanks may have small rounded papillae; neck broad and thick; **small swelling at the back of the head** between the eyespots; pericardium round, a short tapering renopericardium behind it. Parapodia triangular, broadest and separated at the front; edges thick, smooth; broad flaps fold up over the back to form a large anterior siphonal opening and a smaller one at midbody. Rhinophores are relatively short, rolled, tips blunt. Eyes small, behind rhinophores. **A prominent anal papilla at front edge of right parapodium.** Ground color translucent gray. All papillae white. Except for the renopericardium and most of the inner parapodial faces, all surfaces have crowded **dark-to-olive green digestive tract diverticula, orange/red dots, and scattered iridescent blue spots**. Varying concentrations of dots produce different overall colors, e.g. greenish, shades of brown, maroon. Rhinophores are white-spotted distally, apical rim black. Inner parapodial faces have white blotches or pale blue dots and some dark green areas. Head and outer parapodial faces may have white spots. Renopericardium yellowish-white with orange/red dots. **Parapodial edge from middle of front siphon to rear of second siphon forms a thick white band** that may have a narrow dark green margin with transverse white dashes. Eyes black, in unpigmented areas. Inhabits algae on coral reefs or nearby sandy areas. Preferred hosts are *Codium* spp. and *Bryopsis plumosa*, but sometimes *Caulerpa* spp. Usually opens its parapodia when at rest. It does not swim when disturbed. Spawn is an irregularly-shaped gelatinous mass with white eggs. Depth: to ~10 feet. Reported from Southeast Florida, Dry Tortugas, Guadeloupe, Martinique, Carriacou, Brazil, Panama, Costa Rica, Mexico, Jamaica. Rare.

Riviera Beach, Florida

Elysia catulus **(A. A. Gould, 1870)** **eelgrass slug**
Size: to 9 mm Plakobranchidae
ID: Body narrowly ovate; head large, globose, neck slightly constricted; foot narrower than body, front squared, tapers to elongate bluntly pointed tail. Parapodia about two thirds the body length, fold up and almost meet above back; **a prominent frilly-tipped tubercle at inner edge of each parapodium just behind midbody**; back with fine longitudinal folds. Rhinophores short, rolled, broad, tips blunt. **Dorsal pattern of digestive tract has a median longitudinal tube extending back from pericardium, with several side branches**. Ground color green from ingested chlorophyll, but usually much black pigment, especially on head and tail. Several white areas – tips of rhinophores, blotch behind/between rhinophores, marks on pericardium, dots on parapodia, often a diagonal patch from rhinophore's outer edge half-way to dorsal median line. Parapodial tubercles white-dotted, tips white. Inhabits grassy areas in coastal waters, including estuaries; feeds on eelgrasses (*Zostera* spp.). Adults suck sap from flat areas of the grass blades; juveniles hide in leaves' rolled edges. Spawn is a spiral coil of eggs usually attached to the distal one third of an eelgrass blade. Depth: to at least 15 feet. Nova Scotia to South Carolina, Bermuda, Southeast Florida. Catulus is the Latin word for "kitten"– the slug's round head and short "ears" reminded Gould of a kitten's head. *Elysia serca* (see below) is very similar but has a different branching pattern of its digestive tract.

←Guadeloupe Roatan→

Elysia serca **Er. Marcus, 1955** **seagrass elysia**
Size: to 8 mm Plakobranchidae
ID: Viewed from above: broad head, pinched-in neck, body tapers to an elongate tail with a broadly rounded tip. Rhinophores short, rolled, tips rounded or slightly pointed on one side. Parapodia narrow, margin thick, front edges separated and considerably behind head, end before tail. Pericardium large, raised, rounded; renopericardium elongate, tapering, up to half the body length. Foot and parapodia merge without clear demarcation. Eyes prominent, well behind rhinophores. **Dorsal digestive tract has a longitudinal tube extending back from the pericardium's left rear and right rear, each tube with side branches directed outwards**. Greenish-brown to dark brown, rarely blackish; sometimes with **purplish speckles**; patches of white dots on parapodial outer faces and top/front of head; the head is sometimes entirely white. Parapodial margins usually with a prominent **large white patch at midlength**; blackish variants may have a row of white patches. Rhinophores, tail, and pericardium densely white-spotted; **sometimes a purplish blotch just behind the pericardium**. Eyes black, in pale patches. Feeds suctorially on epiphyte-free young leaves of the algae *Halodule wrightii*, *Thalassia testudinum*, and (preferentially) *Halophila engelmanni*. Larvae develop planktotrophically. Depth: tat least 6 feet. Florida, Bahamas, Cuba, U.S. Virgin Islands, Guadeloupe, Barbados, Curacao, Venezuela, Roatan, Brazil.

Roatan

Elysia chlorotica **A. A. Gould, 1870** **emerald elysia**

Size: 20-30 mm, but may reach 60 mm Plakobranchidae

ID: Body elongate and smooth; its prominent neck extends back to its round to ovoid pericardium, which precedes a short and broader renopericardium. Its short pointed oral tentacles extend laterally from the sides of the head. The rhinophores are simple, moderately long, and taper to a point. The parapodia are broad with smooth margins and are usually folded up over the dorsum, but they are often open and flattened; when they are fully open and viewed from above they form an oval having rounded front corners and mildly curved sides tapering to the short triangular tail. Each of the inner parapodial surfaces has a wide longitudinal renopericardial digestive gland vessel that extends to the rear with about 6-9 side branches that also branch near the parapodial margin. Ground color deep emerald green. **The outer parapodial surfaces have small white, light blue, and red speckles**; some of the white speckles coalesce into larger and somewhat regularly distributed white spots; the margins are pale yellow; the vessels are transparent and sometimes speckled with white or red. The head has some smaller white spots. The rhinophores and oral tentacles are white-tipped, on the pericardium is white with red speckles, and the renopericardium is white. It feeds on *Vaucheria* species of algae. When resting it often spreads its parapodia flat on a substrate. The only *Elysia* that can harbor chloroplasts in the cells of its digestive diverticula (and thus remain green) for over a month in the absence of food. Depth: to ~2 feet. Nova Scotia to Florida.

Tampa Bay, Florida

Guadeloupe

← Bermuda Roatan →

Elysia cornigera **T. R. Nuttall, 1989** **horned elysia**

Size: to 8 mm Plakobranchidae

ID: Elongate, stocky, warty-looking (papillate). Parapodia are roughly triangular, broadest at the front, anteriors well-separated; they fold up to form a **prominent vertical mid-dorsal projection;** there is also a smaller posterior one in larger individuals; they taper to form a short rounded tail lobe; edges look "scalloped" due to projecting papillae; rows of papillae on outer surface. **Rhinophores long and knobby-looking**, widen slightly at middle, taper to a point; often point forward like a bull's horns. Renopericardial complex starts near front of parapodia; round pericardium in front of a short renopericardium. Small granules everywhere give it a **white/grayish ground color**. All papillae white. **Green digestive diverticula** numerous on inner parapodial faces; occasionally produce greenish cast on outer faces; diverticula also in sides of head and neck, producing green scrawls/patches. **Small red spots** everywhere, most crowded on the head and parapodial outer faces. Some **larger orange spots** on parapodial outer faces and sides of foot. Rhinophores white with red spots; yellowish-green internally due to diverticula. Pericardium white with red spots. Feeds on the algae *Acetabularia crenulata* and *Acetabularia penicillus*, secreting live chloroplasts and harnessing them for nutrient production via photosynthesis. However, unlike typical "solar-powered slugs" that retain chloroplasts for 3-4 weeks, it digests them quickly, losing their benefits in only ~11 days. Doesn't swim when disturbed; may cover its head with its parapodia. Spawn is a pale yellow gelatinous spiral ribbon. Depth: to ~10 feet. Bermuda, Southern Florida, Bahamas, Cuba, Saba, Guadeloupe,Roatan,Cayman Islands, Jamaica, Yucatan.

↓Virgin Islands

St. Vincent↑

↓ Bonaire

Utila↑

↓St. Vincent

Utila↑

↓Florida "clarki"

Roatan↑

Elysia crispata **Mörch, 1863** — **lettuce slug**

Size: to 150 mm — Plakobranchidae

ID: Elongate, large, smooth. Rhinophores short, wide, rolled. Parapodia broad, margins thin; morphology correlates with habitat: the dark "**clarki ecotype**" is found in **low-light low-flow habitats** (e.g. mangrove lagoons); it has few and shallow ruffles that leave most of the dorsum uncovered; the "**crispata ecotype**" predominant in **high-light high-flow habitats** (patch reefs, tidal channels) has many complex and deep ruffles that cover the entire dorsum. The external coloration is very diverse – creamy-white with white spots and green patches, uniformly dark green, purple with white spots, completely blue. The head is usually green with scattered variably sized white spots and/or iridescent blue specks; the parapodial edge typically has a thin white line, a darker green/gray submarginal band, and a pale yellow/orange line that is sometimes flanked with neon blue but occasionally is just an orange marginal stripe. Rhinophores match the head/back pattern or are paler and unspotted. Foot color is also habitat-dependent: pale cream (crispata); green with pale/white spots (clarki). Inhabits reefs and seagrass beds; usually alone, resting or crawling on rocks, sediment, coral rubble, coral, and algae, often with parapodia splayed. Feeds on a broad range of algae species – *Derbesia tenuissima*, species of *Bryopsis*, *Penicillus*, *Halimeda*, *Acetabularia*, but probably not *Caulerpa* species. Deposits egg masses on vertical flat surfaces of algae as a gelatinous spiral coil with white eggs. Depth: 0-60 feet. Florida, Bahamas, Caribbean, Mexico (Yucatan). Common.

Cayman Brac

Elysia deborahae **Ortea, Espinosa & Moro, 2005**

Size: to 4 mm Plakobranchidae

ID: The body is elongate, 2-3 conical papillae on the flanks; renopericardial complex globose anteriorly, elongate at rear, extends back past the center of the body. Rhinophores elongate, rolled, papillate. Parapodia thick; edges straight, meeting only at midbody, a spherical inclusion at middle rear of inner face. Eyes prominent behind rhinophoral bases. **Body orange-brown with scattered bright blue dots and microscopic yellow granules.** Neck and head white or orange; a strip of blue spots passes between the eyes and extends onto the snout; 2 orange lines extend back from eyes to parapodia and bracket several branched green lines that penetrate the inner parapodial faces. Pericardium pigmented with orange and white; dorsal vessels orange. Lateral papillae white with blue-ringed bases. Parapodia with brown pigment along entire length or in segments; where they meet, each parapodium has a prominent olive-green stripe with a bright blue spot on its front half and white pigment on its rear half. Foot sole orange with bright blue spots at lateral edges and front corners. Rhinophores uniformly reddish-brown, papillae white. Has been found on *Lobophora* brown algae and on *Halimeda* spp. Strongly grips a substrate using a sticky mucous secreted by its foot sole. When it detaches it *swims* by slowly flapping its parapodia. Depth: to 115 feet. Cuba, Cayman Brac, maybe Bahamas. The image from Cuba is from the paper by Ortea et al (2005), so it is definitely *E. deborahae*. Currently, it's the only confirmed *in situ* photo for this species. The Cayman Brac animal resembles it but does not display all of the details mentioned in the description. Moreover, it has very pronounced white marginal patches on the parapodia, which were not mentioned. That animal is either an intraspecific variant or some other species. The uncertainty could be removed by additional anatomical and molecular analyses.

Cuba

Punta Uva, Costa Rica

Elysia diomedea **(Bergh, 1894) Diomedes' sapsucker**

Size: to 35 mm Plakobranchidae

ID: The body is elongate, large, and smooth; the head relatively small; rhinophores short, broad, rolled; parapodia broad, margins thin and ruffled. Ground color olive green, dark on the sides, light on the back. Dorsum and flanks have bright blue streaks and dashes. Parapodial edges are yellow to reddish. Rhinophores externally dark olive green, nearly black, with white longitudinal stripes; internally black. Usually found in shallow beds of *Padina* and *Codium* algae where it feeds by extracting sap from algal cells. Like its congener *Elysia crispata*, which also has very ruffled parapodial edges, it exhibits kleptoplasty – the retention of algal cell chloroplasts for in-body production of essential nutrients via photosynthesis. Depth: shallows to 100 feet. This is a common Eastern Pacific species that was recently recorded from the Caribbean coast of Costa Rica. Whether or not there is an established population remains to be seen.

Elysia entredosaguas **Ortea & Bacallado, 2014**

Tulum elysia

Size: to 65 mm Plakobranchidae

ID: The elongate body tapers to a **short triangular tail**. Broad parapodia begin well-separated well behind the relatively short and rolled rhinophores. The head, parapodia, and rhinophores have scattered rows of **short to long conical papillae**; parapodial faces are sometimes smooth or slightly bumpy, their edges thickened, papillate, and ruffled. Ground color highly variable: **yellow, shades of green, rarely reddish**; probably food-dependent. Papillae white, usually with basal white patches. The head and body are usually peppered with black and brown dots; front edge of head brownish, the short head papillae usually have carmine pigment; often a **white, gray, or tan patch from between rhinophores to pericardium**, usually flanked by white streaks from rhinophores to inner edges of eyes. Parapodial inner faces may have white speckles/papillae; outer faces are variably white-mottled and black-spotted; margin tan to dark brown; **parapodial edge with very thin black line**. Rhinophores tan/lavender/dark brown, tips or entire rear face whitish; rim of pit may be edged with black. Pericardium white to brown. Inhabits shallow beds of *Caulerpa* seagrasses (e.g. *Caulerpa prolifera*), where it sucks out their cell contents using a circular "suction cup" that leaves a grid of circles behind it as it moves. When stressed it exudes a milky white substance. Spawn is a ribbon of capsule-encased orange eggs deposited on a leaf as a spiral coil of up to 5 whorls. Depth: typically < 2 feet. Tulum and Isla Mujeres (Mexico); Banana River and Indian River Lagoon (Florida), Grenada, Belize, Jamaica. This is *Elysia* sp. 5 in the book *Caribbean Sea Slugs* (Valdés et al, 2006).

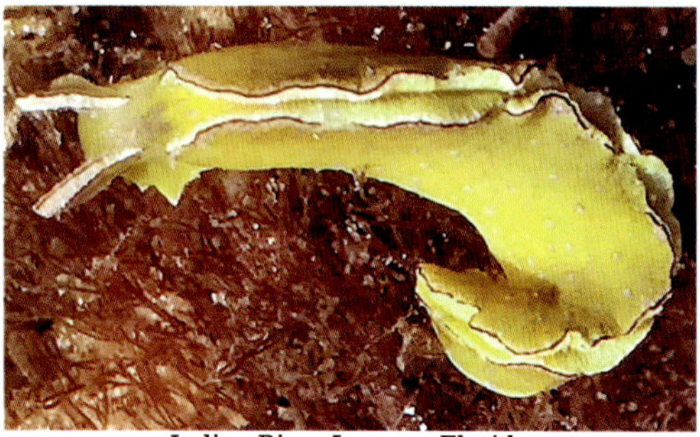

Indian River Lagoon, Florida

Grand Cayman

Guadeloupe

Guadeloupe

Roatan

Cozumel

Elysia ellenae **Ortea, Espinosa & Caballer, 2013** **Ellen's elysia**

Size: to 12 mm Plakobranchidae

ID: Body elongate, slender; distinct head & neck. Parapodia broad, thick; fused at rear, small gap at front through which front of pericardium protrudes; **parapodial margins slightly undulating, flap-like outward extensions produce 3-4 siphonal openings** when parapodia fold up to meet, **parapodial edge thickened and with regularly spaced papillae**. Parapodial inner face and top half of outer face with small dot-like papillae, lower half of outer face with some large papillae. **Rhinophores papillate**, cylindrical, rolled, relatively short, slightly flared apically. **Pericardium large, round, swollen**, a bulge that fills the front siphonal opening; pericardial extension reaches one third of the body length. Ground color usually whitish, sometimes greenish. Papillae white. **Dark green digestive tract diverticula form a variably dense network**, especially on sides of head, body, parapodial inner faces, lower half of parapodial outer faces. Pericardium white, renopericardial extension grayish. Head and pericardium may be crowded with dark brown-green dots. Parapodial outer face uniformly red-dotted; **margin usually dark red** with evenly-spaced yellow spots; submargin faintly iridescent blue. Parapodial inner face with **prominent iridescent light blue submarginal band that is broadest at the siphonal openings**; may also have red dots. Rhinophores with green diverticula on front surface, red blotch/band half-way up, **red patch or arc at base. Pale blue blotch/streak behind red arc at inner edge of eye**, sometimes extending partly onto back of rhinophore. Has been observed on mixed algae and algae-covered rocks; algal food source unknown. When resting, it opens and flattens its parapodia, giving it the appearance of a heart-shaped leaf. Depth: 3-25 feet. Bahamas, Saba, Guadeloupe, Roatan, Cozumel, Grand Cayman. Rare.

Vero Beach, Florida

Puerto Vargas, Costa Rica

Elysia evelinae **Er. Marcus, 1957**

Size: to 8 mm Plakobranchidae

ID: Body short and wide; head long; rhinophores short and wide; eyes well back of the rhinophores; pericardium large and rounded; parapodia short and thick, do not cover the pericardium; tail short and conical; anal papilla dorsally on the right side of the body at the base of the head. Ground color pale cream to brownish gray; head or entire body covered with many red and opaque white spots; sometimes with dark brown to black pigmentation at back of head. Rhinophores are mottled pale cream. Eyes large and black. Usually a dark stripe from each rhinophore to the front of the parapodium, sometimes passing through the eye. A **prominent black ring at base of anal papilla**. Highly cryptic. The only known *Elysia* that feeds suctorially on diatoms. Accordingly, it inhabits shallow, shaded, sheltered areas (e.g. tidal channels) conducive to the formation of dense mats of both benthic diatoms (*Biddulphia* spp.) and epiphytic diatoms attached to *Bryopsis* and *Caulerpa* algae. Populations in Florida have been observed to peak seasonally (spring) when cooler flowing water carrying higher levels of nutrients stimulates the production of diatom mats. Its low reproductive output combined with loss of habitat due to siltation of tidal channels has driven down the Florida population to the point of "very rare", perhaps verging on extinction. Depth: to ~5 feet. Florida (Key Largo, Key Biscayne, Vero Beach), Costa Rica, Brazil. Rare.

↓Bermuda

Roatan↑

↓Guadeloupe

Bonaire↑

Elysia flava **A. E. Verrill, 1901** **white-edge elysia**

Size: to 20 mm Plakobranchidae

ID: Body short, wide, somewhat elevated; head small, about one-fourth body length; tail short, triangular, pointed. Parapodia tall, broad, thick; edges wavy when the animal is still, straight when it is moving; a few small conical papillae on the faces, many along the edges. Rhinophores are short, thick, rolled, smooth, and blunt-tipped. Pericardium short, sends off two main digestive tract vessels (one per side) that arc back almost to the body's rear while branching toward the parapodial edges. Ground color usually translucent yellowish to yellowish-orange. Head and rhinophores have a few scattered white spots. Each parapodial edge has a thin white stripe; all papillae white; parapodial faces usually greenish with gray to black blotches due to visible branches (in adults) of the digestive tract. Eyes prominently black. No published data are available regarding host algae or larval development in the tropical West Atlantic. In the Mediterranean it is suspected to feed on *Cladophora* or *Bryopsis* algae. Usually under rocks during the day. Depth: to at least 10 feet. Amphiatlantic. Bermuda, Southeast Florida, Antilles, Venezuela to Mexico, Brazil. Indo-Pacific records were recently shown to be for the similar *E. obtusa*. The red animal shown above seems to be a red morph, but confirmation is needed.

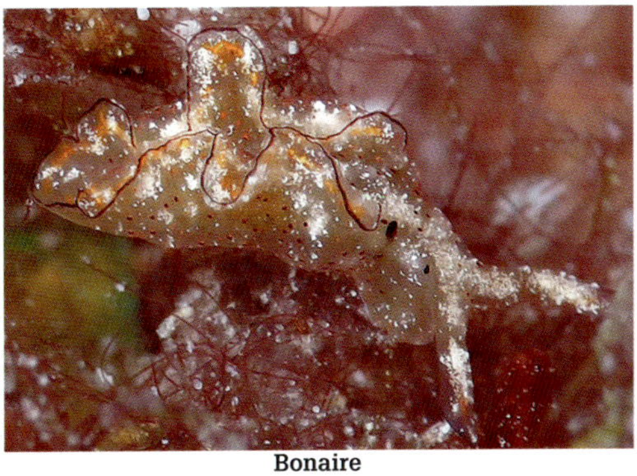

Bonaire

Roatan

Guadeloupe

Curacao

Curacao

Guadeloupe

Elysia jibacoaensis **Ortea, Moro, Caballer & Espinosa, 2011** **Jibacoa elysia**

Size: to 12 mm Plakobranchidae

ID: Elongate, slender, small conical rounded papillae on sides. Rhinophores long, rolled, taper somewhat to apex, a few scattered low papillae. Parapodia long; two pairs of flaps fold up over back to form siphonal openings – one tall and wide at midlength, one short and narrow at rear; parapodial edges very thin, weakly scalloped with similarly-sized well-spaced small papillae. Pericardium oval. Tail short, rounded, an extension of the parapodia. Ground color **pale olive green**; sometimes a pale orange cast; **reddish spots regularly distributed everywhere** except on rhinophores. Sides of body with white dots and low, conical, rounded white papillae. An elongate narrow triangular white patch extends back from between rhinophores, flanked by red dots. Pericardium also has white dots and patches. External parapodial faces variously white-blotched, more so dorsally where they may also have orange-brown highlights; internal faces paler green with white and red spots. Parapodial edge papillae are clear with a white internal granule, either on or flanking a **thin black marginal line** above a broader **fragmented orange line**. Rhinophores have a few light green papillae and a large white spot with orange tinges at front center; **entire edge of unrolled rhinophore with a narrow black margin**, making the rolled rhinophore appear to have a thin vertical black line and a black apex. It advances by repetitively distending the front part of its foot beyond its head and then retracting it, while simultaneously opening and closing its parapodia. When at rest it unfolds its parapodia, shrugs back its rhinophores, and flattens itself. It does not swim but may "hang" from the water's surface with its tail bent up and forward. Known to associate with *Penicillus* spp. green algae, presumably feeding on them. Depth: to at least 50 feet. Southeast Florida, Bermuda, Cuba, Les Saintes (Guadeloupe), Martinique, Bonaire, Curacao, Roatan.

←Roatan Bermuda→

Florida

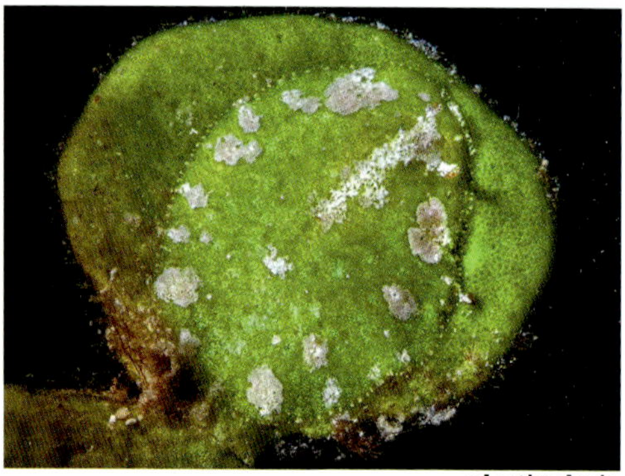

Elysia marcusi **(Ev. Marcus, 1972)** **plastic elysia**

Size: to 5 mm Plakobranchidae

ID: Body smooth, long and thin when moving, but **can change shape** to circular or oval when resting. Head smoothly rounded at front; neck slightly narrowed. Tail short and roundly pointed. Parapodia fused onto back forming a **noticeable median line** from just in front of the tail to somewhat behind the rhinophores where it meets a **furrow that curves diagonally to the right**; the furrow meets the side perpendicularly and continues straight down to the foot. Rhinophores are simple, flattened, fully retractable. Small eyes behind the rhinophores, widely-spaced, more "longish" than round. Light to dark green, either uniformly or with scattered white blotches. Often a prominent **elongate white patch at rear of midline groove**. Color variations: a white patch on head between/behind the rhinophores, scattered iridescent dark blue spots, clusters of white patches along the dorsal midline, body edge with whitish-blue spots or short transverse bars. **Rhinophores white everywhere or just near the apex**. Eyes black. Inhabits and feeds on *Halimeda opuntia* in shallow shaded areas (Florida); also seen on clumps of *Halimeda gorveaui* under ledges and in shaded stands of dead coral. Copulation is head to tail for pairs, head to tail in a circle for threesomes. Deposits a gelatinous coil containing white or yellow egg capsules with similarly-colored blobs of extra-capsular yolk attached to them. Larvae hatch in ~14 days. Depth: to at least 60 feet. Florida Keys, Bahamas, Bermuda, Cuba, Jamaica, Costa Rica, Roatan, Mexico. (See the similar *Bosellia mimetica*.)

Roatan

Elysia **cf.** ***marcusi*** **(Ev. Marcus, 1972)**

Size: to 5 mm Plakobranchidae

ID: This may be *Elysia marcusi* but the rhinophores are long, slender, and rolled, instead of somewhat short and flat as pictured in Krug et al (2016). The description of *E. marcusi* in that paper is the most reliable one currently available, based on an examination of 152 specimens from the Bahamas, Jamaica, and Florida.

Bermuda

Elysia nisbeti **T. E. Thompson, 1977**

Size: to 10 mm

Plakobranchidae

ID: Very elongate, slender, tapering to a sharp point at rear, covered with small rounded papillae. Parapodia narrow, front edges well separated; form 2 siphonal openings when folded up; outer surfaces with longitudinal rows of variously sized low papillae. Parapodial margins thick, fleshy, with conspicuous blister-like swellings along their inner edges. Rhinophores moderately long, rolled, papillate, tips truncate. Ground color rich green. Papillae and parapodial marginal blisters are white. Lower sides are green, shading to a pale yellowish-green foot and a reddish-brown parapodial margin. **Top of head has a Y-shaped pinkish blotch** formed by a broad median line extending forward from the parapodia and splitting into two branches that continue up the backs of the rhinophores almost to their white tips. It has been found on the green alga *Caulerpa verticillata* among mangroves and on a species of Rhodophyta red algae. Does not swim. Sometimes arches its rhinophores back over its dorsum and holds them in that position. Depth: 15-65 feet. Jamaica, Bermuda.

Utila

Little Cayman

feeding; Roatan

4 mm juvenile; Roatan

↓Grand Cayman

Roatan↑

Elysia pratensis **Ortea & Espinosa, 1996**

lined elysia

Size: to 60 mm

Plakobranchidae

ID: Elongate, narrow; head squarish; tail short, triangular, tip rounded. Parapodia somewhat narrow, usually folded up to meet over the back, fronts well separated, large swellings (papillae) straddling their edges. Pericardium small; renopericardial extension as long as body. Rhinophores smooth, rolled, sometimes a few papillae distally; usually inclined forward. Ground color green to yellowish-green to brownish. **Thin crowded white stripes along length of each parapodial outer face** that continue onto sides of head; similar stripes from rear of pericardium to front of head, including rhinophores. **Parapodial edge with thin black line and white swellings**; **outer faces with long white papillae on the white stripes**. Rhinophores with a yellowish-orange cast distally, its "roll" with black edges. Inhabits seagrass beds; usually feeds on blades of *Rhipocephalus phoenix*. It extracts the green chloroplasts from the blade edges and tips, turning them into white-lined mimics of itself. Spawn is a spiral gelatinous coil with an internal flat orange ribbon of yolk weaving its way through the coil and touching each egg capsule. Depth: to 40 feet. Southern Florida, Bahamas, Turks and Caicos, Cuba, Saint Barthélemy, Guadeloupe, Honduras, Belize, Cozumel, Bay of Campeche, Jamaica, Cayman Islands.

St. Vincent

↓Florida Grand Cayman↑

↓Florida Guadeloupe↑

↓Roatan Grand Cayman↑

Elysia ornata **(Swainson, 1840)** **ornate elysia**

Size: to 55 mm Plakobranchidae

ID: Body elongate and smooth. Parapodia long, broad, merge to produce a pointed tail; margins slightly undulating, flap-like outward extensions produce three siphonal openings when the parapodia fold up and meet – a high one at midbody, lower ones ahead and behind. Pericardium small, renopericardial extension short. Rhinophores are short, smooth, rolled, taper to a blunt tip, usually point forward and obliquely out. The front of the foot is rounded, its corners often with recurved tentacles. Ground color usually translucent olive green, occasionally pale green to cream. **Many round black and/or red spots and white dots** on the head, back, and parapodia (both surfaces). The parapodia have a **narrow marginal black band** and a broader more diffuse **yellow/orange submarginal band**. The dorsum has a broad median patch of clear tissue along its entire length, flanked by deep-set white reproductive glands. **Rhinophores are whitish on the proximal half to two-thirds of the rear face; apex and roll edges black; orange to red elsewhere.** Renopericardial complex whitish; some black spots, red/orange dots, and brown flecks. Inhabits shallow reefs, blending in extremely well with green algae that it feeds on (e.g. *Bryopsis plumosa*). Its overall shade of green depends on its food source and digestive gland saturation level – bright green when satiated, very pale green and sometimes tinged with yellow or orange when hungry. Deposits transparent egg capsules in a cylindrical spiral coil (~15 mm across), with 1-6 white eggs per capsule, and regularly spaced large white blobs of extra-capsular yolk along the upper face of the egg mass. Depth: 0-60 feet. Amphiatlantic (reported from Canary Islands and Azores). Bermuda, Florida, Bahamas, Turks and Caicos, Antilles, Venezuela to Bay of Campeche, and Brazil, Common.

Roatan

Grand Cayman

Bocas del Toro, Panama

Guadeloupe

Riviera Beach, Florida

Riviera Beach, Florida

Grand Cayman

Elysia subornata **A. E. Verrill, 1901** **brownlined elysia**

Size: to 70 mm Plakobranchidae

ID: Body elongate, tapers to **short triangular tail**. Parapodia broad, anteriors well separated and well behind rhinophores. Pericardium small, round; renopericardium extends far back as a long tapering tube; ten or more branching renopericardial vessels prominent on upper inner face of each parapodium. Rhinophores relatively short, rolled, papillate. Head, parapodia, and rhinophores with a few to many **short or long conical papillae**; the papillae more abundant and prominent in large individuals, may be indistinct in juveniles. Parapodia may have smooth faces and thickened, ruffled, papillate edges. Ground color highly variable, probably dependent on food type and state of digestion: yellow, olive, dark green, greenish-tan, rarely reddish. **Papillae white**, usually with basal white patches. Head and body peppered with black/brown dots; front edge of head brownish; often a **white, gray, or tan patch from between rhinophores to pericardium**, usually flanked by white streaks from rhinophores to inner edges of eyes. Parapodial inner face often white-speckled, outer face variably white-mottled; **margin tan to dark brown**, sometimes white speckled; **edge with thin black line**. Rhinophores tan/lavender/dark brown, tips white-blotched. Pericardium white to brown, renopericardium whitish/greenish, vessels clear to whitish. Inhabits shallow reefs, grass flats, and mangrove fringes; blends in extremely well with its green algae food source (*Caulerpa* spp.). Doesn't swim when alarmed. Deposits egg capsules in a continuous jelly-filled spiral coil with a bright orange ribbon of yolk winding through it. Depth: 0-15 feet. Bermuda, Florida, Bahamas, most Antilles, Brazil, Panama, to Mexico, Common.

Roatan

Guadeloupe

Key Largo, Florida

Guadeloupe

Elysia papillosa **A. E. Verrill, 1901** **papillose elysia**

Size: to 35 mm Plakobranchidae

ID: Body elongate, tapers to a narrow pointed tail; head large, sometimes a few small papillae, usually 1-2 large ones between the eyes; neck long. Front foot corners are short triangular projections. Rhinophores are elongate, stout, rolled, apices truncate and wide; papillate. Parapodia are relatively low, their anteriors well-separated; do not cover pericardial complex; both faces are papillate, the margin lined with small rounded bumps and a row of protruding three-pointed flaps that (in dorsal view) form two wave-like edges that produce siphonal openings when they meet over the back. The pericardium is near the front of the body, round, with low papillae; the narrower and straight renopericardial extension extends almost to midbody. Its color is predominantly pale green, but ranges from white to tan to olive green. **All papillae are white**. The head is paler green to white on the sides. A whitish/brownish broad median stripe from the front to the pericardium may be flanked by darker streaks that pass through the eyes. The parapodial outer faces have black/brown dots and flecks; the margins may be tan to dark brown; the **edge bumps and protrusions white**; inner faces and back usually mottled with light to dark green, white patches, **iridescent blue/green/orange flecks**, turquoise spots, and brown or black speckles. Pericardium white, brown-spotted; the clear renopericardial extension spotted with orange-brown, faintly speckled with white to light blue. Rhinophores white to tan; a broad (usually oblique) **orange-brown band at or below midlength,** sometimes a fainter one closer to the tip; both bands often obscured by papillae. Has been found on *Caulerpa* spp. and *Halimeda opuntia* algae but its preferred host/food are *Penicillus* spp. **When alarmed it swims by flapping its parapodial margins**. Depth: usually to 15 feet, but as deep as 60 feet. Bermuda, Florida, Bahamas, most Antilles, Venezuela to Yucatan. Common to locally abundant.

St. Vincent

with copepod eggs; Roatan

Riviera Beach, Florida

with copepod eggs; Grand Cayman

Riviera Beach, Florida

juvenile; Grand Cayman

Elysia pawliki **Krug, Vendetti & Á. Valdés, 2016** — **Pawlik's elysia**

Size: to 30 mm — Plakobranchidae

ID: Body elongate, somewhat broad, densely papillate. The rhinophores are relatively short, rolled, blunt, with elongate papillae. The parapodia, when opened and flattened, usually show three laterally-projecting side-flaps – a small front one, large middle one, intermediate rear one – producing a stylized "cross" as wide as a half body length; with the parapodia folded up and touching the flaps produce correspondingly-sized siphonal openings; large individuals may have 2-3 additional flaps at the rear; **thin elongate papillae cover the parapodial outer face and form a row along the entire parapodial edge**. Pericardium large and rounded; the thick renopericardial extension reaches slightly beyond midbody. Eyes very small, behind rhinophoral bases. **All papillae white**. The body and parapodia are usually yellowish-green with brown patches to overall brownish; dark green after feeding. Digestive gland diverticula are visible in the head, body, and parapodia. Long conical papillae on the head and high on the parapodia usually emerge from white patches. The parapodial margin sometimes looks fluffy-white, often with a **narrow brown line** along its inner/outer edges. Head, rhinophores, pericardium, and sides of parapodia are brown-dotted. There are occasional purplish blotches on the parapodia, notably **a large purplish blotch on the outer face near the margin of the front flap**. Front of head with white, tan, and pink patches. Rhinophores white to whitish-brown; sometimes pink/brown spots and white patches; green diverticula are visible at bases of inner surface. Pericardium has a whitish-pink dorsal patch. Eyes black. Lives/feeds on the sea grape alga *Caulerpa racemosa*. The purple eggs of a parasitic copepod are sometimes beside its pericardium. When alarmed it releases mucus but **does not swim**. Depth: to at least 15 feet. Bermuda, Florida, Bahamas, Guadeloupe, St. Vincent, Venezuela, Costa Rica, Roatan, Grand Cayman.

Riviera Beach, Florida

Guadeloupe

Roatan

Bermuda

Riviera Beach, Florida

Guadeloupe

with spawn; Riviera Beach, Florida

Elysia velutinus **Pruvot-Fol, 1947** **velvet elysia**

Size: to 40 mm Plakobranchidae

ID: Body elongate, slender, rear blunt or tapers to short triangular tail. Parapodia smooth or papillate, thick, narrow, begin well behind head; edge slightly undulated, flaps fold up to form an open-ended tunnel that **peaks prominently at midback** with a small siphonal opening. Pericardium large, round. Rhinophores short to moderately long, cylindrical, rolled, distal half with small papillae. **Ground color light to dark green; sometimes yellow**. All papillae white. Scattered **white-to-tan spots and patches; Y-shaped white head and neck marking** (may be tan) formed by patch between pericardium and rhinophoral bases. Rhinophores with **green or brown "band" or mottling near base**, whitish above. Parapodial outer face with white/tan speckles/spots/patches, sometimes with small round white papillae; margin usually with larger irregularly-broad white patches (often papillate); edge may be white or tan. **Back, pericardium, and parapodial inner face with iridescent bluish-green dots**. Usually on/eating *Halimeda* spp. but large individuals are sometimes on other algae. Spawn is an egg coil containing white egg capsules and a bright orange ribbon of extra-capsular yolk that meanders through the mass, touching each capsule. **Bluish-green eggs of a parasitic copepod sometimes attached to back**. Does not swim when alarmed. Depth: 2-100 feet, but is usually less than 10 feet. Bermuda, Florida, Bahamas, Puerto Rico, Saint Martin, Guadeloupe, Barbados, Curacao, Panama to Quintana Roo, Grand Cayman, Jamaica, Brazil. Formerly named *Elysia tuca*. Some of the pictured individuals show considerable spotting with iridescent blue on the head and outer parapodial faces, which is inconsistent with existing descriptions. However, this is likely a previously undocumented intraspecific color variation.

Elysia zemi **Krug, Vendetti & Á. Valdés, 2016** — **orangeleaf elysia**

Size: to 16 mm — Plakobranchidae

ID: Body elongate, narrows to a blunt or slightly pointed rear. Parapodia broad; **3 pairs of rounded flap-like outward expansions that form siphonal openings** when folded up over the back – a small front one over the pericardium, a large midbody one, a posterior large or smaller one; gives animal a leaf-like appearance when parapodia are fully open. Parapodial outer face with a few variable-length conical papillae; both faces with many small rounded bumps; **edge papillae elongate and hair-like**. Body, head, and rhinophoral papillae thin and branched. Rhinophores cylindrical, rolled, usually inclined slightly forward. Pericardium round, renopericardial extension extends to or beyond midbody. Eyes very small, behind the rhinophores. Overall color **greenish-yellow**, **or shades of orange/red/brown**; **papillae white**. Head may have patches of white/pink/gray on top and sides. A grayish-purple patch between the rhinophores; another starts between the eyes, broadens rearward and slightly narrows to near the pericardium. **Pericardium with a white blotch**; renopericardial extension greenish, with white to purplish patches behind it. Parapodial "bumps" are dark reddish-brown, sometimes with white apices; edge papillae brownish and white-spotted; **parapodial margin tan, submarginal band brown/dark-orange/red**. Rhinophores white-spotted. It has been observed near and on *Caulerpa racemosa* algae. Depth: 3-50 feet. Guadeloupe, Martinique, Petit Nevis (St. Vincent and the Grenadines), Venezuela, Roatan, Grand Cayman. Rarely reported.

Guadeloupe

Key Largo, Florida

Roatan

Delray Beach, Florida

Bermuda

juvenile; Roatan

Elysia zuleicae Ortea & Espinosa, 2002

Zuleica's elysia

Plakobranchidae

Size: to 17 mm

ID: Body elongate, the sides smooth to variously papillate. Usually an elongate pointed tail; less usual in juveniles. Parapodia thin, the edges slightly undulating; a row of large papillae along the margin, often in peaked clusters of 3 or 5. Pericardium small and rounded; the renopericardial extension wider, reaches back about one-fourth body length. Rhinophores are cylindrical, long, rolled, tips truncate and slightly flared; some rounded papillae. Ground color olive to dark green to greenish-brown. The head often has an elongate white/brown patch from the rhinophores to the front of the parapodia; the patch may have a yellow/orange smudge. The **back and parapodial inner faces have brown flecks and iridescent blue to white speckles**; **the outer faces are white-specked**; the margin is white to brown; large adults may have a purplish marginal band and a white/brown submarginal band; **some (e.g. juveniles) with a thin black edge-stripe flanked by thick white bands. Parapodial orange streaks also occur. The renopericardial area is white to black with dark speckles; its emanating vessels are white with iridescent blue highlights**. Rhinophores are white to brown to brownish-purple with distal white patches and white papillae. The **tail, when present, usually matches the color of the parapodial margin: if black it unambiguously distinguishes this species from similar-looking ones**. It is usually found on, and eating, the fan alga *Udotea flabellum*. Its spawn is a coiled egg mass containing a white ribbon of extra-capsular yolk that meanders past the egg capsules. **When disturbed it swims vigorously by flapping its parapodia**. Depth: 15-65 feet. Bermuda, South and Southeast Florida, Bahamas, southeast Cuba, Guadeloupe, Panama, Costa Rica, Roatan, Mexico, Grand Cayman.

The identities of the following *Elysia* spp. are uncertain. They all have similarities with known species but also either lack some of their distinctive features or have additional features that have not been described. Such a mix of known and unknown features could be the result of previously unrecognized intraspecies variability, but could also represent the presence of an undescribed species. Future studies should resolve the uncertainties.

Belize

***Elysia* sp. 1**

Size: ~15 mm Plakobranchidae

ID: Body elongate; parapodia broad, elevated, margins convex and unruffled, edges thick and relatively smooth; external faces minutely papillate; rhinophores long, rolled, tips truncate. Parapodial external faces brownish-yellow to pale green, margins brown to reddish-brown; tiny papillae white but may also have groups of larger white, green, to bluish branched papillae. Rhinophores are brown to reddish-brown, with a few white papillae, apices white. Two median white to bluish blotches on the head – a large one on the front of the head and a smaller one between the rhinophores and the eyes. Found crawling on a hard substrate. Depth: to 15 feet.

all images from Roatan

***Elysia* sp. 2**

Size: up to 10 mm Plakobranchidae

ID: Body elongate, tapers to a pointed tail. The parapodia broad, elevated, their front edges well separated; margins are thick and scalloped with some vertical undulations and low rounded papillae; the outer face is moderately smooth with a few variably sized papillae. Rhinophores are long, rolled, tips truncate, a few tiny papillae distally. The ground color is yellowish-orange with minute brownish specks; the papillae white. Small white blotches on the back behind the head. Parapodial edge broadly white or a thin brown line; outer face anteriorly green with a broad white submarginal blotch. Rhinophores brownish with white papillae distally. Eyes black, in pale areas. Crawling on various algae. Depth: to 60 feet.

all photos from Destin, Florida

***Elysia* sp. 3**
Size: ~8 mm Plakobranchidae
ID: This has the color pattern and most of the diagnostic features of *Elysia cornigera* – a short body, long rhinophores, parapodia that peak halfway along the body, parapodial margins somewhat papillate, largely whitish with numerous red spots and several larger orange spots, the digestive diverticula greenish. However, there are some discrepancies: the rhinophoral tips are squared-off instead of pointed; the rhinophores lack noticeable papillae; the pictured animals have tails that are elongate and pointed instead of very short to non-existent. Nevertheless, this is probably a variant of *E. cornigera*.

Saba

***Elysia* sp. 4**
Size: ~8 mm Plakobranchidae
ID: Body elongate and slender; the neck elongate. All surfaces are prominently granular-looking. Rhinophores long, cylindrical, taper slightly to a rounded apex. Parapodia begin considerably behind the rhinophores, fronts well-separated, lateral undulations form 2 pairs of vertical projections – a high one at midbody, a low one further back – each pair flanking an elongate siphonal opening; parapodia taper to a broadly triangular "tail". Head, rhinophores, and parapodial margins are densely covered with white granules; some red/orange dots on the parapodia; a small blue patch on the upper rear of the neck. Found among algae and hydroids on a coral reef. Depth: ~10 feet. The parapodial structure and red/orange dots suggest an *Elysia cornigera* (q.v.) but this animal is proportionally slimmer, especially in the neck region, and the rhinophores are not noticeably "bumpy" with papillae.

***Elysia* sp. 5**
Size: ~8 mm Plakobranchidae
ID: The body is elongate; each parapodium arches to a high peak at midbody; the parapodia and rhinophores are bumpy with small papillae; the parapodia have thick margins and their edges have broad-based conical papillae; the rhinophores are short, rolled, and have truncate tips. The upper half of each parapodial outer face is pinkish-orange with white mottling and a few small purple blotches; the lower half is grayish with more numerous and larger purple blotches; the margin is greenish and the edge papillae are white. Head spotted with brown and white, eyes in greenish patches. Rhinophores are greenish with white papillae, and have a wide brown band at midlength. Found on rubble. Depth: ~10 feet.

Puerto Rico

Elysia sp. 6

Size: ~4 mm Plakobranchidae

ID: Body elongate; tail broadly triangular, the tip rounded. Most surfaces with small rounded tubercles, large on rhinophores. Parapodia broad, edges smooth, undulating margins create three siphonal openings, middle highest. Rhinophores long, rolled, tips truncate. Ground color greenish. Body, parapodia, and the tail with many orange dots, some of them blue. Rhinophores, top of head, upper parapodia, much of the tail, and all tubercles opaque white. The parapodial edge has a thin dark orange-brown line. The tail and the inner parapodial faces with some orange blotches. Found on fuzzy algae attached to the side of a large rock. Depth: ~12 feet. So far it has been reported only from Southeast Florida. This is probably a variant (juvenile?) of *Elysia jibacoaensis*.

4 mm; Pompano Beach, Florida

Colon Island, Panama

Elysia sp. 7

Size: ~10 mm Plakobranchidae

ID: Body long and slender; tail very short. All surfaces with small rounded tubercles. Parapodia are long, low, held open; margin/edge with a few prominently larger conical tubercles. Rhinophores are long, rolled, tips truncate. Ground color is pale green with a dense distribution of black dots. Papillae are white. Sides of the head dark; a median light brown patch behind rhinophores. Parapodia mottled dark green to blackish below; the margin pale brown with dark spots. Dorsum between the parapodia ground color with white and blue small patches. Rhinophores pale brown with a broad dark brown band at midlength. On encrusting green algae in shallow water.

Grand Cayman

Guadeloupe

Grand Cayman

Elysia sp. 8

Size: ~5 mm Plakobranchidae

ID: Body elongate, slender; rhinophores long, cylindrical, rolled, tips flared; parapodia begin well back of the rhinophores, widely separated anteriorly, long, broad, convex (no undulations), taper to form a short triangular tail. A few small rounded tubercles on the rhinophores, parapodial outer faces, and along parapodial edges. Eyes prominent, widely spaced, set back from rhinophoral bases. The ground color is grayish-white. Body and head slightly greenish, densely covered with white speckles. Parapodia with white tubercles and a submarginal row of reddish-brown spots (visible from either face). The rhinophores have a reddish brown distal third and white tubercles. Eyes black. Found crawling on algae leaves. Depth: to 56 feet.

Guadeloupe

***Elysia* sp. 9**
Size: ~9 mm Plakobranchidae
ID: The body is elongate and slender. The rhinophores are short, stout, conical, and rolled; apices truncate and constricted; with small rounded tubercles on their distal half. Parapodia widely separated in front, long and broad, their edges thin, a midlength undulation creates a siphon; they taper to a broadly triangular very short tail; there are some small tubercles along their edges and some on the faces. Pale greenish with a scattering of reddish-brown dots; all tubercles are white. Rhinophores are reddish-brown on the distal half. Found crawling on algae leaves at a depth of ~15 feet.

***Elysia* sp. 10**
Size: ~5 mm Plakobranchidae
ID: Body narrow, elongate; rhinophores relatively short, rolled, apices truncate; parapodia moderately broad with thick margins, anteriors widely separated, taper to a short broadly triangular rounded tail. Some small rounded papillae on the rhinophores, some also concentrated along the parapodial margins. Eyes large, well behind the rhinophores. Pale green, crowded with reddish-brown dots; the papillae white. Rhinophores are brownish with some reddish-brown spots and whitish tips. Parapodial margins are pale with some brown areas. Indistinct white Y-shaped patch on the head – its stem median from the front of parapodia to between the eyes, its forks extending to the bases of the rhinophores. Eyes black in pale areas. Found in samples of *Caulerpa* collected from shallows with coral rubble, seagrasses, and other algae. Depth: less than 3 feet.

Bocas del Toro, Panama

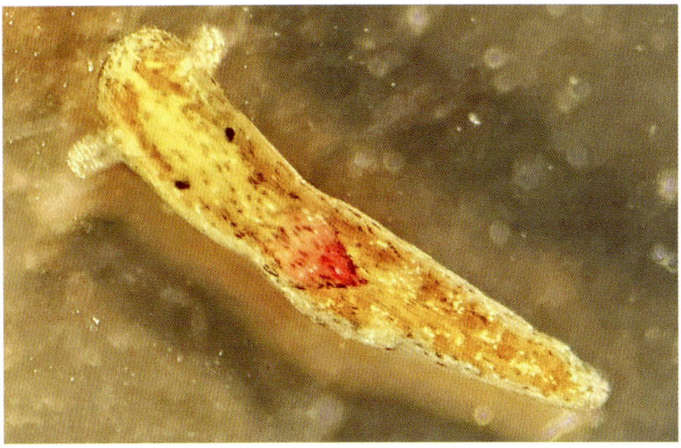

Vero Beach, Florida

***Elysia* sp. 11**
Size: ~8 mm Plakobranchidae
ID: Short, moderately wide; head long, front rounded; **rhinophores short**, wide; eyes well back of rhinophores; parapodia short, narrow, begin well-separated at midlength. Tail short, broadly rounded. Yellowish. Upper surface with numerous reddish-brown dots and patches, and small white papillae; flanks less marked. Each side of back with a narrow reddish-brown irregular streak that starts in front of the rhinophore, grazes its inner base, passes through the black eye, and continues toward the rear. The triangular opening at the front of the parapodia frames a **median red patch**. The rhinophores have a patch of white papillae distally. Its external morphology is similar to that of *Elysia evelinae*, which was also found in the same location. This animal, however, does not have a black ring on its top right side. Shallow.

***Elysia* sp. 12**
Size: ~10 mm Plakobranchidae
ID: This has the same morphology as *Elysia ellenae* and also some of its color pattern: a brick-red spot at the midlength on the rhinophores; the inner parapodial face green, the outer face with many minute red spots, the parapodial margin with a hint of red spotting, the head white. However, the more conclusive color features of a blue patch and a curved red patch in front of the eyes are not visible. In addition, the animal has a greenish pericardium and parapodial outer faces with a greenish base color, as well as many iridescent blue specks low on the outer face of the parapodium (none of which has been previously mentioned in the scientific literature). Nevertheless, it is probably *Elysia ellenae*. It was found at Puerto Rico.

Aquadilla, Puerto Rico

Key Largo, Florida

Riviera Beach, Florida

Martinique

Guadeloupe

Guadeloupe

Thuridilla malaquita Ortea & Buske, 2014 **malachite slug**

Size: to 40 mm Plakobranchidae

ID: Body very long and slender with many variably sized tiny warts distributed randomly over its surface. Parapodia are usually folded up, the wavy edges vertical and in contact, the outer surface with a submarginal row of large well-spaced conical warts. Rhinophores are rolled into a nearly cylindrical hollow tube, as long as or longer than a body width; the front of the foot is broadly rounded, visible between the front of the mantle and the bases of the rhinophores; the rear of the foot does not protrude as a tail. Eyes are very small, behind the bases of the rhinophores. The ground color of the mantle is malachite green ("malaquita") to yellowish-green to pale-brown to purplish (smallest juveniles), with **white warts**. The back is whitish with orange patches; **parapodia with a broad pale orange edge** and a broader **white submarginal band containing yellowish warts**. Rhinophores are greenish with white flecks below, purple tips, orange and white patches between; orange spot at each inner base. Anterior of foot whitish with some orange patches above, orange below, the rest ground color. Eyes black. **A distinctive black wedge originates behind each eye and widens posteriorly**. Found on rocky bottoms, with and without algae. It is so slender that, when crawling, it can almost reverse direction by turning its front half in the new direction of travel while its rear half remains attached to the substrate. Depth: to ~20 feet. Southern Florida, Saba, Guadeloupe, Martinique, Grenada, Venezuela, Colombia, Costa Rica. Rare.

Guadeloupe

↑Roatan juvenile, Roatan↓

Thuridilla mazda Ortea & Espinosa, 2000 **mazda slug**
Size: to 17 mm Plakobranchidae
ID: Body elongate, slender, tapers to a short pointed tail; head and neck rather long. **Parapodia broad, reach the tip of the tail, edges slightly sinuous** with ~6-7 undulations; **usually folded up to meet medianly above the back**. Rhinophores short, broad-based, distinctly separated, usually inclined forward and sideways. Eyes are small, behind the rhinophoral bases. **Ground color dark reddish-brown to black. Parapodial margin narrow, white to pale yellow to orange; submarginal with a row of small and large half-moon-shaped patches that project downwards**, the largest ones corresponding to valleys in the edge undulations. **Color patterns of the half-moon patches can vary:** four crescent-shaped stripes – yellow, orange, blue, orange (top to bottom); or white, orange, black, and blue; or pale yellow followed by orange. The interior (dorsal) parapodial surface is black with blue lines posteriorly and some orange spots near the middle/front. **A broad white median stripe on the neck and the back of the head, usually ending with an orange spot between the rhinophores. The rhinophores are dark, the rear face with a white stripe and a bluish spot near its top. White pigment on the head, neck, and rhinophores merges to form a prominent "Y" when viewed from above**. Edge of foot orange. The tail has a white median stripe blue tip. Eyes are black, at the edges of the white head stripe. Little is known about the ecology of this species, but members of this family are known to feed on green algae and diatoms. It has been found on algae, bare rocks, and poorly vegetated reefs. Depth: 0-65 feet. Amphiatlantic (recently found in The Azores). Southeast Florida, northern Cuba, Bahamas, Saba, Guadeloupe, Costa Rica, Belize, Roatan, Mexico, Grand Cayman.

←Cozumel Grand Cayman→

Roatan

Thuridilla picta (A. E. Verrill, 1901) **painted elysia**
Size: to 16 mm Plakobranchidae
ID: Body elongate, slender, smooth, tapers to a point. Parapodia broad, smooth, usually folded up over the back, reach the tip of the tail; **edges are very sinuous with** ~6 large undulations and a few small ones. Rhinophores are long, cylindrical, rolled, their bases almost merged; usually inclined forward and sideways. Ground color black. There are red/orange, blue, and yellowish spots on the sides of the head. **The parapodia have a broad red/orange margin**, and an offset **submarginal row of quadrangular blotches** that extend down to the mid-side nested between downward extensions of large undulations; the blotches are **whitish to yellow, their sides and tops bright blue**, the yellow and blue often overlapping slightly to produce a diffuse greenish band. Below the blotches there is a scattering of whitish/greenish/yellow spots, sizes decreasing ventrally. A **whitish/yellow median stripe runs forward from a blue spot at the back of the neck onto the bottom half of each rhinophore**, the top half continuing with a narrow blue band, a narrow black band, a wide red/orange band, and a blue-spotted black tip. It feeds on green algae in areas of reef rubble and sand near shallow patch reefs. It can swim by flapping its large parapodia. Spawn consists of orange eggs within a gelatinous matrix in the form of a ribbon wound into a tight spiral of ~4 whorls. Depth: 0-35 feet. Amphiatlantic. Bermuda, Southeast Florida, Bahamas, east Cuba, Guadeloupe, Martinique, St. Vincent, Curacao, Colombia, Roatan, Cozumel, Bay of Campeche, Jamaica, Cayman Islands.

Family Volvatellidae Pilsbry, 1895

A family of small bubblesnails with external shells, consisting of about two dozen species in the genera *Ascobulla* and *Volvatella*. They are distributed globally in temperate to tropical waters and both genera are represented in the tropical West Atlantic.

Ascobulla has a bilobed and deeply furrowed cephalic shield, but *Volvatella* has oral tentacles and rhinophores. Body oval, radula uniseriate, teeth long with denticulate edges, ascus present. Shell external, bubble-shaped to cylindrical to pyriform, fragile, coiled, large enough for the entire animal to retract into; its yellowish to yellowish-green periostracum is superimposed on a rigid brittle white calcareous layer; aperture broad, its posterior extended as a siphonal canal and a variable-length spout.

They inhabit shallow tidal waters, feeding on species of *Caulerpa* green algae either by extracting nutrients from their protruding stalks or by burrowing into the sand to remove them from small root-like structures (rhizoids) that anchor the plant to the substrate.

Bocas del Toro, Panama

Ascobulla ulla **(Er. Marcus & Ev. Marcus, 1970)**
rice-grain bubblesnail

Size: to 8 mm Volvatellidae

ID: Animal's cephalic shield large, with a deep median furrow. Shell fragile, cylindrical but slightly pyriform, wider at the rear; the apex is truncate and shallowly concave, the protoconch is within a cavity; outer lip long, almost straight; it contains most of the animal except for its cephalic shield. The animal is translucent white with opaque white spots, its viscera are visible through its translucent white to yellowish shell. It is found in shallows burrowing into the sand and mud sediment near *Caulerpa* spp. algae, usually *Caulerpa racemosa*. When disturbed it exudes a white milky fluid from its posterior. Found in shallows to ~6 feet. Bermuda, Florida, Bahamas, Grand Turk, St. Croix, Brazil, Aruba, Venezuela, Belize, Mexico, Cayman Islands.

Bermuda

Bermuda

Volvatella bermudae **K. B. Clark, 1982**

Size: to 10 mm Volvatellidae

ID: Body oval; head small, flat, usually inside the shell margin. Rhinophores and oral tentacles are present. The rhinophores are conical but usually contracted as low crests. The mantle has highly branched diverticula of the digestive gland. The shell is inflated, fragile, has a broad aperture, its periostracum is thick and strong; it can completely contain the body of the animal. The mantle is green, the body and foot white. The shell is translucent whitish, its periostracum yellowish-green. It feeds on the rhizoids or vesicles of *Caulerpa racemosa* alga that is usually anchored on vertical rock faces and mangrove roots in areas of strong water circulation. It retains symbiotic chloroplasts for energy production. Depth: to ~10 feet. Bermuda, Florida Keys, Belize.

Family ONCHIDIIDAE Rafinesque, 1815

Except for the Arctic and Antarctic extremes, this is a worldwide family consisting of about 80 species in 13 genera of small (10-50 mm) air-breathing marine slugs (2 species are terrestrial only) with lung-like organs instead of gills, and the vast majority without any shell in the adult stage. Their evolutionary shell loss has been compensated for by a tough leathery mantle and good camouflage. They develop in sea water and inhabit intertidal areas like rocky seashores and marine or (rarely) brackish mangrove forests. There, during low tide, they graze in the open air on films of microalgae and bacteria that coat rocks, roots, sand, and mud. If submerged during high tide, many species can still survive for a few hours by using evolved backup respiratory mechanisms: some species burrow into the soft substrate where they trap an air bubble to breathe from; others can absorb small amounts of oxygen through their mantle; yet others breathe via small branching accessory gill-like appendages on their backs.

For many decades the taxonomy of these slugs was in chaos, making them notoriously difficult to identify, especially from external features alone. However, in the past 15 years it has been revised entirely, one genus at a time, based on a great deal of specimen collection and the subsequent delineation of genera and species based on anatomies and genetic sequences (mitochondrial and nuclear).

In the Tropical West Atlantic the only records are for species of the genus *Onchidella*: *O. floridana*, *O. armadilla*, *O. incisa*, and *O. indolens*.

← Bermuda →

Onchidella floridana **(Dall, 1885)** **Florida onchidella**
Size: to 20 mm Onchidiidae
ID: Body elongate-oval, slightly convex; mantle much broader than the foot sole; the mantle's velvety dorsal surface is densely covered with low, small, rounded tubercles, and its ventral surface is smooth; there are two prominent, narrow, retractile, cephalic tentacles that protrude forward from beneath the mantle's front margin; its eyes at the tips of the tentacles. It is uniformly slate-gray to slate-blue to grayish-green dorsally; lighter ventrally. It's an intertidal and sheltered-bay species that hides in crevices within ironshore and between/within shells of dead oysters at high tide, emerging at low tide to graze among algae. Prefers to feed when it's overcast, calm, and dry. Its lung, which is adapted for breathing air, is at the rear end of the body where it opens via a contractile pore. Hermaphroditic – a female genital opening at the rear of the body, a male organ opening near the right cephalic tentacle. Depth: intertidal. Bermuda, South Carolina to Florida, Bahamas, Cuba, Trinidad and Tobago, Colombia, Costa Rica. Abundant in Southwest Florida.

Miami, Florida

Onchidella incisa **(Quoy & Gaimard, 1832)**
Size: to 30 mm Onchidiidae
ID: Body elongate-oval and convex. The mantle has many large rounded smooth tubercles with variably sized smaller ones between them; its edges are crowded with laterally-projecting tentacle-like conical papillae; there are two slender, retractile, cephalic tentacles that protrude from under its front margin. The ground color can be yellowish, grayish, dark shades of green, to almost black. Mantle margins are usually yellowish; edge papillae yellow. The large tubercles are paler than the mantle, typically yellowish, cream, pale brown; their bases often being centered in similarly-colored circular patches. The body is yellowish-cream to pale orange ventrally. Inhabits upper intertidal zone, on/under rocks and pebbles on sand and silt. Its lung (adapted for breathing air) is at the rear end of the body and opens via a contractile pore. Southeast and Southwest Atlantic. Found in the Gulf of Mexico, Bermuda, Dominican Republic, Guadeloupe, Colombia, Brazil.

all photos from Espirito Santo, Brazil

Family Tylodinidae Gray, 1847

A family of a few species in the single genus *Tylodina* distributed circumglobally in temperate to tropical waters. So far, only one species, *Tylodina americana*, has been photographed in the tropical West Atlantic.

Body oval, smooth; mantle smooth, thin, its front and rear entire; oral veil narrow, overhangs mouth, no papillae; mouth with a protrusible buccal mass; two rolled oral tentacles by the oral veil, slightly connected to each other; rhinophores short, well separated, rolled; foot with a large flat sole, smooth upper surface, no anterior vertical cleft; gill well back on body's right rear quadrant and attached to it for the front half of its length; rachis smooth; anus above rear end of gill membrane, not flush with the body; a sperm groove leads from the non-protrusible penis to the genital aperture at the base of the right oral tentacle; the radula is broad, with a rachidian tooth and simple laterals with weak cusps, sometimes with a small denticle at the base of the inner face.

The shell is external, small to medium-sized, circular to oval, smooth, dorsally central, elevated with a conical apex (the protoconch is often inconspicuous); and adherent periostracum is present, often forming prominent radial/circular ridges, easily detachable, extends well beyond the edge of the shell; large enough to fully cover the animal's body.

Feed primarily on sponges. Shell lengths are typically less than 40 mm.

Tylodina americana Dall, 1890

American tylodina

Size: to 10 mm — Tylodinidae

ID: The external shell is disk shaped, slightly convex, with a prominent conical central apex, with up to 20 fine radial ridges; the periostracum considerably overhangs the shell. The body is oval, smooth, a bit larger than the shell; the obscured mantle is thin, smooth, with a serrated edge; the foot has a large flat sole; rhinophores are short, smooth, rolled, well separated, and protrude slightly from under the shell; the eyes are between and in front of the bases of the rhinophores; oral tentacles are broad, rolled, grooved externally, at the corners of a narrow oral veil that overhangs the mouth; the gill is short, bipinnate, along the middle third of the right side, with its rear half detached. The animal is yellow, its shell's periostracum reddish to reddish-brown. It feeds on sponges in the family Aplysinellidae. The egg mass is a spirally-wound yellow ribbon bearing cream-white eggs, attached flat to a substratum. Found in depths of 60-480 feet. South Carolina, Bermuda, Florida, Texas, Cuba. Rare.

all photos from Florida

Family Umbraculidae Dall, 1889 (1827)

It consists of the genera *Spiricella* and *Umbraculum*, the former with two species known only from fossils and the latter with a single species, *Umbraculum umbraculum*, with a circumglobal distribution in temperate to tropical waters.

Umbraculum: Body oval to roundish, top truncated; mantle thin, margins serrate, covered with large soft tubercles; rhinophores anteriorly median, separated, tubular, retractile, with a basal swelling; eyes between and below rhinophoral bases; rolled oral tentacles at base of foot, no oral veil between them; foot circular, muscular, rather high, front with a deep vertical cleft, upper surface with smooth tubercles; gill along groove below mantle from left front corner of body around the front and rearward along the right side for most of the body length; gill plumes tripinnate; anus on a prominent papilla at right rear of body behind gill; non-retractable penis is an elongate lobe in a median vertical cleft below the rhinophores and above the mouth; genital opening on right side of penial base; radula very broad, no rachidian tooth but up to 800 laterals per side for each of up to 130 rows. Shell external, small to medium, circular to oval, somewhat flattened but with an off-center conical apex, protoconch sinistrally coiled; sometimes calcified; covered with a sturdy periostracum that inhibits colonization by fouling organisms; too small to cover the entire body. Feed on sponges in moderate depths but have been dredged from >1200 feet. Lengths typically 20-30 mm but up to 280 mm.

Bonaire

←Brazil

with spawn; Bonaire→

←Guadeloupe←

Umbraculum umbraculum **([Lightfoot], 1786)** **umbrella slug**

Size: to 280 mm Umbraculidae

ID: See above for general morphological details. High, broad, almost circular, flat-topped, widens downwards. **Apical shell central**, **umbrella-shaped**; apex low, conical, covers ~50% of top; usually concealed by an **algal mat** encrusted with plant growth, sponges, anemones, dirt, other detritus. A narrow opening beneath it provides a view of the smooth mantle with its serrated edge, and allows **the rhinophores and the large penis slightly below them** to protrude. Below them is the **foot covered by smooth tubercles;** near its base behind a vertical slit and straight down from the rhinophores are its mouth and two pairs of oral tentacles. The ground color is uniformly whitish, orange, yellow, brown, to black. Its algal mat is white/gray/brownish; the tubercles usually ground color, often paler to whitish. It burrows into sand and mud bottoms and hides under rocks, in crevices and caves. Feeds on sponges. The shell's top surface initially has a beard-like periostracum that erodes to enable colonization by plants and animals. Its rhinophores sense prey but also siphon oxygen-rich water toward its gill. Spawn is a convoluted pink/red/orange ribbon-like spiral coil deposited on a hard substrate. Depth: 0-260 feet. Bermuda, Florida to Texas, Antilles, Venezuela to Roatan, south to Brazil. Uncommon.

References

Alvim, J. & Pimenta, Á. D. (2015). Taxonomic review of *Berthella* and *Berthellina* (Gastropoda: Pleurobranchoidea) from Brazil, with description of two new species. Zoologia. 32(6): 497-531.

Alvim, J. & Pimenta, Á. D. (2016). Comparative Morphology and Redescription of *Pleurobranchus* Species (Gastropoda, Pleurobranchoidea) from Brazil. Zoological Studies. 55: doi:10.6620/ZS.2016.55-15.

Alvim, J., Simone, L. R. L. & Pimenta, Á. D. (2014). Taxonomic review of the genus *Pleurobranchaea* (Gastropoda: Pleurobranchoidea) from Brazil, with description of a new species. Journal of Molluscan Studies. 80: 604-623.

Baba, K. (1959). The genus *Petalifera* and a new species, *P. ramosa*, from Japan. Publications of the Seto marine Biological Laboratory. 7(3): 43-44.

Baez, D. P.; Ardila, N.; Valdés, Á.; Acero, A. (2011). Taxonomy and phylogeny of *Armina* (Gastropoda: Nudibranchia: Arminidae) from the Atlantic and eastern Pacific. Journal of the Marine Biological Association of the United Kingdom. 91(5), 1107-1121.

Bergh, 1877 (Nudibranchia: Polyceridae: Nembrothinae) in the Atlantic Ocean. Proceedings of the California Academy of Sciences. 54(22): 381-392

Buske, Y. (2018). Inventaire des Mollusques "Euthyneura" ou Limaces de Mer de la Martinique. Rendu à la DEAL Martinique. 1-138.

Caballer, M. & Ortea, J. (2007) Nueva especie del género *Hermaea* Lovén, 1844 (Mollusca: Sacoglossa), de la costa norte de La Habana, Cuba. Avicennia 19: 127-132.

Caballer, M. & Ortea, J. (2012) Description of a new species of *Hypselodoris* (Gastropoda: Nudibranchia: Chromodorididae) from Venezuela. Revista de la Academia Canaria de Ciencias 23(3): 93-106. [Journal issue for 2011; published April 2012]

Caballer, M. & Ortea, J. (2014). A new sibling species of *Notobryon* (Gastropoda, Nudibranchia) from the Caribbean Sea. Journal of the Marine Biological Association of the United Kingdom. 94(7): 1465-1470.

Caballer, M. & Ortea, J. (2015). Finishing to untangling the taxonomic knot: new species of the genus *Bulbaeolidia* Carmona, Pola, Gosliner & Cervera, 2013 (Mollusca: Aeolidiidae) from the Caribbean and Galapagos. Revista de la Academia Canaria de Ciencias. 27: 113-123.

Caballer, M. & Ortea, J. (2019 [nomenclatural availability: 2018]). A rare new species of *Corambe* Bergh, 1869 (Mollusca: Gastropoda: Nudibranchia) from the Caribbean Sea, with a review of the taxonomical history of the Corambidae Bergh, 1871. Marine Biodiversity. 49(3): 1371-1383.

Caballer, M.; Ortea, J.; Espinosa, J. (2001). A new species of *Eubranchus* Forbes, 1834. Avicennia suplemento 4 (Moluscos del Caribe de Costa Rica): 55-56.

Camacho-Garcia, Y. E.; Gosliner, T. M. (2008). Systematic revision of *Jorunna* Bergh, 1876 (Nudibranchia: Discodorididae) with a morphological phylogenetic analysis. Journal of Molluscan Studies. 74 (2): 143-181.

Camacho-Garcia, Y. E.; Marta Pola, M.; Carmona, L.; Padula, V.; Villani, G.; Cervara, J. L. (2014). Diversity and distribution of the heterobranch sea slug fauna on the Caribbean of Costa Rica. Cah. Biol. Mar. 55: 109-127.

Camacho-García, Y. E.; Ornelas-Gatdula, E.; Gosliner, T. M.; Valdés, Á. (2013). Phylogeny of the family Aglajidae (Pilsbry, 1895) (Heterobranchia Cephalaspidea) inferred from mtDNA and nDNA. Molecular Phylogenetics and Evolution. 71: 113-126.

Carmona, L.; Gosliner, T. M.; Pola, M.; Cervera, J. L. (2011). A molecular approach to the phylogenetic status of the aeolid genus *Babakina* Roller, 1973 (Nudibranchia). Journal of Molluscan Studies. 77(4): 417-422.

Carmona, L.; Lei, B.R.; Pola, M.; Gosliner, T. M., Valdés, Á.; Cervera, J. L. (2014). Untangling the *Spurilla neapolitana* (Delle Chiaje, 1841) species complex: a review of the genus *Spurilla* Bergh, 1864 (Mollusca: Nudibranchia: Aeolidiidae). Zoological Journal of the Linnean Society, 170(1): 132–154.

Carmona, L.; Malaquias, M. A. E.; Gosliner, T. M.; Pola, M.; Cervera, J. L. (2011). Amphi-Atlantic distributions and cryptic species in sacoglossan sea slugs. Journal of Molluscan Studies, 77: 401-412

Carmona, L.; Pola, M.; Gosliner, T. M.; Cervera, J. L. (2014). Review of *Baeolidia*, the largest genus of Aeolidiidae (Mollusca: Nudibranchia), with the description of five new species. Zootaxa, 3802 (4): 477–514.

Carmona, L.; Pola, M.; Gosliner, T. M.; Cervera, J. L. (2017). Integrative taxonomy and biogeography of the genus *Bulbaeolidia* (Nudibranchia: Aeolidida). Journal of Molluscan Studies. 83(4): 440-450.

Carmona, L.; Pola, M.; Gosliner, T. M.; Cervera, J. L. (2013). A Tale That Morphology Fails to Tell: A Molecular Phylogeny of Aeolidiidae (Aeolidida, Nudibranchia, Gastropoda). PLOS ONE, 8(5): 1-13.

Carmona, L.; Pola, M.; Gosliner, T. M.; Cervera, J. L. (2014). The Atlantic-Mediterranean genus *Berghia* Trinchese, 1877 (Nudibranchia: Aeolidiidae): taxonomic review and phylogenetic analysis. Journal of Molluscan Studies. 80(5): 482-498.

Carmona, L.; Pola, M.; Gosliner, T. M.; Cervera, J. L. (2014). The end of a long controversy: systematics of the genus *Limenandra* (Mollusca: Nudibranchia: Aeolidiidae). Helgoland Marine Research, 68: 37-48.

Carmona, L.; Bhave, V.; Salunkhe, R.; Pola, M.; Cervera, J. L. (2014). Systematic review of Anteaeolidiella (Mollusca, Nudibranchia, Aeolidiidae) based on morphological and molecular data, with a description of three new species. Zoological Journal of the Linnean Society, 171: 108-132.

Chan, J. M. & Gosliner, T. M. (2006). Description of a new species of *Thordisa* (Nudibranchia: Discodorididae) from Panamá. Proceedings of the California Academy of Sciences (4) 57(35): 981-990.

Churchill, C. K. C.; Valdés, Á.; Ó Foighil, D. (2014). Molecular and morphological systematics of neustonic nudibranchs (Mollusca: Gastropoda: Glaucidae: *Glaucus*), with descriptions of three new cryptic species. Invertebrate Systematics. 28: 174–195.

Crescini, R.; De Sisto, M.; Villalba, W. (2013). A new species of nudibranch of the genus *Learchis* (Gastropoda: Heterobranchia: Facelinidae) from the tropical western Atlantic Ocean. American Malacological Bulletin 31(2): 339-341.

Cunha, C. M. (2011) A new species of *Acteon* (Opisthobranchia: Acteonidae) from Northeast Brazil. Zoologia 28(2): 229-232.

DaCosta, S.; Padula, V.; Schrödl, M. (2010) A new species of *Hypselodoris* and a redescription of *Hypselodoris picta lajensis* (Nudibranchia: Chromodorididae) from Brazil. The Veliger 51(1):15-25.

Dayrat, B. (2006). A taxonomic revision of *Paradoris* sea slugs (Mollusca, Gastropoda, Nudibranchia, Doridina). Zoological Journal of the Linnean Society 147: 125-238.

Dayrat, B. (2009). Review of the current knowledge of the systematics of Onchidiidae (Mollusca: Gastropoda: Pulmonata) with a checklist of nominal species. Zootaxa. 2068: 1-26.

Dayrat, B. (2010). A monographic revision of basal discodorid sea slugs (Gastropoda, Opisthobranchia, Nudibranchia, Doridina). Proceedings of the California Academy of Sciences. Series 4, vol. 61, suppl. I, 1-403, 382 figs.

de Jong, K. M. & Coomans, H. E. (1988). Marine Gastropods from Curaçao, Aruba and Bonaire. Studies on the Fauna of Curaçao and Other Caribbean Islands: No. 214.

Domínguez, M.; García, F. J.; Troncoso, J. S. (2006). Some aspects of the family Chromodorididae (Opisthobranchia: Nudibranchia) from Brazil, with description of a new species. Scientia Marina. 70(4), 621-634.

Dominguez, M.; Troncoso, J. S.; García, F. J. (2008). The family Aeolidiidae Gray, 1827 (Gastropoda Opisthobranchia) from Brazil, with a description of a new species belonging to the genus *Berghia* Trinchese, 1877. Zoological Journal of the Linnean Society. 153(2): 349-368.

Edmunds, M.; Just, H. (1983). Eolid nudibranchiate Mollusca from Barbados. Journal of Molluscan Studies. 49: 185-203.

Espinoza, E.; DuPont, A.; Valdés, Á. (2013) A tropical Atlantic species of *Melibe* Rang, 1829 (Mollusca, Nudibranchia, Tethyiidae). ZooKeys 316: 55-66.

Espinoza, E.; DuPont, A.; Valdés, Á. (2014). Molecular data reveal an undescribed cryptic species of *Costasiella* Pruvot-Fol, 1951 (Euthyneura: Sacoglossa: Limapontidae) in the Bahamas. American Malacological Bulletin. 32(2): 173-182.

Espinosa, J.; Ortea, J.; Caballer, M.; Moro, L. (2006) Criaturas del paisaje sumergido de Cuba. Avicennia: Revists de Biodiversidad Tropical, Suplemento 5.

Espinosa, J. & Ortea, J. (2001). Mollusca of Costa Rica Caribbean Sea: from Cahuita to Gandoca. Avicennia, Suplemento 4: 1-77.

Espinosa, J.; Ortea, J. (2004). Nuevas species de moluscos gasteropodos marinos (Mollusca: Gastropoda) de las Bahamas, Cuba y el Mar Caribe de Costa Rica. Rev. Acad. Canar. Cienc, XV (Nums. 3-4), 207-216.

Espinosa, J.; Ortea, J.; Diez García, Y. (2017). Nuevas especies y nuevos registros de moluscos gasterópodos (Mollusca: Gastropoda) marinos de la región oriental de Cuba. Avicennia. 21: 59-67.

Feliciano, K.; Malaquias, M. A. E.; Stout, C.; Brenzinger, B;, Gosliner, T. M.; Valdés, Á. (2021). Molecular and morphological analyses reveal pseudocryptic diversity in *Micromelo undatus* (Bruguière, 1792) (Gastropoda: Heterobranchia: Aplustridae). Systematics and Biodiversity. DOI: 10.1080/14772000.2021.1939458: 1-25.

Filho, H. G.; Paulay, G.; Krug, P.J. (2019). Eggs sunny-side up: A new species of *Olea*, an unusual oophagous sea slug (Gastropoda: Heterobranchia: Sacoglossa), from the western Atlantic. Zootaxa 4614 (3): 541–565.

Garcia, F. J. & Troncoso, J. S. (2003) Two unknown species of Mollusca Gastropoda from Archipelago Fernando de Noronha (Brazil), with description of a new species belonging to the genus *Phidiana* Gray, 1850 and a new record of *Dendrodoris senegalensis* Bouchet, 1975. Scientia Marina 67(2): 159-166.

García-Méndez, K.; Padula, V.; Valdés, Á. (2022). Integrative systematics of the genus *Dondice* Marcus, 1958 (Gastropoda, Nudibranchia, Myrrhinidae) in the Western Atlantic. Marine Biodiversity. 52(5).

Ghanimi, H., Schrödl, M., Goddard, J. H. R.; Ballesteros, M.; Gosliner, T. M.; Buske, Y.; Valdés, Á. (2020). Stargazing under the sea: molecular and morphological data reveal a constellation of species in the *Berthella stellata* (Risso, 1826) species complex (Mollusca, Heterobranchia, Pleurobranchidae). *Marine Biodiversity*. 50(1).

Gilmer, R. W. & Harbison, G. R. (1986). Morphology and field behavior of pteropod molluscs: feeding methods in the families Cavoliniidae, Limacinidae and Peraclididae (Gastropoda: Thecosomata). Marine Biology. 91(1): 47-57.

Goodheart, J.; Camacho-García, Y.; Padula, V.; Schrödl, M.; Cervera, J. L.; Gosliner, T. M.; Valdés, Á. (2015). Systematics and biogeography of *Pleurobranchus* Cuvier, 1804, sea slugs (Heterobranchia: Nudipleura: Pleurobranchidae). *Zoological Journal of the Linnean Society*. 174: 322-362.

Goodheart, J. A.; Ellingson, R. A.; Vital, X. G.; Galvão Filho, H. C.; McCarthy, J. B.; Medrano, S. M.; Bhave, V. J.; García-Méndez, K.; Jiménez, L. M.; López, G.; Hoover, C. A.; Awbrey, J. D.; De Jesús, J. M.; Gowacki, W.; Krug, P. J.; Valdés, Á. (2016). Identification guide to the heterobranch sea slugs (Mollusca: Gastropoda) from Bocas del Toro, Panama. Marine Biodiversity Records. 9: 56.

Gosliner, T. M. (1980). Systematics and phylogeny of the Aglajidae (Opisthobranchia: Mollusca). Zoological Journal of the Linnean Society. 66: 325-360.

Gosliner, T. M. (1987). Review of the nudibranch genus *Melibe* (Opisthobranchia: Dendronotacea) with descriptions of two new species. The Veliger. 29(4): 400-414. [1 April 1987].

Gosliner, T. M. (1989). Revision of the Gastropteridae (Opisthobranchia: Cephalaspidea) with descriptions of a new genus and six new species. The Veliger. 32(4): 333-381.

Gosliner, T. M. (1994). New records of Flabellinidae (Opisthobranchia: Aeolidacea) from the tropical Americas, with descriptions of two new species. Proceedings of the California Academy of Sciences, series 4. 48(9):171-183.

Gosliner, T. M. & Fahey, S. J. (2008). Systematics of *Trapania* (Mollusca: Nudibranchia: Goniodorididae) with descriptions of 16 new species. Systematics and Biodiversity. 6(1): 53-98.

Gosliner, T. M.; Ghiselin, M. T. (1987). A new species of *Tritonia* (Opisthobranchia: Gastropoda) from the Caribbean Sea. Bulletin of Marine Science. 40(3): 428-436.

Gosliner, T. M.; González-Duarte, M. M.; Cervera, J. L. (2007). Revision of the systematics of *Babakina* Roller, 1973 (Mollusca: Opisthobranchia) with the description of a new species and a phylogenetic analysis. Zoological Journal of the Linnean Society. 151, 671–689.

Gosliner, T. M. & Johnson, S. (1994). Review of the genus *Hallaxa* (Nudibranchia: Actinocyclidae) with descriptions of nine new species. The Veliger. 37(2): 155-191.

Gosliner, T. M.; Kuzirian, A. M. (1990). Two new species of Flabellinidae (Opisthobranchia: Aeolidacea) from Baja California. Proceedings of the California Academy of Sciences. 4, 47(1):1-15.

Gosliner, T. M.; Smith, V. G. (2003). Systematic review and phylogenetic analysis of the nudibranch genus *Melibe* (Opisthobranchia: Dendronotacea) with descriptions of three new species. Proceedings of the California Academy of Sciences. 4(54): 302-356.

Gowacki, W. A.; Bell, S. S.; Pierce, S. K. (2016). Correcting taxonomic confusion for the kleptoplastic sea slug *Elysia papillosa* Verrill (1901). Amer. Malac. Bull. 34(2): 85–91.

Grune, S.; Crescini, R.; De Sisto, M.; Velásquez, M.; Villalba, W. (2014). Opisthobranchs of the National Park Laguna de La Restinga, Margarita Island, Venezuela. Amici Molluscarum 22(2): 25-35.

Gutierrez, M. C.; Ortea, J.; Rivero, N.; Tucker, C.T.; Malaquias, M. A. E.; Narciso, S. (2015). The opisthobranch gastropods (Mollusca: Heterobranchia) from Venezuela:an annotated and illustrated inventory of species. Zootaaxa 4034 (2): 201-256.

Hamann, J. C.; Farmer, W. M. (1988). Two new species of *Plocamopherus* from the western warm water Atlantic. The Veliger. 31: 68–74.

Henryco, B.; de Meirelles, C. A. O.; García-Méndez, K.; Camacho-García, Y.; Valdés, Á.; Schrödl, M.; Padula, V. (2025). The lady of the rings: Integrative systematics of *Nanuca* Er. Marcus, 1957 (Mollusca: Nudibranchia: Myrrhinidae) reveals a hidden diversity. Zoological Journal of the Linnean Society. 205(1): zlaf085: 1-32.

Hewitt, S. J.; Picton, B. E.; DuPont, A.; Zahner, T. M.; Salvador, R. B. (2021). New records of marine molluscs from Saba, Caribbean Netherlands. Basteria. 85 (1): 59–72. Leiden. [7 May 2021].

Hewitt, S. J.; van Leeuwen, S. (2017). Marine molluscs from the island of St. Eustatius, Eastern Caribbean: Results of the Statia Marine Biodiversity Expedition 2015. Vita Malacologica 16: 44-80.

Jensen, K. R. (1980). *Oxynoe azuropunctata*, n. sp., a new sacoglossan from the Florida Keys (Mollusca: Opisthobranchia). Journal of Molluscan Studies. 46 (3): 282–292.

Johnson, R. F. (2011) Breaking family ties: taxon sampling and molecular phylogeny of chromodorid nudibranchs (Mollusca, Gastropoda). Zoologica Scripta 40(2): 137-157.

Johnson, R. F. & Gosliner, T. M. (2012) Traditional taxonomic groupings mask evolutionary history: A molecular phylogeny and new classification of the chromodorid nudibranchs. PLoS ONE 7(4): e33479.

Korshunova, T.; Fletcher, K.; Martynov, A. (2025). The endless forms are the most differentiated – how taxonomic pseudo-optimization masked natural diversity and evolution: the nudibranch case. Zoological Journal of the Linnean Society. 204(4): zlaf057: 1-93.

Krug, P. J.; Berriman, J. S.; Valdés, Á. (2018). Phylogenetic systematics of the shelled sea slug genus *Oxynoe* Rafinesque, 1814 (Heterobranchia: Sacoglossa), with integrative descriptions of seven new species. Invertebrate Systematics. 32(4): 950-1003.

Krug, P. J.; Vendetti, J. E.; Valdés, Á. (2016). Molecular and morphological systematics of *Elysia* Risso, 1818 (Heterobranchia: Sacoglossa) from the Caribbean region. Zootaxa. 4148(1): 1-137.

Krug, P. J.; Wong, N. L. W. S.; Medina, M. R.; Gosliner, T. M.; Valdés, Á. (2018). Cryptic speciation yields remarkable mimics: A new genus of sea slugs that masquerade as toxic algae (*Caulerpa* spp.). Zoologica Scripta. 47(6): 699-713.

Laetz, E.; Christa, G.; Händeler, K.; Wägele, H. (2014). The *Cylindrobulla* /*Ascobulla* complex–unraveling problems in identification and adding to *Cylindrobulla* diversity (Gastropoda, Heterobranchia, Sacoglossa) by describing a new species. Zootaxa. 3893(3): 339-362.

Lalli, C. M. & Gilmer, R. W. (1989). Pelagic Snails: The Biology of Holoplanktonic Gastropod Mollusks. Stanford Univ. Press, Stanford, California.259 pp.

Lima, P. O. V. & Simone, L. R. L. (2015). Anatomical review of *Doris verrucosa* and redescription of *Doris januarii* (Gastropoda, Nudibranchia) based on comparative morphology. Journal of the Marine Biological Association of the United Kingdom. DOI: 10.1017/S0025315415000296I.

Malaquias, M. A. E. (2014). New data on the heterobranch gastropods ('opisthobranchs') for the Bahamas (tropical western Atlantic Ocean). Marine Biodiversity Records. doi:10.1017/S175526721400030X; Vol. 7; e27; 201 January.

Malaquias, M. A. E. & Reid, D. G. (2008). Systematic revision of the living species of Bullidae (Mollusca: Gastropoda: Cephalaspidea), with a molecular phylogenetic analysis. Zoological Journal of the Linnean Society. 153(3): 453-543.

Marcus, Er. (1955). Opisthobranchia from Brazil. Boletim da Faculdade de Filosofia, Ciências e Letras da Universidade de São Paulo (Zoologia). 20: 89-261.

Marcus, Er. (1957). On Opisthobranchia from Brazil (2). Journal of the Linnean Society of London (Zoology). 43 (292): 390-486.

Marcus, Er. & Marcus, Ev. (1970). Opisthobranchs from Curaçao and faunistically related regions. Studies on the Fauna of Curaçao and other Caribbean Islands. 33 (1): 1-129.

Marcus, Ev. (1971 ["1970"]). Opisthobranchs from northern Brazil. Bulletin of Marine Science. 20(4): 922-951.

Marcus, Ev. (1972). On some Opisthobranchs from Florida. Bulletin of Marine Science. 22 (2): 284-308.

Marcus, Ev. (1976). Opisthobranchia von Santa Marta, Colombia. Studies on the Neotropical Fauna & Environment. 11 (1-2): 119-150.

Marcus, Ev. (1978). The western Atlantic species of *Onchidella* (Pulmonata). Sarsia. 63: 221-224.

Marcus, Ev. (1979). The Atlantic species of *Onchidella* (Gastropoda Pulmonata) part 2. Boletim de Zoologia (Sao Paulo). 4(4): 1-38.

Marcus, Ev. (1980). Review of Western Atlantic Elysiidae (Opisthobranchia Ascoglossa) with a description of a new *Elysia* species. Bulletin of Marine Science. 30 (1): 54-79.

Marcus, Ev. (1984). The western Atlantic warm water Notaspidea (Gastropoda, Opisthobranchia). Parte 2. Boletim de Zoologia, Universidade de São Paulo. 8: 43-76.

Marcus, Ev. & Hughes, H. P. I. (1974). Opisthobranch mollusks from Barbados. Bulletin of Marine Science. 24(3): 498-532.

Marcus, Ev. & Marcus, Er. (1960). Opisthobranchs from American Atlantic warm waters. Bulletin of Marine Science of the Gulf and Caribbean. 10: 129-203.

Marcus, Ev. & Marcus, Er. (1962). Opisthobranchs from Florida and the Virgin Islands. Bulletin of Marine Science. 12 (3): 450-488.

Marcus, Ev. & Marcus, Er. (1963). Opistobranchs from the Lesser Antilles. Studies on the Fauna of Curaçao and Other Caribbean Islands: no 79. 19(1): 1-76.

Marcus, Ev. & Marcus, Er. (1967). American Opisthobranch Mollusks Part I, Tropical American opisthobranchs; Part II, Opisthobranchs from the Gulf of California. Studies in Tropical Oceanography. 6: 1–256.

Marcus, Ev. & Marcus, Er. (1967). Some opisthobranchs from Sapelo Island, Georgia, U.S.A. Malacologia. 6(1-2):199-222.

Martinez, E. (1996). On *Petalifera petalifera* (Rang, 1828) (Gastropoda: Opisthobranchia): new anatomical and geographical data. Journal of Molluscan Studies 62(2): 243-250.

Martínez, E. & Ortea, J. (1999). Redescription of *Atys macandrewii* E.A. Smith, 1872, an amphiatlantic cephalaspidean. American Malacological Bulletin 14(2): 133-138.

McCarthy, J. B.; Krug, P. J.; Valdés, Á. (2017). Integrative systematics of *Placida cremoniana* (Trinchese, 1892) (Gastropoda, Heterobranchia, Sacoglossa) reveals multiple pseudocryptic species. Marine Biodiversity 49, 357–371.

McDonald, G. R. & Nybakken, J. W. (1997). A worldwide review of the food of nudibranch mollusks. I. Introduction and the suborder Arminacea. The Veliger. 40(2): 157-159.

McDonald, G. R. & Nybakken, J. W. (1999). A worldwide review of the food of nudibranch mollusks. II. The suborder Dendronotacea. The Veliger. 42(1).

Medina, M.; Collins, T.; Walsh, P. J. (2005). Phylogeny of sea hares in the *Aplysia* clade based on mitochondrial DNA sequence data. Bulletin of Marine Science. 76(3): 691-698.

Medrano, S.; Krug, P. J.; Gosliner, T. M.; Kumar, A. B.; Valdés, Á. (2018). Systematics of *Polybranchia* Pease, 1860 (Mollusca: Gastropoda: Sacoglossa) based on molecular and morphological data. Zoological Journal of the Linnean Society. 186(1): 76-115.

Mikkelsen, P. M. (1998). *Cylindrobulla* and *Ascobulla* in the western Atlantic (Gastropoda, Opisthobranchia, Sacoglossa): Systematic review, description of a new species, and phylogenetic reanalysis. Zoologica Scripta. 27(1): 49-71.

Millen, S. V. & Gosliner, T. M. (1985). Four new species of dorid nudibranchs belonging to the genus *Aldisa* (Mollusca: Opisthobranchia), with a revision of the genus. Zoological Journal of the Linnean Society. 84: 195-233.

Millen, S. V. & Hamann, J. C. (1992). A new genus and species of Facelinidae (Opisthobranchia: Aeolidacea) from the Caribbean Sea. The Veliger. 35(3): 205-214.

Millen, S. V. & Hamann, J. C. (2006). A new nudibranch species, genus *Flabellina* (Opisthobranchia: Aeolidacea) from the Caribbean with redescriptions of *F. verta* (Marcus 1970), and *F. dushia* (Marcus and Marcus, 1963). Proceedings of the California Academy of Sciences (4). 57(30): 925-936.

Moro, L. & Ortea, J. (2015). New taxa of seaslugs of the Canary Islands and CapeVerde Islands (Mollusca: Heterobranchia). Vieraea 43: 21-86.

Muniain, C.; Ardila, N. E.; Cervera, J. L. (2007). *Pleurobranchaea inconspicua* Bergh, 1897 (Opisthobranchia: Pleurobranchidae): Redescription and distribution from Argentina and Colombia. Bonner zoologische Beiträge 55: 291-300.

Nuttall, T. R. (1989) A new *Elysia* (Opisthobranchia: Ascoglossa) from the Florida Keys. The Veliger 32(3): 302-307.

Ortea, J. (2001). El género *Doto* Oken, 1815 (Mollusca: Nudibranchia) en el mer Caribe: Historia natural y descripción de nuevas especies. Avicennia. Supplement 3: 1-46.

Ortea, J. (2010) Tres nuevas especies de *Doto* Oken, 1815 (Mollusca: Nudibranchia), colectadas en el Caribe de Costa Rica y en las Islas Galapagos. Revista de la Academia Canaria de Ciencias 21(3-4): 109-115.

Ortea, J. (2013) Notas en Opistobranchia (Mollusca, Gastropoda): *Doto chica* Marcus y Marcus, 1970, un nombre para dos especies. *Revista de la Academia Canaria de Ciencias*, 25: 83-86.

Ortea, J. (2016). Una segunda especie atlántica del género *Hoplodoris* Bergh, 1880 (Mollusca: Nudibranchia) colectada en Martinica, Antillas Menores. Revista de la Academia Canaria de Ciencias. 28: 201-208.

Ortea, J. (2018). El alzamiento de especies, origen de sinonimias en el género *Elysia* Risso, 1818 (Mollusca: Sacoglossa) en el Mar Caribe. Avicennia. 23: 57-62.

Ortea, J. (2020). Nuevos datos sobre el género *Cylindrobulla* P. Fischer, 1857 en el Mar Caribe (Mollusca: Sacoglossa). Revista de la Academia Canaria de Ciencias. 32: 55-66.

Ortea, J. (2022). Una nueva especie del género *Doto* Oken, 1815,(Gastropoda: Heterobranchia), de la isla de Martinica dedicada al Concejo de Cangas del Narcea, Asturias. Avicennia. 30: 1-6.

Ortea, J.; Bacallado, J. J. (2007). Description de una nueva especie de *Hypselodoris* Stimpson,1855 (Mollusca: Nudibranchia: Chromodorididae) Nombrada en homenaje a Olga Ucelay Sabina. Rev. Acad. Canar. Cienc. 18(4), 53-60.

Ortea, J. & Buske, Y. (2014). Una nueva especie de *Thuridilla* Bergh, 1872 (Mollusca: Sacoglossa) de la isla de Martinica, Antillas Menores. Revista de la Real Academia Canaria de Ciencias. 26: 189-194.

Ortea, J. & Caballer, M. (2002). New data about the genus *Eubranchus* Forbes, 1838 (Mollusca: Nudibranchia) in warm waters of West Atlantic. Avicennia 15: 77-90.

Ortea, J. & Caballer, M. (2007) Nueva especie de *Doto* Oken, 1815 (Mollusca: Cladobranchia), nombrada en honor de Torrelavega, ciudad hermanada con La Habana. Avicennia 19: 121-126.

Ortea, J.; Caballer, M.; Espinosa, J. (2003) Nuevos Aeolidaceos (Mollusca: Gastropoda) de Costa Rica. Avicennia 16: 129-142.

Ortea, J.; Caballer, M.; Moro, L. (2001). Una nueva especie del género *Noumea* Risbec, 1928 (Mollusca: Nudibranchia) del Caribe arrecifal de Costa Rica, descrita con motivo del Quinto Centenario de su descubrimiento. Avicennia 14: 1-6.

Ortea, J.; Caballer, M.; Moro, L. (2004). Dos aeolidaceos con ceratas rojos de la region macaronesica y el mar Caribe (Mollusca: Nudibranchia). Vieraea 32: 83-96.

Ortea, J.; Caballer, M.; Moro, L.; Espinosa, J. (2014). What the shell tells in Aglajidae: A new genus for *Aglaja felis* (Opisthobranchia: Cephalaspidea). Revista de la Academia Canaria de Ciencias. 26: 83-119.

Ortea, J. & Espinosa, J. (1998) Dos nuevas especies de Moluscos marinos (Mollusca: Gastropoda) recolectadas en los subarchipiélagos Jardines del Rey y Jardines de la Reina, descritas en honor de la Reyes de España por su primera visita a Cuba. Avicennia 8-9: 1-6.

Ortea, J. & Espinosa, J. (2000). Nueva especie del género *Janolus* Bergh, 1884 (Mollusca: Nudibranchia) de Cuba y Costa Rica. Avicennia 12/13: 79-83.

Ortea, J. & Espinosa, J. (2000). New species of the genus *Thuridilla* Bergh, 1872 (Mollusca: Sacoglossa) from Cuba and Costa Rica. Avicennia. 12/13: 87-90.

Ortea, J. & Espinosa, J. (2000). Nueva especie del género *Okenia* Menke, 1830 (Mollusca: Nudibranchia) de Cuba. Avicennia. 12 (13) pp. 84-86.

Ortea, J. & Espinosa, J. (2001). A new species of *Philinopsis* Pease, 1860. Moluscos del mar Caribe de Costa Rica, supl.4, 41-44.

Ortea, J. & Espinosa, J. P. (2001). Descripción de una nueva especie de Moluscos del Mar Caribe de Costa Rica: desde Cahuita hasta Gandoca. Avicennia. Supplement 4: 41-42.

Ortea, J. & Espinosa, J. (2001) Descripción de una nueva especie de *Dendrodoris* Ehrenberg, 1831. In: Espinosa J. & Ortea J., Moluscos del Mar Caribe de Costa Rica: desde Cahuita hasta Gandoca. Avicennia Suppl. 4: 52-53.

Ortea, J. & Espinosa, J. (2002). Nuevas especies del género *Elysia* Risso, 1818 (Mollusca: Sacoglossa) con caracteres singulares. Avicennia 15: 129-140.

Ortea, J.; Espinosa, J.; Caballer, M. (2005 ["2004"]). Nuevos taxones y registros de la familia Polyceridae (Mollusca: Nudibranchia) en las costas de Cuba. Avicennia 17: 101-106.

Ortea, J.; Espinosa, J.; Caballer, M.; Buske, Y. (2012). Initial inventory of the seaslugs (Opisthobranchia and Sacoglossa) from the expedition Karubentos, held in May 2012 in Guadeloupe (Lesser Antilles, Caribbean Sea). Revista de la Academia Canaria de Ciencias 24: 153-182.

Ortea, J.; Espinosa, J.; Camacho, Y. (1999). Especies del género *Polycera* Cuvier, 1816 (Mollusca: Nudibranchia) recolectadas en la epifauna de algas rojas del Caribe de Costa Rica y Cuba. Avicennia 10-11: 157-164.

Ortea, J.; Espinosa, J.; Moro, L. (2009). Descripcion de una nueva especie de Nuevos taxones de babosas marinas de las islas Canarias y de Cabo Verde (Mollusca: Heterobranchia) A. Adams, 1850 (Mollusca: Opisthobranchia: Cephalaspidea) de las costas cubanas en el Golfo de México. Revista de la Academia Canaria de Ciencias 20(4): 29-34. [September 2009].

Ortea, J.; Espinosa, J.; Moro, L. (2009). Descripcion de una nueva especie del género *Kankelibranchus* Ortea, Espinosa & Caballer, 2004 (Mollusca: Nudibranchia: Polyceridae) de la Peninsula de Guanahacabibes, costa oeste de Cuba. Revista de la Academia Canaria de Ciencias 20(4): 35-43.

Ortea, J.; Espinosa, J.; Moro, L.; Caballer, M.; Bacallado, J. J. (2012). Notas en Opisthobranchia (Mollusca, Gastropoda) 5: Sobre el uso de la concha interna como caracter sistematico de primer orden en el inventario de las especies atlanticas de la familia Aglajidae (Mollusca: Cephalaspidea). Revista de la Academia Canaria de Ciencias. 24: 183-195.

Ortea, J.; Luque, A. A.; Templado, J. (1990). Contributions to the knoweldge of the genus *Aegires* Lovén, 1844 (Opisthobranchia: Doridoidea: Aegiretidae) in the North Atlantic, with descriptions of two new species. Journal of Molluscan Studies 56 (3): 333-338

Ortea, J. & Moro, L. (2016). Nuevos datos sobre el género *Jorunna* Bergh, 1876 (Mollusca: Heterobranchia: Discodorididae) en la Macaronesia y el mar Caribe. Vieraea. 44: 25-52.

Ortea, J.; Moro, L.; Bacallado, J. J. (2017). Nota sobre dos especies del género *Elysia* Risso, 1818 (Mollusca: Sacoglossa) en las islas de Cabo Verde. Revista de la Academia Canaria de Ciencias. 29: 21-30.

Ortea, J.; Moro, L.; Bacallado, J. J.; Caballer, M. (2014). Música y Naturaleza: Descripción de dos especies nuevas de babosas marinas (Mollusca: Gastropoda) colectadas entre dos aguas, Algeciras y Cancún, nombradas en honor de Paco de Lucía y su obra. Rev. Acad. Canar. Cienc., Vol. XXVI, 281-292.

Ortea, J.; Moro, L.; Caballer, M.; Espinosa, J. (2011). Avances en el inventario del género *Elysia* Risso, 1818 (Mollusca: Sacoglossa) en el Mar Caribe: redescripción de *Elysia purchoni* Thompson, 1977 y descripción de tres nuevas especies. Revista de la Academia Canaria de Ciencias 22(3): 199-213. [October 2011].

Ortea, J.; Moro, L.; Espinosa, J. (2007 ["2006"]). Descripción de dos nuevas especies de Pease, 1860 (Mollusca: Opisthobranchia: Cephalaspidea) de Cuba y Bahamas con comentarios sobre las especies atlánticas del género. Revista de la Academia Canaria de Ciencias 18(3-4): 33-52. [Nominal issue for 2006; published August 2007].

Ortea, J.; Moro, L.; Espinosa, J. (2009). The genus *Okenia* Menke, 1830 (Mollusca: Nudibranchia) in Canary Islands, with remarks on Okenia zoobotryon (Smallwood,1910), a species in permanent controversy. Vieraea 37:7583.

Ortea, J.; Moro, L.; Espinosa, J. (2015). Estudio de un grupo de especies caribeñas enmascaradas en el nombre *Aegires sublaevis* Odhner, 1932 (Mollusca: Nudibranchia) utilizando técnicas tradicionales. Revista de la Academia Canaria de Ciencias. 27: 243-258.

Ortea, J.; Moro, L.; Espinosa, J. (2017). El género *Lapinura* (Marcus & Marcus, 1963) (Mollusca: Runcinacea) en el Atlántico, con la descripción de nuevas especies de las islas de Cabo Verde y Costa Rica. Avicennia. 21: 11-18.

Ortea, J. & Templado, J. (1989). Una nueva especie de *Cyerce* Bergh, 1871 (Opisthobranchia: Ascoglossa) de la isla de Cuba. Iberus. 8(1): 11-14.

Ortea, J.; Valdés, Á.; Espinosa, J. (1994). North Atlantic Nudibranchs of the *Chromodoris clenchi* Color Group (Opisthobranchia: Chromodorididae). J. Moll. Stud. 60: 237-248.

Ornelas-Gatdula, E.; Camacho-García, Y.; Schrödl, M.; Padula, V.; Hooker, Y.; Gosliner, T. M.; Valdés, Á. (2012). Molecular systematics of the '*Navanax aenigmaticus*' species complex (Mollusca, Cephalaspidea): coming full circle. Zoologica Scripta. 41(4): 374-385.

Ornelas-Gatdula, E.; Dupont, A.; Valdés, Á. (2011). The tail tells the tale: taxonomy and biogeography of some Atlantic *Chelidonura* (Gastropoda: Cephalaspidea: Aglajidae) inferred from nuclear and mitochondrial gene data. Zoological Journal of the Linnean Society. 163, 1077-1095.

Ornelas-Gatdula, E. & Valdés, Á. (2012). Two cryptic and sympatric species of (Cephalaspidea: Aglajidae) in the Bahamas distinguished using molecular and anatomical data. Journal of Molluscan Studies. 78(4): 313-320.

Ortigosa, D.; Lemus-Santana, E.; Simoes, N. (2015). New records of "opisthobranchs" (Gastropoda: Heterobranchia) from Arrecife Alacranes National Park, Yucatan, Mexico. Marine Biodiversity Records. Vol. 8, 1-18.

Ortigosa, D. & Valdés, Á. (2012). A new species of *Felimare* (formerly *Mexichromis*) (Gastropoda: Opisthobranchia: Chromodorididae) from the Yucatan Peninsula, Mexico. The Nautilus. 126(3), 98-104.

Oskars T.R., Bouchet P. & Malaquias M.A. (2015). A new phylogeny of the Cephalaspidea (Gastropoda: Heterobranchia) based on expanded taxon sampling and gene markers. Molecular Phylogenetics and Evolution. 89: 130-150.

Padula, V.; Araújo, A. K.; Matthews-Cascon, H.; Schrödl, M. (2014). Is the Mediterranean nudibranch *Cratena peregrina* (Gmelin, 1791) present on the Brazilian coast? Integrative species delimitation and description of *Cratena minor* n. sp. Journal of Molluscan Studies. 80: 575-584.

Padula, V.; Bahia, J.; Stöger, I.; Camacho-García, Y.; Malaquias, M. A. E.; Cervera, J.,L.; Schrödl, M. (2016). A test of color-based taxonomy in nudibranchs: Molecular phylogeny and species delimitation of the *Felimida clenchi* (Mollusca: Chromodorididae) species complex. Molecular Phylogenetics and Evolution. 103: 215-229.

Padula, V. & Valdés, Á. (2012). Phylogeny and biogeography of *Paradoris* (Nudibranchia, Discodorididae), with the description of a new species from tyhe Caribbean sea. The Veliger, 51(3): 165-176. [16 November.

Pola, M.; Cervera, J. L.; Gosliner, T. M. (2006). Taxonomic revision and phylogenetic analysis of the genus *Tambja* Burn, 1962 (Mollusca, Nudibranchia, Polyceridae). Zoologica Scripta 35(5):491-530.

Pola, M.; Cervera, J. L.; Gosliner, T. M. (2005). A new species of *Tambja* (Nudibranchia: Polyceridae: Nembrothinae) from southern Brazil. Journal of the Marine Biological Association of the United Kingdom, 85(4): 979-984.

Pola, M.; Rudman, W. B.; Gosliner, T. M. (2009). Systematics and preliminary phylogeny of Bornellidae (Mollusca: Nudibranchia: Dendronotina) based on morphological characters with description of four new species. Zootaxa. 1975: 1-57.

Pola, M.; Cervera, J. L.; Gosliner, T. M. (2008). Description of the first *Roboastra* species (Nudibranchia, Polyceridae, Nembrothinae) from the Western Atlantic. Bulletin of Marine Science. 83(2): 391-399.

Pola, M. (2014). The identity of *Okenia zoobotryon* (Smallwood, 1910) (Nudibranchia: Goniodorididae): Redescription and proposed designation of a neotype. American Malacological Bulletin. 33(1): 1-6.

Pola, M.; Cervera, J. L.; Gosliner, T. M. (2005). Review of the systematics of the genus *Roboastra* Bergh, 1877 (Nudibranchia, Polyceridae, Nembrothinae) with the description of a new species from the Galápagos Islands. Zoological Journal of the Linnean Society, 144(2): 167-189.

Redfern, C. (2001). Bahamian Seashells: A Thousand Species from Abaco, Bahamas. Bahamian Seashells, Boca Raton, Forida, ix + 280 pp., 124 pls.

Redfern, C. (2013). Bahamian Seashells: 1161 Species from Abaco, Bahamas. Bahamian Seashells.com Inc., Boca Raton, Forida, 501 pp., 3600 color photos, 170 b/w photos.

Rudman, W.; Sea Slug Forum. https://www.seaslugforum.net

Rudman, W. B. (1974). A comparison of *Chelidonura*, *Navanax* and *Aglaja* with other Genera of the Aglajidae (Opisthobranchia: Gastropoda). Zoological Journal of the Linnean Society. 54: 185-212.

Saad, L. O.; Magenta, C.; Colpo, K. D.; Valdés, Á. (2014). It's not what it looks like: molecular data fails to substantiate morphological differences in two sea hares (Mollusca, Heterobranchia, Aplysiidae) from southern Brazil.

Sales, L.; Migotto, A. E.; Baroni, S.; Cunha, C. M. (2019). Taxonomic reassessment and redescription of *Okenia polycerelloides* (Ortea & Bouchet, 1983) (Nudibranchia: Gonio,dorididae) based on morphological and molecular data. Marine Biodiversity; https://doi.org/10.1007/s12526-019-00972-w.

Sanvicente-Añorve, L.; Solis-Weiss, V.; Ortigosa, J.; Hermoso-Salazar, M.; Lemus-Santana, E. (2012). Opisthobranch fauna from the National Park Arrecife Alacranes, southern Gulf of Mexico. Cah. Biol. Mar. 53: 447-460.

Schrödl, M. & Millen, S. V. (2001). Revision of the nudibranch gastropod genus *Tyrinna* Bergh, 1898 (Doridoidea: Chromodorididae). Journal of Natural History 35(8): 1143-1171.

Silva, F. V.; Azevedo, V. M.; Matthews-Cascon, H. (2014). A new species of *Tritonia* (Opisthobranchia: Nudibranchia: Tritoniidae) from the tropical South Atlantic Ocean. Journal of the Marine Biological Association of the United Kingdom. 94(3): 579–585.

Troncoso, J. S.; Garcia, F. J.; Urgorri, V. (1998). Anatomical data on a rare *Hypselodoris picta* (Schultz, 1836) (Gastropoda, Doridacea) from the coast of Brazil with description of a new subspecies. Bulletin of Marine Science 63 (1): 133-141.

Thompson, T. E. (1977). Jamaican opisthobranch molluscs I. Journal of Molluscan Studies. 43: 93-140.

Thompson, T. E. (1980). Jamaican Opisthobranch Molluscs II. Journal of Molluscan Studies. 46, 74-99.

Thompson, T. E. (1977). The taxonomic status of two Bermudan opisthobranchs. Journal of Molluscan Studies, 43: 217-222.

Thompson, T. E. (1981). Redescription of a rare North Atlantic Doridacean Nudibranch, *Aegires sublaevis* Odhner. The Veliger 23 (4): 315-316.

Valdés, Á. (2002). A phylogenetic analysis and systematic revision of the cryptobranch dorids (Mollusca, Nudibranchia, Anthobranchia). Zoological Journal of the Linnean Society. 136: 535-636.

Valdés, Á. (2002). How many species of *Hexabranchus* (Opisthobranchia: Dorididae) are there ?. Molluscan Research. 22(3): 289-301.

Valdés, Á (2004). Tropical western Atlantic species of *Diaulula* Bergh, 1878 (Mollusca, Nudibranchia), with the description of a new species. Contributions in Science (Los Angeles). (501):1-7.

Valdés, Á. (2005). A new species of *Aeolidiella* Bergh, 1867 (Mollusca: Nudibranchia: Aeolidiidae) from the Florida Keys, USA. The Veliger 47(3): 218-223.

Valdés, Á & Afonso, C. M. L. (2003). Synonymy of *Cadlina burnayi* Ortea, from the Cape Verde Islands (Opisthobranchia: Chromodorididae) with *Tyrinna evelinae* (Marcus, 1958). The Veliger. (2): 176-179.

Valdés, Á.; Breslau, E.; Padula, V.; Schrödl, M.; Camacho, Y.; Malaquias, M. A. E.; Alexander, J.; Bottomley, M.; Vital, X. G.; Hooker, Y.; Gosliner, T. M. (2018). Molecular and morphological systematics of *Dolabrifera* Gray, 1847 (Mollusca: Gastropoda: Heterobranchia:Aplysiomorpha).Zoological Journal of the Linnean Society, 184, 31-65.

Valdés, Á. & Campillo, O. A. (2004). Systematics of pelagic aeolid nudibranchs of the family Glaucidae (Mollusca, Gastropoda). Bulletin of Marine Science 75(3): 381–389.

Valdés, Á. & Hamann, J. (2008). Two new species of *Doriopsilla* from the tropical Western Atlantic with remarks on Cariopsillidae Ortea & Espinosa, 2005. The Veliger. 50(3): 210-218.

Valdés, Á.; Hamann, J.; Behrens, D. W.; DuPont, A. (2006). Caribbean Sea Slugs: A field guide to the Opisthobranch Mollusks from the Tropical Northwestern Atlantic. Sea Challengers Natural History Books, Gig Harbour, Washington: 289 pp.

Valdés, Á.; Medrano, S.; Bhave, V. (2016). A new species of *Cuthona* Alder and Hancock, 1855 (Gastropoda: Heterobranchia: Nudibranchia: Tergipedidae) from the Caribbean Sea. The Nautilus. 130(2): 72-78.

Valdés, Á. & Ortea, J. (1996). Review of the family Phyllidiidae in the Atlantic Ocean. Amer. Malac. Bull. 13(1/2).

Valdés, Á. & Ortea, J. (1997). Review of the genus *Doriopsilla* Bergh, 1880 (Gastropoda, Nudibranchia) in the Atlantic Ocean. The Veliger. 40(3): 240-254.

Valdés, Á.; Ortea, J.; Ávila, C.; Ballesteros, M. (1996). Review of the genus *Dendrodoris* Ehrenberg, 1831 (Gastropoda: Nudibranchia) in the Atlantic Ocean. Journal of Molluscan Studies. 62(1): 1-31.

Wägele, H.; Vonnemann, V.; Rudman, W. B. (2006). *Umbraculum umbraculum* (Lightfoot, 1786) (Gastropoda, Opisthobranchia, Tylodinoidea) and the synonymy of *U. mediterraneum* (Lamarck, 1812). Records of the Western Australian Museum Supplement. 69, 69-82.

Zamora-Silva, A.; Malaquias, M. A. E. (2018 [nomenclatural availability: 2017]). Molecular phylogeny of the Aglajidae head-shield sea slugs (Heterobranchia: Cephalaspidea): new evolutionary lineages revealed and proposal of a new classification. Zoological Journal of the Linnean Society. 183(1): 1-51.

Zamora-Silva, A.; Ortigosa, D. (2012). Nuevos registros de opistobranquios en el Parque Nacional Sistema Arrecifal Veracruzano, México. Revista mexicana de biodiversidad, 83: 359 - 369.

Glossary

acleioproctic	anus that is not enclosed within or associated with a ceratal arch (see cleioproctic); usually on/near the dorsal midline
acoel	A small (< 15 mm) soft-bodied unsegmented marine worm in the order Acoela. The name 'acoel' refers to the lack of a coelom, the cavity that holds the digestive tract and other visceral organs.
aeolid	a nudibranch in the superfamily Aeolidioidea
aglajid	a sea slug in the family Aglajidae
aplysiid	a sea slug in the family Aplysiidae; i.e. a seahare
aposematism	use of bright colors as a warning sign of toxicity or unpalatability
amphiatlantic	inhabits the east and west sides of the Atlantic Ocean
annulate	in reference to a rhinophore – having a series of rings or bands along its stalk
antero-	a prefix meaning "at the front"
aperture	the opening or gape of a shell
apex (pl. apices)	peak, tip
arborescent	branched, tree-like
ascus	a pouch in the buccal cavity of sacoglossans for storing worn out radular teeth
auriculate	having ear-shaped parts or appendages; ear-like
autotomize	to discard a body part, usually as a distraction for self defense
benthic	associated with or occurring at the bottom of a body of water
biconical	shaped like two cones joined at their bases
bifid	cleft or split into two parts or lobes
bifurcate	two-branched
bilabiate	two-lipped
bipinnate	describes a stalk with unipinnate appendages (here, in reference to gill plumes)
branchial	related to the gills; e.g. branchial plumes, leaves, pocket, sheath, etc.
bryozoans	Small colonial aquatic invertebrates belonging to the phylum Bryozoa. Colonies can be flat encrusting sheets on rocks, shells, algae, etc., or branching lace-like structures ranging in size from a few millimeters to several meters across.
buccal	related to the mouth
buccal mass	a muscular feeding structure between the oral tube and esophagus which contains and controls the radula
buccal pump	a powerful sucking organ that some species use to pull food into their mouths.
caenosarc	the shared, living, soft tissue that connects the individual members (zooids) of a colonial invertebrate
cardiac	pertaining to the heart
caruncle	an elevated ridge or lobe between or anterior to the rhinophores in some nudibranchs; presumed to have a sensory function
caryophyllidium	a sensory apparatus consisting of a tubercle encircled by spicules (pl. caryophyllidia)
cephalic	pertaining to the head.
cephalic shield	a flat broad fleshy plate covering the head and often a portion of the body behind it.
cerata (sing. ceras)	finger-like respiratory dorsal appendages configured as a row of clusters along each side of the back; the cerata within a cluster are usually arranged to form arches or oblique lines; digestive gland branches may extend into the cerata
cilia	microscopic hair-like surface structures (e.g. on rhinophores, oral tentacles, foot) that have several functions: locomotion by coordinated "beating"; sensory perception for locating food, a mate, or a predator; food ingestion by creating currents to direct floating particles to the mouth
circumtropical	globally distributed in tropical regions
circumglobal	globally distributed
cleioproctic	anus on the right side of the body within the arch of the frontmost ceratal cluster

cnidosac	the sac at the tip of each ceras, where nematocysts are stored for defense
columella	the central axial pillar within the coil of a gastropod shell
congener	a member of the same genus
conspecific	the same species
copepod	a small (0.5-2.0 mm) translucent teardrop-shaped marine crustacean belonging to the class Copepoda; very abundant in the water column, often the dominant form of zooplankton in the ocean
crenulated	having an edge or margin that is finely notched, scalloped, or wavy
cryptic species	genetically distinct species having nearly identical external appearances; their differences have usually been attributed to intraspecies variability, but genetic studies are now often pin-pointing subtle external characters that allow for reliable identifications
cryptobranch	a dorid that can retract its gill into a protective cavity
cyanobacterium	a microscopic, single-celled organism belonging to the domain Bacteria; defined by its ability to perform photosynthesis; often incorrectly called "blue-green algae"
cytosol	the liquid portion of the cytoplasm within a cell
denticle	a small tooth-like projection (denticulate = with denticles)
dentitheca	Historically, a term that refers to a slug's entire "used tooth repository", not just the storage pouch within it called the "ascus". However, in modern usage the two terms are commonly synonymized.
digitiform	finger-like
dinoflagellate	From: dino (Gr.) = whirling; flagellum (L.) = whip. A single-celled marine organism belonging to the phylum Dinoflagellata. It has two whip-like tails: a longitudinal one providing forward thrust while the other wraps around the cell in a groove and causes it to spin. The base of many tropical food webs (e.g. diet for larval stages of many slugs) and for important symbiotic interactions.
distal	away from a structure's point of attachment
diverticula (sing. diverticulum)	Pouch and tube-like structures that branch off from the digestive tract (primarily the digestive gland) and run throughout the body and may extend into many of the external projections (e.g. cerata, rhinophores). They are the sites where food digestion and nutrient absorption and storage take place.
dorid	any sea slug belonging to the order Doridida
dorso-lateral	on top and on the side
dorso-ventral	in the top-to-bottom direction, such as from the mantle to the foot
dorsum	top surface, back
echinoids	A collective term for marine invertebrates belonging to the class Echinoidea. It includes sea urchins, sand dollars, sea biscuits.
ecotype	a distinct form or population within a species that has genetically adapted to specific local environmental conditions but still remains within that species
eelgrasses	The common name for species of marine flowering plants belonging to the genus *Zostera*. They form extensive and dense underwater meadows in shallow, protected coastal areas, bays, and estuaries.
epibenthic	living on or just above the seafloor
epifauna	organisms that live on top of a substrate, either permanently attached or mobile
epipelagic zone	the ocean layer extending from the surface to ~660 feet, where sunlight can support photosynthesis
epiphyte	an organism that lives on the surface of another living organism without feeding on it
eversible	capable of being pushed out of its normal position, often to perform a function like grasping or probing, and then quickly withdrawn into its original position
exhalant	describes the outflow of water (the opposite of inhalant)
flexions	motions involving the bending of a body or its appendages
foramen	in an anatomical context – a natural opening, hole, or passage found in bone, cartilage, or other hard or semi-rigid structure

foraminifera	A large phylum of single-celled organisms found throughout marine and freshwater environments and distinguished by their shells, which range in size from microscopic to a few centimeters. The shells have pores called "foramina" through which thread-like appendages extend for the purposes of feeding, locomotion, or anchoring. Small foraminifera are common prey for planktonic larval stages of sea slugs.
fusiform	spindle-shaped (tapering gradually toward both ends)
gill	the respiratory organ
globose	globe-shaped
headshield	cephalic shield
herbivorous	feeding on plant material
hermaphrodite	an individual having functional male and female sex organs and able to produce both sperm and eggs
heterostrophic	a condition of a shell where the protoconch whorls are coiled in the opposite direction from that of the adult whorls
holoplankton	marine organisms that spend their entire lives as plankton
hydroids	Small invertebrates (often colonial) in the class Hydrozoa within the phylum Cnidaria. Most form feather-like structures with thousands of stinging cells (cnidocytes) on rocks, shells, and algae. They are the primary food source for all aeolid nudibranchs, which store undischarged cnidocytes in their cnidosacs and, when required, discharge their nematocysts for defensive purposes.
hydrozoan	any organism in the class Hydrozoa (hydroids, siphonophores, fire coral)
hyponotal	on the underside of the mantle
imperforate	lacking an opening, hole, or perforation; closed off
infauna	organisms that live within soft sediments, i.e. beneath the surface of a substrate
inrolled	rolled inward along its length, creating a slit or tube opening along one side; scroll-like
interhepatic	between the parts of the digestive gland
intertidal zone	region between low and high tide lines which is alternately exposed and submerged
intraspecific	occurring or arising within the bounds of a single species
ironshore	a term used in the Caribbean region denoting rocky, rugged coastal shorelines composed of exposed, dark, highly pitted, eroded limestone
jaws	thick chitinous plates inside the buccal cavity; used for holding prey while feeding
kleptoplasty	a biological phenomenon where an organism sequesters functional chloroplasts from its food source and integrates them into its own tissues in order to utilize these "stolen" organelles to perform photosynthesis for its own metabolic needs
lamella (pl. lamellae)	a thin layered structure designed to maximize surface area for sensory, respiratory, or structural functions
lamellate	having lamellae
lateral teeth	teeth flanking the central tooth on the radula
lecithotrophic	a mode of development in which the larva sustains itself entirely on the yolk (lecithin) reserves within the egg; characterized by a shorter planktonic period, larger eggs (more yolk), and a reduced dispersal ability relative to planktotrophic feeding
littoral zone	The entire near-shore coastal region, which includes the intertidal zone and extends both above and below it with abundant sunlight, oxygen and nutrients.
macroalgae	seaweeds
mamillate	nipple-shaped
mantle	dorsal body wall that covers the visceral mass; in shelled mollusks it builds the shell
mantle cavity	a protected chamber enclosed by a fold of the mantle; it houses vital organs in most gastropod mollusks; absent in nudibranchs and dorids
mantle glands	specialized structures located within mantle tissue that produce and store defensive chemicals; especially important for sea slugs that lack a shell and its physical protection
masticatory	related to, adapted for, or serving the function of chewing or grinding food
median	on the midline

mesopelagic zone	Ocean layer typically from 660 feet to 3300 feet, where there is not enough light to support photosynthesis. Many of its inhabitants engage in a diurnal mass vertical migration, wherein they ascend to the epipelagic zone at night and return before daylight
microalgae	Microscopic single-celled organisms that perform photosynthesis. The "unseen" primary producers that sustain all other life in the epipelagic zone.
morphology	the structural features of an organism
morphometric	related to measurements of form and shape
multipinnate	as used here: any or all of unipinnate, bipinnate, tripinnate, quadripinnate
nematocyst	A small fluid-filled high-pressure capsule inside a stinging cell (cnidocyte) that contains a tightly coiled, barbed, and venom-injecting tubule. They are defensive weapons created primarily inside the tentacles of all cnidarians (anemones, corals, hydroids, etc.). Sea slugs that feed on those animals can extract the cnidocytes from the ingested tissues, store them within their cnidosacs, and deploy the nematocysts as needed for defensive purposes.
notum	top surface of mantle; dorsum
operculum	a calciferous plate that completely or partially closes the aperture through which an animal has withdrawn into its shell
oral tentacles	a pair of sensory tentacles or flaps around the mouth used for exploring the immediate environment and locating food
oral veil	a sensory structure at the anterior of the head above the mouth; typically a pair of flattened lobes or projections
papilla (pl. papillae)	a small finger-like protuberance
papillate/papillose	covered with papillae
parapodia	thin flap-like lateral extensions of the foot that usually fold up and over the body, where they sometimes meet; some functions are physical protection, creation of dorsal siphons for directing water flow past the gills, and swimming when flapped like wings
pedunculate	supported by a stalk, stem, or peduncle
pelagic	occurring in the open ocean; the opposite of benthic
penial	related to the penis
perfoliate	Often used synonymously with the term "lamellate" for describing a rhinophore. However this word draws attention to the fact that the lamellae appear to be "pierced through" by the supporting stalk.
pericardial	pertaining to the pericardium
pericardium	a membranous sac or cavity that completely encloses and protects the heart
periostracum	a shell's organic outermost layer in some mollusks; it can be thin, thick, smooth, fibrous, hairy, horny; it often obscures a shell's color pattern
perisarc	the non-living transparent protective covering that surrounds the living tissue (the coenosarc) of colonial hydroids and some other hydrozoans
phagomimicry	a form of defensive mimicry where an organism has evolved to resemble a body or substance that a predator has either already consumed and found unpalatable, or one that the predator finds unworthy of consumption (e.g. detritus, waste)
phanerobranch	(Greek: phaneros = visible, brankhia = gill), in reference to a dorid having a gill that can contract into a tight cluster of gill leaves just above the surface of the mantle but cannot retract into a pocket beneath it.
phylogenetic	relating to or based on evolutionary history, relationships, and descent
phytoplankton	planktonic organisms such as microscopic algae and cyanobacteria that survive via photosynthesis; basal to the marine food web, producing about 50% of the planet's oxygen
pinnate	a symmetrical feather-like structure in which branches (pinnae) project equally from two opposite sides of a central axis
plankton	a group of organisms that passively drift in the water column usually because they are unable to swim against a current (planktos (Gr.) means "wandering")
planktonic	living in the water column as a passive drifter; their sizes vary greatly from microscopic (e.g. bacteria, viruses) to macroscopic (e.g. jellyfishes, *Sargassum*)

planktotrophic	a mode of development in which the larva enables its development by actively feeding on small organisms floating in the water column (plankton) until its metamorphosis into a juvenile; characterized by small eggs (with little yolk), extended floating and feeding time, and high dispersal ability
pleuroproctic	anus on the right side of the body below the ceratal clusters
plicate	having folds, ridges, or creases (plicae)
-pod (pl. -poda)	Greek for foot (feet), in terms related to foot structures; e.g. pseudo**pod**, Ptero**poda**
polycerid	any sea slug belonging to the family Polyceridae
post- / (postero-)	a prefix meaning after or behind / (at the rear)
propodium / (propodial)	frontmost section of the foot / (related to the frontmost section of the foot)
protandrous	matures first as a functional male and later changes sex to become a functional female
protoconch	the initial larval shell that is formed while a gastropod mollusk is still an embryo or a free-swimming larva
proximal	close to a structure's point of attachment (opposite of distal)
pseudobranch	a "false gill"; used here specifically for *Doto* species in reference to the tentacle-like appendages at the inner bases of some of their cerata
pseudoconch	a "false shell"; in reference to Gymnosomata it is a small, ephemeral, often lightly calcified structure that may temporarily form around the internal organs to replace a discarded protoconch
pseudofeces	in Pteropoda, the undigested material and rejected particles that are bound together by mucus and expelled without having passed through the digestive tract and anus
pustulate	covered with small blister-like or pimple-like bumps (pustules)
pyriform	pear-shaped
quadrate	with reference to a biological structure it often implies a blocky or squarish appearance
rachidian tooth	the central tooth in each transverse row of teeth on the radula
rachis (pl. rachises)	in reference to a gill consisting of leaves (plumes) it is the axis of the gill leaf
radula (pl. radulae)	a flexible ribbon-like extendable structure within the buccal cavity having many transverse crowded rows of tiny chitinous teeth; feeding occurs via scraping, tearing, puncturing, and drawing in the prey (e.g. sponges, corals, other slugs); it is self-renewing: worn-out anterior teeth are shed and new posterior ones are moved forward; the tooth shape/configuration is a useful tool for species identification and classification
renopericardium	the complex formed by the adjacent and interconnected kidney (nephridium) and the pericardium
resorb	to absorb and assimilate back into the body mass after its purpose has been fulfilled or it is no longer needed
rhinophores	a pair of tentacle-like chemo-sensory organs on the head or at the front of the mantle
rhizoid	a root-like structure on non-vascular plants (like algae) for anchoring to a substrate
Sargassum	a genus of large buoyant brown seaweed typically found floating in massive mats on the surface of the open ocean
seagrasses	flowering plants evolved to live fully submerged in marine or estuarine environments; unlike seaweeds they have true roots, stems, and produce flowers and seeds
sesquiterpenes	Natural organic compounds found in sponges and algae that are produced as secondary metabolites to act as powerful chemical defenses. Nudibranchs that feed on sponges do not digest sesquiterpenes but sequester them in mantle tissues or specialized glands for defensive purposes.
sidegill	a gill located on the right side of the body between the foot and the edge of the mantle
siphonal	pertaining to a siphon
speciose	consisting of many species
spicule	a small glass-like needle or rod embedded in mantle tissue to give it reinforcement
spiculose/spiculated	possessing spicules
spire	the set of whorls of a gastropod shell excluding the final one (the body whorl)
stellate	star-like

striae	fine parallel grooves on the surface of a shell
stylet	A small, hard, sharp, needle-like or dagger-like structure that is used by an organism primarily for piercing, stabbing, or injecting material. Slugs may have a penial stylet for reproduction or a feeding stylet (radula) for feeding.
subapical	just below the apex
submarginal	situated just inside, beneath, or near the edge/margin
sublittoral zone	the part of the littoral zone from the low-tide line to the edge of the continental shelf
subtidal zone	same as sublittoral zone
suctorial	an adjective describing a specialized feeding method using a sucking mechanism
suture	a line or groove on the surface of a spiral shell at the junction of two adjacent whorls
taxon (pl. taxa)	any unit of classification used in the scientific system of taxonomy
teleoconch	The portion of a shell secreted and grown by a mollusk after it leaves the larval stage. It begins where the protoconch ends.
temperate zone	a zone characterized by moderate temperatures and distinct seasons
tetrapinnate	describes a stalk with tripinnate appendages (here, in reference to gill plumes)
Thalassia	a genus of seagrasses that form large dense meadows in shallow, tropical and subtropical coastal waters; commonly referred to as "turtle-grass"
trilobate	three-lobed
tripinnate	describes a stalk with bipinnate appendages (here, in reference to gill plumes)
triseriate	arranged in three series or rows
tubercle	a small, rounded, surface protuberance with a base diameter greater than or equal to its height
tunicate	sea squirt, ascidian
umbilicate	having a small, central depression or pit (umbilicus)
uncinate	hook-shaped
unipinnate	describes a stalk with unbranched appendages (here, in reference to gill plumes)
uniseriate	having a single row
utricle	A small, flask- or club-shaped cell that forms the outer layer of tissue in certain large siphonous green algae. Utricles are the targets for feeding sap-sucking slugs.
velar	related to the oral veil
veliger	the second, usually free-swimming larval stage of many aquatic mollusks, named for possessing a **velum** that it uses for swimming and feeding
velum	Usually a pair of large ciliated lobes that extend from the head of a veliger larva. It enables locomotion through a water column via the beating of its surface cilia, which also create a current to sweep microscopic food particles (e.g. phytoplankton) into the larva's mouth. The velum is lost as the veliger transitions into its bottom-dwelling adult form (metamorphosis).
ventral	pertaining to the underside
ventrum	the underside
visceral mass (viscera)	cluster of internal organs (e.g. stomach, intestines, digestive glands, gonads, kidney, gill)
wingplate	parapodium of a pteropod, used exclusively for swimming in the open ocean
zoanthids	Small, colonial, anemone-like, sessile invertebrates belonging to the Order Zoantharia, found primarily in tropical and subtropical marine habitats, often forming dense, colorful mats that cover rocks, dead coral, and sponges. Preyed on and camouflaged by slugs that have evolved immunity to their nematocysts and toxins.
zooplankton	animal-like planktonic organisms (e.g. copepods, krill, some slugs, larval stages of many animals) that feed on phytoplankton and other zooplankton, thereby providing sustenance for higher levels of the food chain
zooxanthellae	single-celled photosynthetic algae that live symbiotically with numerous marine invertebrates, including some sea slugs

Scientific Name Index

Common Name Index

Photo Credits (by page)

Vertical position: T=top, U=upper middle, M=middle, L=lower middle, B=bottom
Horizontal position: l=left, m=middle, r=right

1: Bill Frank (Ml), Conquiliologistas do Brazil (Br), Marlo F. Krisberg (T)
2: Daphne Tsapalas (Br), Ellen Muller (Tr), Everett Turner (Mr), Linda Ianniello (Tl), Sandra Edwards (Ml)
3: Anne DuPont (Tl), Ben Eddy (Ml), Luke Foster (Mr), Markus Niemeyer (Tr), Mickey Charteris (Bl)
4: Alev Ozten (Mr), Anne DuPont (Tl), Everett Turner (Br), Linda Ianniello (Bl/Ll), Mickey Charteris (Tr/Ul)
5: Anne DuPont (Tr), Cynthia Abgarian (Ur), Keri Wilk (Ml), Mélodie Caussat (Tm), Pauline Walsh Jacobson (Bl), Sandra Edwards (Tl)
6: Keri Wilk (Tl), Mauricio Riquelme (Mr), Patricia Horie (Ml), Pauline Walsh Jacobson (Tr), Phil Croy (Bl/Br)
7: André de Molenaar (Br), Anne DuPont (Bl), Ariane Dimitris (Mr), Keri Wilk (Ml), Ray Haberman (Tl)
8: Frank Krasovec (Tl), John Fowler (Ml), Keri Wilk (Bl), Mickey Charteris (Mr), Nuch Uthairat (Br)
9: Anne DuPont (Ur), Jack Israel (Lr), Kris Wilk (Tr), Linda Ianniello (Tl), Mélodie Caussat (Bl/Ml), Tom Poff (Br)
10: Robert Aguilar (Smithsonian) (Bl/Br), Sandra Edwards (Tl), Terence Gosliner (Tr)
11: Jacob Weston (B), Marlo F. Krisberg (T), Prof. Carlos Sigmaringa & PhD Alexandre Pimenta (M)
12: Marlo F. Krisberg (Br/Tr)
13: Alain Goyeau (Tr), Anne DuPont (Lr), Ellen Muller (Ll), Julie Mellinger (Bl/Tl), Laetitia Rossignol (Ul), Pauline Walsh Jacobson (Br), Terence Gosliner (Ur)
14: Deneb Ortigosa (Bl), Mélodie Caussat (Tl), Terence Gosliner (Ml/Mr)
15: Angel Valdes (Tl), Linda Ianniello (Ml), Marlo F. Krisberg (Tr), Sandra Edwards (Mr), Xavier Salvador (Bl/Br)
16: Anne DuPont (Ml/Mr), Carole Assier de Pompignan (Bl), Cherie-Anne Henderson Dam (Tl), Marlo F. Krisberg (Br), Terence Gosliner (Tr)
17: Anne DuPont (Bl), Everett Turner (Ml), Markus Niemeyer (Tr), Marlo F. Krisberg (Br), Mélodie Caussat (Tl), Terence Gosliner (Mr)
18: Everett Turner (Mr), Frank Krasovec (Tr), Gustav Paulay (FMNH) (Bl), Marlo F. Krisberg (Br/Ml), Mickey Charteris (Tl)
19: Courtesy of Jesus Ortea (Tl), Kerry Lewis (Br), Smithsonian Marine Station (Tr), Yan Buske (Bl/Mr)
20: Anne DuPont (Bl/Br), Cynthia Abgarian (Tr), Mickey Charteris (Tl)
21: Cherie-Anne Henderson Dam (Ml), Keri Wilk (Br/Mr), Kerry Lewis (Tr), Kierstin Zetterberg (Tl), Sandra Edwards (Bl)
22: David Kirsh (Bl/Tr), Manuel Caballer (Br/Tl)
23: Ben Eddy (Mr/Tr), Bill Frank (Bl), Phil Liff-Grieff (Tl), Robert Aguilar (Smithsonian) (Br), Terence Gosliner (Ml)
24: Courtesy of Jesus Ortea (Bl/Ml/Tl), Courtesy of Jesus Ortea et al (Mr), Leon Altoff (Br), Manuel Caballer (Tr)
25: Courtesy of Jesus Ortea (Mr), Terence Gosliner (Bl/Br/Ml/Tl/Tr)
27: Linda Ianniello (Bl/Bm/Br/Ml/Mm/Mr)
28: Linda Ianniello (Bl/Br/Tm/Tr), Susan Mears (Bm), Suzan Meldonian (Tl)
29: Alicia Hermosillo McKowen (Tl), Linda Ianniello (Br/Mr/Tr), Susan Mears (Ml), Suzan Meldonian (Bl)
30: Linda Ianniello (Bl/Bm/Br/Tl), Suzan Meldonian (Tm/Tr)
31: Karen Osborn (Br), Linda Ianniello (Ml/Tr), Senem Çağlar and Bilal Ozturk (Tl), Steven Kovacs (Bl), Susan Mears (Mr)
32: Linda Ianniello (Bl/Br/Tl/Tr)
33: Deborah Devers (Bl), Linda Ianniello (Br/Ml/Mr/Tl/Tr)
34: Linda Ianniello (Bl/Br/Mr/Tl/Tr), Suzan Meldonian (Ml)
35: Deborah Devers (Tl/Tr), Linda Ianniello (Bm/Br/Tm), Susan Mears (Bl)
36: Alain Goyeau (Br), Alicia Hermosillo McKowen (Tr), Anne DuPont (Bl), Flavio Mendes (Ml), Julie Mellinger (Mr), Linda Ianniello (Tl)
37: Francisco Riquelme Tapia (Tl/Tr), Jorge Lopez (Br), Mickey Charteris (Bl/Mr), Ned DeLoach (Ml)
38: Anderson Rabello Pereira (Ur), Andy Martinez (Tl/Tr), Anne DuPont (Bl/Lr), Frank Krasovec (Br), Mickey Charteris (Ml)
39: Ariane Dimitris (Ml/Tl), Cynthia Abgarian (Mr), Everett Turner (Br/Tr), Mélodie Caussat (Bl)
40: Anne DuPont (Tl/Tr), Ariane Dimitris (Ml), Mélodie Caussat (B), Mickey Charteris (Mr)
41: Anne DuPont (Tr), Carole Assier de Pompignan (Tl), Julie Mellinger (Bl), Mickey Charteris (Ml), Sandra Edwards (Mr), Suzan Meldonian (Br)
42: Alain Goyeau (Ml), Jessica Pate (Tr), Laetitia Rossignol (Bl), Linda Ianniello (Br/Tl), Sandra Edwards (Mr)
43: Alain Goyeau (Bl), Cynthia Abgarian (Br), Everett Turner (Tl), Keri Wilk (Tr), Laetitia Rossignol (Mr), Mickey Charteris (Ml)
44: Julie Mellinger (Br/Tl), Julie Piron (Mr), Mickey Charteris (Tr), Rino Jackson (Bl), Terence Gosliner (Ml)
45: Alain Goyeau (Tl), Anne DuPont (Tr), Mélodie Caussat (Mr), Mickey Charteris (Bl), Ray Haberman (Br)
46: Ariane Dimitris (Br/Tr), Julie Mellinger (Ml), Mickey Charteris (Bl/Mr/Tl)
47: Ariane Dimitris (Tl/Tr), Betsey Putnam Hansen (Bl), Lucas Pescador Barcelos (Br), Nicholas Curtis (M)
48: Cherie-Anne Henderson Dam (Mr), Keri Wilk (Ml), Ray Haberman (Tl/Tr)
49: Carole Assier de Pompignan (Tr), Julie Mellinger (Tl), Laetitia Rossignol (Mr), Linda Ianniello (Bl/Br/Ml)
50: Everett Turner (M), Mickey Charteris (Br), Tricia O'Malley (Bl), Yan Buske (T)
51: Alicia Hermosillo McKowen (Br), Ariane Dimitris (Ml), Markus Niemeyer (Mr), Sandra Edwards (T), Yan Buske (Bl)
52: Jorge Mayorga (Bl/Br), Linda Ianniello (Ml), Sandra Edwards (Mr/Tl), Terence Gosliner (Tr)
53: Courtesy of Jesus Ortea (B), Manuel Caballer (M/T)
54: Alicia Hermosillo McKowen (Bl/Br), Mélodie Caussat (Tr), Mickey Charteris (Tl), Ray Haberman (Ll/Ul)
55: Andrew McKinlay (Tr), Chris Venus (Br), Erik Daniel Erikson (Ml), Keri Wilk (Tl), Leah Marcus (Bl), Terence Zahner (Mr)
56: Ariane Dimitris (Bl), Brenda Hill (Ml), Matt Sullivan (Tr), Mickey Charteris (Mr), Ray Haberman (Tl)
57: Alain Goyeau (Ml), Alasdair Dunlap-Smith (Tl/Tm), Keri Wilk (Mr), Lee Richter (Tr), Tim Kary (Br), Yan Buske (Bl)
58: Joe Tomoleoni (Tl), Karen Oliver (Bl), Leila Kim (Ml/Mr), Markus Niemeyer (Ml (inset)/Tr), Nestor Ardila (Br)
59: Alain Goyeau (Tl), Daniel Pinelli (Mr), Deneb Ortigosa (Bl/Br), Julie Mellinger (Ml), Yan Buske (Tr)
60: David Sanchez (Ll/Ul), Gwenaelle Barbotteau (Bl), Jason Phillip (Lr), Keri Wilk (Br/Tl), Linda Ianniello (Tr), Pauline Walsh Jacobson (Ur)
61: Daniel Prairie (Tr), Dewalt Kotze (Lr), Evelyn Caporali (Ur), Lureen Ferretti (Tl), Mickey Charteris (Bl), Paul Young (Br), Sandra Edwards (Ml)
62: Ariane Dimitris (Ml), Dianna Burkholder (Mr), Evelyn Caporali (Bl), Linda Ianniello (Br), Sandra Edwards (Tl), Svetlana Speransky (Tr)
63: Alain Goyeau (Ll), Armando de Luca Junior (Br), Kadu Pinheiro (Bl), Keri Wilk (Tr/Ur), Mauricio Riquelme (Lr), Terence Zahner (Tl)

64: Dave Holladay (Ml), Keri Wilk (Tl/Tr), Mickey Charteris (Br/Mr), Rino Jackson (Bl)
65: Bernard Picton (Br), Mike Krampf (Bl/Tr), Ray Haberman (Ml/Mr), Terence Zahner (Tl)
66: Ariane Dimitris (Bl), Dewalt Kotze (Tr), Evelyn Caporali (Tl), Linda Ianniello (Br), Lyndsey Emery (Ml), Raul Fernandez (Mr)
67: Cynthia Abgarian (Tr), Jason Phillip (Br), Keri Wilk (Lr/Tl/Ul), Linda Ianniello (Bl), Svetlana Speransky (Ur), Terence Zahner (Ll)
68: Ariane Dimitris (Ml/Tm), Brian Gortney (Tr), Dewalt Kotze (Mm), Manuel Caballer (Bl/Br), Pauline Walsh Jacobson (Mr), Scott McDuff (Tl)
69: Bernard Picton (Ml/Mr), Kadu Pinheiro (Bl/Br), Leslie Harris (LACM) (T)
70: Manuel Caballer (B), Mike G. Rutherford (T), Terence Gosliner (L/Ul/Ur)
71: Ben Eddy (Ml), Jessica Riederer (Mm/Mr/Tr), Luke Foster (Br), Ned DeLoach (Tl), Terence Gosliner (Bl)
72: Cynthia Abgarian (Ml), Daniel Prairie (Bl/Mr), Jason Phillip (Tl), Keri Wilk (Tr), Sandra Edwards (Br)
73: Anne DuPont (Tm), Ariane Dimitris (Bl), David Sanchez (Br), Everett Turner (Tr), Gabriel Jensen (Mr), Linda Ianniello (Tl)
74: Anne DuPont (Br), Ariane Dimitris (Ml), Linda Ianniello (Bl/Tr), Robin Gwen Agarwal (Tl)
75: Alain Goyeau (Tl), Alicia Hermosillo McKowen (Ml), Ellen Muller (Tr), Emanuel Goyco (Mr), Ray Haberman (Bl), Yan Buske (Br)
76: Daniel Prairie (Mr/Tr), David Sanchez (Tl), Dewalt Kotze (Ml), Nancy Sheridan (Florida Fish and Wildlife) (Bl/Br)
77: Anne DuPont (Tl), Emanuel Goyco (Ml/Mr/Tr), Yan Buske (B)
78: Flavio Mendes (Ll), João Luís Carraro (Mr), Jorge Mayorga (Ul), Nancy Sheridan (Florida Fish and Wildlife) (Br), Nestor Ardila (Tl), S. Kaicher (Bl), Vinicius Padula (Tr)
79: Linda Ianniello (M), Mickey Charteris (B), Rino Jackson (Tl/Tr)
80: Alicia Hermosillo McKowen (L), Ariane Dimitris (T), Frank Krasovec (Ul), Jeff Hamann (B), Seth Pikal (Ur)
81: Athila Bertoncini (Bl), Craig Hoover (Mr), Flavio Mendes (Br), Nelsy Rivero (Ml), Peter Vermeul (T)
82: Ariane Dimitris (B/Mr/Tl/Tr), Daniel Prairie (Ml)
83: Carlo Magenta Cunha (Bl/Tl/Tr), Linda Ianniello (Br)
84: Alain Goyeau (Tl/Tr), Florida Fish & Wildlife Conservation Commission, photos by Janessa Cobb (B/Ml/Mr)
85: Ellen Muller (Lr), Jeff Goddard (Br), Keri Wilk (Tr), Mickey Charteris (Bl), Ray Haberman (Ul)
86: Alicia Hermosillo McKowen (Bl), Francisco Garcia & Jesus Troncoso (U), Greg McFall (NOAA) (Br), Jim Garin (Ll/Lr), Leopoldo Moro Abad (T)
87: Anne DuPont (Bl), David Sanchez (Br), Mickey Charteris (L/Ul), Nick Kowalske (T), Rino Jackson (Ur)
88: Emanuel Goyco (Bl/Br/Tl), Enrique Carnivali (Tr), Jason Phillip (Lr/Ur)
89: Ariane Dimitris (Ml), Evelyn Caporali (Br), Mauricio Riquelme (Mr), Mickey Charteris (Bl/Tl/Tr)
90: Alicia Hermosillo McKowen (Bl), Mickey Charteris (Br/Tl/Tr)
91: Ariane Dimitris (Tl), Mélodie Caussat (Tr), Mickey Charteris (Br), Ray Haberman (Ul), Terence Gosliner (L/Ur), Yan Buske (Bl)
92: Alain Goyeau (Tl), Jeff Goddard (Bl/Br/Ml/Mr), Ray Haberman (Tr)
93: Alicia Hermosillo McKowen (Br), Jeff Goddard (M/Tl/Tr), Mickey Charteris (Bl)
94: Ben Eddy (B/Ml/Mr), Mickey Charteris (Tl/Tr)
95: Bob Blauvelt (Br), David Sanchez (Ml), Linda Ianniello (Tr), Mélodie Caussat (Tl), Tom Poff (Bl), Yan Buske (Mr)
96: Benjamin Wilson (B/Ll/Lr), Everett Turner (Tr/Ul), Mauricio Riquelme (Tl)
97: Linda Ianniello (Mr), Mickey Charteris (Bl/Br), Ray Haberman (Tl/Tr)
98: Anne DuPont (Tl), Ariane Dimitris (Br), Craig Hoover (Tr), Mickey Charteris (Mr), Ray Haberman (Ml), Terence Gosliner (Bl)
99: Anne DuPont (Ml), Craig Hoover (Bl), Daniel Prairie (Br), Julie Mellinger (Mr), Mickey Charteris (Tl/Tr)
100: Anne DuPont (M), Brenda Hill (Tl), Emanuel Goyco (Bl), Markus Niemeyer (Br), Pauline Walsh Jacobson (Tr)
101: Bob Blauvelt (B), Flavio Mendes (Ml/Mr), Mauricio Riquelme (Tl), Yan Buske (Tr)
102: Craig Hoover (T), Mauricio Riquelme (Bl), Mickey Charteris (Ml/Mr), Rino Jackson (Br)
103: Mauricio Riquelme (Tl), Mickey Charteris (B/Ml/Mr/Tr)
104: Alicia Hermosillo McKowen (Br), Jeff Goddard (Ml), Kerry Lewis (Bl/Mr), Linda Ianniello (Tl), Sandra Edwards (Tr)
105: Kerry Lewis (Tl), Maria De las Alas (Tr), Mauricio Riquelme (B), Oliver Vincent (Ml/Mr)
106: Angel Valdes (B), Courtesy of Jesus Ortea et al (Ml/Mr), Kerry Lewis (Tl), Yan Buske (Tr)
107: Judy Townsend (Tl), Pauline Walsh Jacobson (Ul), Ray Haberman (Br/Lr/Tr)
108: Alicia Hermosillo McKowen (Bl/Br/Ll), Linda Ianniello (Tr), Suzan Meldonian (Tl), Terence Gosliner (Ur)
109: Jose H. Leal (Tl), Judy Townsend (Ml), Mickey Charteris (Bl), Nestor Ardila (Mr), Randall Spangler (Tr), Ray Haberman (Br)
110: Alain Goyeau (Ur), Everett Turner (Br/Ul), Mickey Charteris (Tl/Tr), Nestor Ardila (Ll), Rino Jackson (Bl), Terence Gosliner (Lr)
111: Courtesy of Jesus Ortea et al (Bl/Br/Mr), Ray Haberman (Tl)
112: Jeff Goddard (Lr), Kris Wilk (Ul), Mickey Charteris (Br/Tr), Ray Haberman (Tl), Rino Jackson (Bl), Terence Gosliner (Ll)
113: Bernard Picton (Bl/Br), Keri Wilk (Mr), Nancy Sheridan (Florida Fish and Wildlife) (Ml), Svetlana Speransky (Tl/Tr)
114: Mickey Charteris (Bl/Br/Ml/Mr/Tl/Tr)
115: Alicia Hermosillo McKowen (Ml/Mr), Don Zoecklein (Br), Flavio Mendes (Bl), Jeff Hamann (T)
116: Ariane Dimitris (Tl), Mickey Charteris (Ml/Mr), Ray Haberman (Bl/Br/Tr)
117: Ray Haberman (Bl/Br/Ml/Mr/Tl/Tr)
118: Laura Kushner (Tl/Tr), Mickey Charteris (Bl/Br/Ml/Mr)
119: Alain Goyeau (Lr), Cynthia Abgarian (Ll/Ul), Everett Turner (Br/Ur), Mickey Charteris (Bl), Ray Haberman (T)
120: Ariane Dimitris (Bl/Br), David Sanchez (Tr), Mickey Charteris (Ml), Ray Haberman (Mr/Tl)
121: Cynthia Abgarian (Bl), Ray Haberman (Br/Tl/Tr)
122: Mélodie Caussat (Tr), Mickey Charteris (Ml), Ray Haberman (Bl/Br/Mr/Tl)
123: Alicia Hermosillo McKowen (Br), Flavio Mendes (Bl), Jorge Mayorga (Mr), Ray Haberman (Ml/Tl/Tr)
124: Cynthia Abgarian (Mr), Mélodie Caussat (Bl), Mickey Charteris (Tr), Ray Haberman (Ml), Rino Jackson (Br)
125: André de Molenaar (Bl), Angel Valdes (T), Ellen Muller (Ml/Mr), Tricia O'Malley (Br)
126: Alain Goyeau (Ml), Ben Eddy (Mr), Jeff Goddard (Br/Tl/Tr), Luke Foster (Bl)
127: Angel Valdes (T), Mickey Charteris (Br), Ray Haberman (Ml), Terence Gosliner (Bl), Yan Buske (Mr)
128: Amy Tripp (Bl/Br/Ml/Mr), Flavio Mendes (Tl), Kadu Pinheiro (Tr)
129: Alain Goyeau (Tl), Julie Mellinger (Tr), Mickey Charteris (Mr), Nuch Uthairat (Br), Rino Jackson (Bl/Ml)

130: Alain Goyeau (B), Ariane Dimitris (Tr/U), Julie Mellinger (Ll/Lr), Pol Bosch (Tl)
131: Anne DuPont (Mr), Ben Eddy (B), Evelyn Caporali (Ml), Jose Nunez (T)
132: Daniel Prairie (Lr), Evelyn Caporali (Tl/Tr), Jack Israel (U), Linda Ianniello (B/Ll)
133: Alvaro Migotto (Tl/Tr), Arturo Telle (Ll), Jose Juan Calderin Peñate (B), Linda Ianniello (U), Pablo Samper Mendez (Lr)
134: Alicia Hermosillo McKowen (U), David Sanchez (Tl), Dianna Burkholder (Ll), Pol Bosch (B), Svetlana Speransky (Lr), Terence Zahner (Tr)
135: David Sanchez (Bl), Linda Ianniello (Bm/Tl), Markus Niemeyer (Br), Melissa Meadows (Ml/Mr), Sandra Edwards (Tr)
136: Chuck Catlett (Br), Cynthia Abgarian (Bl), Dianna Burkholder (Tl/Tr/U), Ellen Muller (Ll/Lr)
137: Everett Turner (Ml/Mr), Frank Krasovec (Bl/Tr), Keri Wilk (Tl), Mickey Charteris (Br)
138: Alicia Hermosillo McKowen (B/Tl/Tr), Mélodie Caussat (Ml/Mr)
139: Ariane Dimitris (Tl/Tr), Dewalt Kotze (Ml), Evelyn Caporali (Mr), Gwenaelle Barbotteau (Bl), Mélodie Caussat (Br)
140: Ariane Dimitris (B), Emanuel Goyco (Ml), Ray Haberman (Tl/Tr), Scott Chandler (Mr)
141: Carlo Magenta Cunha (Tl/Tr), Luis Fernando Cassino (M), Ross Gundersen (B)
142: Keri Wilk (Tl/Tr), Mickey Charteris (L/U), Paula Sills (Bl/Br)
143: Cynthia Abgarian (Ml/Tl), Mickey Charteris (Bl/Br/Tr), Tommy Nezik (Mr)
144: Alicia Hermosillo McKowen (U), Everett Turner (T), Mélodie Caussat (Bl/Br/L)
145: Ariane Dimitris (Bl/Br/Tl/Tr)
146: Alicia Hermosillo McKowen (Tl), Ariane Dimitris (Tr), Heather Turek (Bl), Linda Ianniello (Br/Ml), Tom Poff (Mr)
147: Alicia Hermosillo McKowen (M/Tl/Tr), Deborah Devers (Br), Tom Poff (Bl)
148: Brian Ricker (Tl/Tr), David Sanchez (B), Douglas Klug (M)
149: Alicia Hermosillo McKowen (Ll), Carol Cox (Bl/Br), Joshua Stone (Lr), Mélodie Caussat (Tl/Tr/Ur)
150: Alicia Hermosillo McKowen (Tr), Dianna Burkholder (Ur), Evelyn Caporali (Br), Laura Kushner (Lr), Linda Ianniello (Ll), Svetlana Speransky (Tl)
151: Alain Goyeau (Bl), Alicia Hermosillo McKowen (Tr), David Sanchez (Ul), Dewalt Kotze (Ur), Dianna Burkholder (Tl), Ray Haberman (Ll)
152: Alicia Hermosillo McKowen (Tl/Tr), Ariane Dimitris (Bl), Svetlana Speransky (Br), Terence Gosliner (Ml)
153: Carlo Magenta Cunha (Tr), Jessica Link (Mr), Kadu Pinheiro (Tl), Rodrigo Canalli (Bl/Br/Ml)
154: Alicia Hermosillo McKowen (Bl), Kadu Pinheiro (Tl/Tr), Paul Humann (Mr), Robert Cox (Br)
155: Alicia Hermosillo McKowen (Bl/Br), Evelyn Caporali (Mr), Joyce Burek (T), Linda Ianniello (Ml)
156: Alicia Hermosillo McKowen (Bl/Br), Athila Bertoncini (Tl), Carlo Magenta Cunha (Tr), Kadu Pinheiro (M)
157: Gustav Paulay (FMNH) (B), Judy Townsend (Mr), Julie Mellinger (Tr), Mélodie Caussat (Tl), Terence Gosliner (Ml)
158: Alain Goyeau (Ml), Cedric Pau (Br), Hart Kerbel (Mr), Linda Ianniello (Tl), Mélodie Caussat (Tr), Rino Jackson (Bl)
159: Ariane Dimitris (Mr/Tl/Tr), Hans Bertsch (Ml), Stephen C. Kempf (B)
160: Carol Cox (Mr), Julie Mellinger (Ml), Mélodie Caussat (Bl), Nestor Ardila (Tr), Robin Gwen Agarwal (Br), Yan Buske (Tl)
161: Jack Mackay (Br/Lr), Linda Ianniello (Tr), Mickey Charteris (Bl/Ll), Sandra Edwards (Tl/Ur)
162: Alicia Hermosillo McKowen (Br/Tr), Ariane Dimitris (Bl/Mr/Tl), Sandra Edwards (Ml)
163: André de Molenaar (Ml), Ariane Dimitris (Br), Keri Wilk (Tl), Linda Ianniello (Mr), Mélodie Caussat (Tr), Svetlana Speransky (Bl)
164: Don Lewis (Tl), Ellen Muller (B/Ml/Mr), Keri Wilk (Tr)
165: Linda Ianniello (Tl/Tr), Mélodie Caussat (Ml), Sandra Edwards (B), Terence Zahner (Mr)
166: Everett Turner (Tl), Markus Niemeyer (Ml), Mélodie Caussat (Bl), Pauline Walsh jacobson (Br), Pol Bosch (Mr)
167: Alain Goyeau (Bl), Alicia Hermosillo McKowen (Mr), Carol Cox (Ml), Flavio Mendes (Br), Mickey Charteris (Tl), Sandra Edwards (Tr)
168: Julie Mellinger (Bl), Mickey Charteris (Br/Tl/Tr), Pauline Walsh Jacobson (Mr), Sandra Edwards (Ml)
169: Julie Mellinger (Br), Keri Wilk (Tr), Mélodie Caussat (Mr), Mickey Charteris (Tl), Phil Allman (Ml), Sandra Edwards (Bl)
170: Alain Goyeau (Mr), Alev Ozten (Tl), Ellen Muller (B/Ml/Tr)
171: Alex Parry (Tr), Greg Graeber (Br), Linda Ianniello (Tl), Stephanie Taylor (Bl)
172: Ariane Dimitris (Br), Cynthia Abgarian (Tl/Tr), Everett Turner (Mr), Mélodie Caussat (Bl), Tracey Winholt (Ml)
173: Alain Goyeau (Bl), Enrique Carnivali (Br), Keri Wilk (Ml), Mélodie Caussat (Tr), Mickey Charteris (Tl)
174: Dianna Burkholder (Tl), Ellen Muller (Ml), Keri Wilk (Mr), Linda Ianniello (Tr), Pauline Walsh Jacobson (Bl), Robert Stansfield (Br)
175: Ariane Dimitris (Tl), Linda Ianniello (Tr), Matt Sullivan (Ul), Rino Jackson (Ll/Lr), Sandra Edwards (B)
176: David Sanchez (T), Laetitia Rossignol (Bl/Br)
177: Linda Ianniello (M), Ray Haberman (Tl), Svetlana Speransky (Tr), Terence Zahner (Bl/Br)
178: Dianna Burkholder (Ml/Mr), Kennedy Hanson, Apalachicola NERR (Florida DEP) (Tl/Tr), Sandra Edwards (B)
179: Anne DuPont (Bl), Linda Ianniello (Br/Mr/Tl), Mélodie Caussat (Ml), Mickey Charteris (Tr)
180: Linda Ianniello (Bl), Mickey Charteris (Tl/Tr), Sandra Edwards (Br/M)
181: Dianna Burkholder (Bl), Linda Ianniello (Ml/Tl), Robert Stansfield (Mr/Tr), Sandra Edwards (Br)
182: Alain Goyeau (Br), Cynthia Abgarian (T), Linda Ianniello (Mr), Mauricio Riquelme (Bl), Sandra Edwards (Ml)
183: Everett Turner (Ml/Mr), Mickey Charteris (B/T)
184: Alicia Hermosillo McKowen (Mr/Tl/Tr), David Sanchez (Br), Mélodie Caussat (Bl)
185: Linda Ianniello (Tr), Pauline Walsh Jacobson (Bl/Br), Rino Jackson (Tl)
186: Cherie-Anne Henderson Dam (Tl/Tr), Dewalt Kotze (B), Frank Krasovec (Ml/Mr)
187: Alvaro Migotto (B), Courtesy of Jesus Ortea (Tl), Manuel Caballer (Tr)
188: Magali Marquez-Ramos (Ml/Mr), Pauline Walsh Jacobson (Tl/Tr), Smithsonian Marine Station (B)
189: Anne DuPont (Tl/Tr), Dianna Burkholder (B/Mr), Evelyn Caporali (Ml)
190: Alain Goyeau (Br), Alicia Hermosillo McKowen (Tr), Ariane Dimitris (Mr), Linda Ianniello (Tl), Mélodie Caussat (Bl), Pauline Walsh Jacobson (Ml)
191: Daniel Prairie (Br), Dianna Burkholder (Tl), Linda Ianniello (Tr), Shawn Turek (Bl)
192: Alicia Hermosillo McKowen (Ml), Ariane Dimitris (Mr), Courtesy of Jesus Ortea (Bl/Br), Mickey Charteris (T)
193: Elizabeth Pelkey (Ml), Frank Krasovec (Mr), Shawn Turek (Bl), Yan Buske (Br)

194: Cynthia Abgarian (Mr), Mickey Charteris (Br/Ml), Rino Jackson (Bl), Vance Solseth (Tl/Tr)
195: Evelyn Caporali (Bl/Br), Frank Krasovec (Mr/Tr), Linda Ianniello (Ml), Pauline Walsh Jacobson (Tl)
196: Linda Ianniello (Br), Mickey Charteris (Bl/Ml/Mr/Tl/Tr)
197: Linda Ianniello (Bl/Br), Mickey Charteris (T), Pauline Walsh Jacobson (Ml/Mr)
198: Alicia Hermosillo McKowen (Tl), Cynthia Abgarian (Ml/Mr), Yan Buske (B/Tr)
199: Carol Cox (Tr), Ray Haberman (Bl/Br/Ml/Mr), Yan Buske (Tl)
200: Alain Goyeau (Ml/Mr), Magali Marquez-Ramos (Br/T), Svetlana Speransky (Bl)
201: Alfe Villan Bolos (Bl), Dianna Burkholder (Ml/Mr), Mélodie Caussat (T), Tracy Candish (Br)
202: Alain Goyeau (Ml), Cherie-Anne Henderson Dam (T), Dianna Burkholder (Br), Lazaro Coto (Mr), Markus Niemeyer (Bl)
203: Alain Goyeau (Tl), Robert Gundy (Tr), Sandra Edwards (Bl/Br/Ml/Mr)
204: Evelyn Caporali (Bl/Mr), Everett Turner (Tl), Markus Niemeyer (Tr), Sandra Edwards (Br), Shawn Turek (Ml)
205: Linda Ianniello (B/Tl/Tr), Rino Jackson (Ml/Mr)
206: Dianna Burkholder (Bl/Br), Linda Ianniello (Tl/Tr), Mélodie Caussat (Ml/Mr)
207: Alain Goyeau (Bl), April McCormack (Ml), John Kalimnios (Tr), Linda Ianniello (Br/Tl), Markus Niemeyer (Mr)
208: Alain Goyeau (Tr), Evelyn Caporali (Br), Linda Ianniello (Tl), Magali Marquez-Ramos (Mr), Mickey Charteris (Bl), Tracey Winholt (Ml)
209: Ariane Dimitris (Br), Courtesy of Roberta Crescini (Tl/Tm/Tr), Daniel Prairie (Mr), Gwenaelle Barbotteau (Bl), Terence Zahner (Ml)
210: Ellen Muller (Ur), Everett Turner (Tr), Julie Mellinger (Ll), Keri Wilk (Tl), Laetitia Rossignol (B/Lr)
211: Linda Ianniello (Tl), Pauline Walsh Jacobson (Bl/Br/Mr), Tom Poff (Ul)
212: Ben Eddy (Tr), Carlo Magenta Cunha (Bl), Dan Schofield (M), Francisco Garcia & Jesus Troncoso (Br), Markus Niemeyer (Tl)
213: Cynthia Abgarian (Bl/Mr), Keri Wilk (Br), Ray Haberman (Tl)
214: Craig Hoover (Ll), Dewalt Kotze (Ur), Linda Ianniello (Br/Tr/Ul), Mélodie Caussat (Lr), Mickey Charteris (Bl), Sandra Edwards (Tl)
215: Mickey Charteris (Lr/Tl/Tr/U), Robert Stansfield (B), Tracey Winholt (Ll)
216: Alicia Hermosillo McKowen (Br), David Sanchez (Ml/Tl/Tr), Linda Ianniello (Mr), Sandra Edwards (Bl)
217: Cynthia Abgarian (Br/Tr), Everett Turner (Bl), Frank Krasovec (Tl), Marli Wakeling (M)
218: Ariane Dimitris (Tl/Tr), Denisse Pohls (M), Nuch Uthairat (B)
219: Alicia Hermosillo McKowen (Br), Ellen Muller (Bl), Enrique Carnivali (Tr), Terence Zahner (Tl)
220: Daniel Prairie (Mr), Julie Mellinger (Bl), Linda Ianniello (Br), Tom Poff (Tr), Tommy Nezik (Ml)
221: Ariane Dimitris (Tl), Ellen Muller (Tr), Linda Ianniello (Br)
222: Bernard Picton (Mr), David Sanchez (Bl), Ellen Muller (Tr), Linda Ianniello (Br), Terence Zahner (Ml)
223: Cynthia Abgarian (Tl/Tr), Linda Ianniello (Ml), Markus Niemeyer (Br/Mr), Svetlana Speransky (Bl)
224: Linda Ianniello (L/Tr), Pauline Walsh Jacobson (Bl/Tl), Robin Gwen Agarwal (U)
225: Robert Stuke Jr. (T), Sandra Edwards (U), Terence Zahner (Ll/Lr), Tricia O'Malley (B)
226: Angelica Matos (Ml), Anna Torres (Mr), Ariane Dimitris (Tl), Ben Hartley (Tr), Giulia Oliveira (B)
228: Dianna Burkholder (Br), Mickey Charteris (Tl/Tr), Sandra Edwards (Bl)
229: David Sanchez (Tr), Diane Leazenby (Br), Linda Ianniello (Bl/Ml), Svetlana Speransky (Mr/Tl)
230: Alvaro Migotto (B), Dewalt Kotze (Tr), Pauline Walsh Jacobson (Ml/Mr/Tl)
231: Alicia Hermosillo McKowen (M/Tl/Tr), Dianna Burkholder (Bl/Br)
232: Dewalt Kotze (Br), Frank Krasovec (Ml/Tl), Keri Wilk (Mr), Mickey Charteris (Tr), Tracey Winholt (Bl)
233: Craig Hoover (B), Dianna Burkholder (Ul/Ur), Kerry Lewis (Tl), Linda Ianniello (Lr), Sandra Edwards (Tr), Susan Mears (Ll)
234: Ariane Dimitris (B/M), Rino Jackson (T)
235: Ariane Dimitris (Tr/Ur), Linda Ianniello (Lr/Ul), Mélodie Caussat (Ll), Mickey Charteris (Tl), Svetlana Speransky (B)
236: Ariane Dimitris (Ml), Brenda Hill (Bl), Linda Ianniello (Br/Tr), Sandra Edwards (Tl)
237: Linda Ianniello (Tr), Pauline Walsh Jacobson (Br), Robert Stuke Jr. (Bl), Robin Gwen Agarwal (Mr), Shawn Turek (Ml), Svetlana Speransky (Tl)
238: Cynthia Abgarian (Br/Ml), Jessica Riederer (Mr), Keri Wilk (Tl), Mickey Charteris (Bl), Terence Gosliner (Tr)
239: Anne DuPont (Bl), Ariane Dimitris (Br), Linda Ianniello (Mr), Mickey Charteris (Tl), Terence Gosliner (Ml), Yan Buske (Tr)
240: Ariane Dimitris (Mr), Bob Blauvelt (B/Ml), Markus Niemeyer (Tr), Pauline Walsh Jacobson (Tl)
241: Dianna Burkholder (Br), Elizabeth Pelkey (Bl), Kara Birkenmayer (U), Linda Ianniello (Lr), Mickey Charteris (Tl/Tr), Yan Buske (Ll)
242: Evelyn Caporali (Ur), Everett Turner (Lr), Gaby Carias Tucker, Isla des Aves, Venezuela (B), Judy Townsend (Ul), Linda Ianniello (Tl/Tr)
243: April McCormack (Bl), Courtesy of Jesus Ortea (T), Enrique Carnivali (Lr), Jeff Hamann (Ul), Linda Ianniello (Ur), Maria De las Alas (Br), Mélodie Caussat (Ll)
244: Alvaro Altenkirch (Ul), Evelyn Caporali (Ll), Julie Mellinger (Tl/Tr), Laetitia Rossignol (Ur), Mélodie Caussat (B), Tommy Nezik (Lr)
245: Linda Ianniello (Bl/Br/T), Mickey Charteris (Ml)
246: Deborah Devers (Bl), Linda Ianniello (Br/Tl/Tm/Tr)
247: Hanan Atia (Tl), Linda Ianniello (Br/Tr), Suzan Meldonian (Bl/Ml)
248: John Kalimnios (Bl/Mr), Kris Wilk (Ml), Ryan Nelson (T)
249: Cherie-Anne Henderson Dam (Tr), Cynthia Abgarian (Ml), Douglas Kahle (Br), Kerry Lewis (Mr), Patric Lengacher (Bl), Robert Stansfield (Tl)
250: Alain Goyeau (Mr), Alfe Villan Bolos (Tr), Cynthia Abgarian (Ml), Mélodie Caussat (B/Tl)
251: Ariane Dimitris (Br), Julie Mellinger (M), Mélodie Caussat (Tr), Mickey Charteris (Tl), Sandra Edwards (Bl)
252: Felipe De Vasconcelos Silva (B/M/Tl/Tr)
253: Alicia Hermosillo McKowen (Br), Cynthia Abgarian (Ll), Everett Turner (Ur), Linda Ianniello (Ul), Mickey Charteris (Bl/Lr), Stephane Hourdez (T)
254: Alain Goyeau (Tr), Dave Holladay (Ml), Jessica Riederer (Mr), Keri Wilk (Tl), Linda Ianniello (Bl/Br)
255: Ben Eddy (Bl/Tr), Julie Mellinger (Ml), Mickey Charteris (Tl), Ray Haberman (Br), Terence Gosliner (Mr)
256: Ariane Dimitris (Bl), Cynthia Abgarian (Br), Ellen Muller (Ml), Everett Turner (Tr), Keri Wilk (Tl), Mélodie Caussat (Mr)
257: Julie Mellinger (B), Linda Ianniello (T)
258: Anne DuPont (Tl/Tr), Julie Mellinger (Bl/Mr), Paul Selden (Br/Ml)
259: Alicia Hermosillo McKowen (M), Claire Jeuffroy (Tl), Ellen Muller (Bl), Yan Buske (Br/Tr)
260: Ariane Dimitris (Ul/Ur), Jorge Mayorga (T), Mélodie Caussat (Br/L), Rino Jackson (Bl)

261: Jeff Goddard (Tl/Ul/Ur), Mickey Charteris (B/Ll/Lr/Tr)
262: Alain Goyeau (Mr), Anne DuPont (Ml), Judy Townsend (Tl), Keri Wilk (Br), Ray Haberman (Bl)
263: Ariane Dimitris (Ml), Bob Blauvelt (Bl), Mickey Charteris (Mr/Tr), Pauline Walsh Jacobson (Br), Ray Haberman (Mm/Tl)
264: Athila Bertoncini (Ml/Mr/Tl/Tr), Linda Ianniello (B)
265: Ariane Dimitris (Bl), Keri Wilk (Br), Ray Haberman (Mr/Tr)
266: Daniel Prairie (Tr), Linda Ianniello (Ml/Mr), Mickey Charteris (B)
267: Dianna Burkholder (Tl/Tr), Evelyn Caporali (U), Jack Israel (B), Mickey Charteris (Ll/Lr)
268: Ariane Dimitris (Bl), Jack Israel (Tr), Julie Mellinger (Br), Mélodie Caussat (Tl), Mickey Charteris (Ml), Pauline Walsh Jacobson (Mr)
269: Alain Goyeau (U), Cherie-Anne Henderson Dam (Bl/Br), Ray Haberman (Tr), Ross Gundersen (Tl), Terence Gosliner (L)
270: Courtesy of Jesus Ortea (B/Tr), David Kirsh (M), Terence Gosliner (Tl)
271: Carlo Magenta Cunha (Tr), Julie Mellinger (Bl/Br), Luke Foster (Ml/Mr), Yan Buske (Tl)
272: Keri Wilk (Tl), Mélodie Caussat (Br), Mickey Charteris (Ml/Mr), Robert Stansfield (Tr), Terence Gosliner (Bl)
273: Dianna Burkholder (Mr), Mickey Charteris (B/Ml/Tl/Tr)
274: Ariane Dimitris (Tr), Ellen Muller (Ml), Mickey Charteris (Br)
275: Ariane Dimitris (Ml/Mr), Michael Middlebrooks (T), Ray Haberman (Bl/Br)
276: Linda Ianniello (Tl/Tr), Mickey Charteris (Ml/Mr), Robert Stansfield (B)
277: Ariane Dimitris (Lr), Dianna Burkholder (Tl), Evelyn Caporali (Tr), Mauricio Riquelme (Ur), Rino Jackson (Ll), Sandra Edwards (B), Sharon Davies (Ul)
278: Courtesy of Jesus Ortea (M), Cynthia Abgarian (Br), Dianna Burkholder (Tl/Tr), Everett Turner (Bl)
279: Christian Jones (Tl), Dianna Burkholder (Lr/Ml), Julie Mellinger (Br), Mickey Charteris (Tr/Ur), Rino Jackson (Bl)
280: Mélodie Caussat (Bl/Br/Tl/Tr), Mickey Charteris (Ml/Mr)
281: Courtesy of Jesus Ortea (T), Mélodie Caussat (Ul/Ur), Ray Haberman (Bl/Br/L)
282: Alfe Villan Bolos (Ml), Deborah Devers (Tr), Julie Mellinger (Mr), Mélodie Caussat (B), Svetlana Speransky (Tl)
283: David Kirsh (Tr), Ross Gundersen (Tl), Terence Gosliner (B)
284: Alain Goyeau (Tl), Mélodie Caussat (B/Tr), Mickey Charteris (Ml/Mr)
285: David Sanchez (Bl), Evelyn Caporali (Br), Linda Ianniello (Ml), Luke Foster (Mr), Markus Niemeyer (Tl), Sandra Edwards (Tr)
286: David Sanchez (Tl), Dewalt Kotze (Br), Dianna Burkholder (Bl/Mr), Evelyn Caporali (Ml), Julie Mellinger (Tr)
287: Dianna Burkholder (Ml/Tr), Evelyn Caporali (Mr), Gustav Paulay (FMNH) (Bl/Br), Sandra Edwards (Tl)
288: Corinne Fuchs (Br), Julie Mellinger (Bl), Lazaro Ruda (Tr), Mickey Charteris (Ml/Mr), Svetlana Speransky (Tl)
289: Daniel Prairie (T), David Sanchez (B), Julie Mellinger (Ll/Lr), Linda Ianniello (U)
290: Alain Goyeau (Br), Julie Mellinger (Tr), Keri Wilk (Tl), Sandra Edwards (Bl)
291: Ben Eddy (Mr), Dianna Burkholder (Bl), Evelyn Caporali (Tr), Mickey Charteris (Tl), Pauline Walsh Jacobson (Ml), Russ Goldenberg (Br)
292: Alain Goyeau (Bl), Ben Eddy (Mr/Tr), Mélodie Caussat (Br), Mickey Charteris (Tl), Rino Jackson (Ml)
293: Alain Goyeau (Mr), Everett Turner (Br), Mélodie Caussat (Tl/Tr), Mickey Charteris (Ml), Tracey Winholt (Bl)
294: Alicia Hermosillo McKowen (Tl), Julie Mellinger (Br), Linda Ianniello (Tr), Sandra Edwards (Ml), Terence Gosliner (Mr), Yan Buske (Bl)
295: Dianna Burkholder (Tl), Evelyn Caporali (Tr), Laura Kushner (U), Mélodie Caussat (Ll), Mickey Charteris (B/Lr)
296: Ben Eddy (Bl), Julie Mellinger (Mr), Mélodie Caussat (Ml), Michael Middlebrooks (T), Mickey Charteris (Br)
297: Keri Wilk (Ll/Lr/Ul/Ur), Lee Richter (Bl), Mickey Charteris (Br), Pauline Walsh Jacobson (Tl)
298: Audrey Smith (B), Courtesy of Jesus Ortea (U), Erik Daniel Erikson (L), Kerry Lewis (Tl/Tr)
299: Everett Turner (Tl), Julie Mellinger (Mr), Mélodie Caussat (Ml/Tr), Mickey Charteris (Bl), Robert Stansfield (Br)
300: Andrea Bernecker (Tr), Ellen Muller (Br), Luke Foster (Ml), Matthew J. Scripter (Tl), Mélodie Caussat (Bl), Rino Jackson (Mr)
301: Ellen Muller (Tl), Martin Nobs (Bl), Mélodie Caussat (Br/Ml), Pol Bosch (Mr), Rino Jackson (Tr)
302: Ariane Dimitris (Ml), Ben Eddy (Tr), Mauricio Riquelme (Br), Mickey Charteris (Bl/Tl), Svetlana Speransky (Mr)
303: Ben Eddy (Tl/Tr), Keri Wilk (Ul), Mickey Charteris (Br), Pol Bosch (Ll), Rino Jackson (Lr), Russ Goldenberg (Bl), Terence Zahner (Ur)
304: Cherie-Anne Henderson Dam (Br/Ur), Keri Wilk (Tr), Laetitia Rossignol (Lr), Nuch Uthairat (Ul), Rino Jackson (Bl), Sandra Edwards (Ll)
305: Everett Turner (Br/Tr), Linda Ianniello (Bl/Mr), Mélodie Caussat (Mm), Mickey Charteris (Tl), Terence Gosliner (Ml)
306: Alain Goyeau (Bm/Ml), Julie Mellinger (Br), Mauricio Riquelme (Tr), Mélodie Caussat (Bl), Mickey Charteris (Tl), Pauline Walsh Jacobson (Mr)
307: Cherie-Anne Henderson Dam (Br), Cynthia Abgarian (Mr), Daniel Prairie (Ml), Linda Ianniello (Bl), Mickey Charteris (Tr)
308: Alain Goyeau (Bl), Ariane Dimitris (Mr), Ben Eddy (Mm), Linda Ianniello (Tl), Mélodie Caussat (Tr), Mickey Charteris (Ml), Sandra Edwards (Br)
309: Dave Holladay (Tl), Laetitia Rossignol (Ml), Mélodie Caussat (Mr/Tr), Mickey Charteris (Bl), Rino Jackson (Br)
310: Ariane Dimitris (Mr), Ben Eddy (Bl), Jack Israel (Tr), Julie Mellinger (Tl), Phil Croy (Ml), Rino Jackson (Br)
311: Cristobal Molina (Bl/L), Jeff Goddard (Tl/Tr/U), Rino Jackson (Br)
312: Alicia Hermosillo McKowen (Tl/Tr/U), Emanuel Goyco (B), Terence Zahner (L)
313: Ariane Dimitris (T), Julie Mellinger (B), Kerry Lewis (Ll/Lr), Terence Gosliner (U)
314: Jorge Mayorga (U), Julie Mellinger (T), Magali Marquez-Ramos (B), Matthew J. Scripter (L)
315: Ariane Dimitris (Tl), Julie Mellinger (Bl), Linda Ianniello (Tr), Mélodie Caussat (Br/Mr), Yan Buske (Ml)
316: Cynthia Abgarian (Lr), Denisse Pohls (Ll), Evelyn Caporali (Tr), Mélodie Caussat (Tl), Ray Haberman (B), Rino Jackson (U)
317: Terence Gosliner (Bl/Br/M/Tl/Tr)
318: Ben Eddy (Tl), Flavio Mendes (B/Ll/Lr), Invert E Base (CC0 1.0) (U), Luke Foster (Tr)
319: Alicia Hermosillo McKowen (Br/T), Larry Harris (Bl)
320: Alain Goyeau (Bl), Athila Bertoncini (M), Ellen Muller (Br/Tr), Peter de Graaf (Tl)

Uncredited images were taken by the author.

Photo Credits (by photographer)

Vertical position: T=top, U=upper middle, M=middle, L=lower middle, B=bottom
Horizontal position: l=left, m=middle, r=right

Cynthia Abgarian: 5Ur, 20Tr, 39Mr, 43Br, 67Tr, 72Ml, 119Ll/Ul, 121Bl, 124Mr, 136Bl, 143Ml/Tl, 172Tl/Tr, 182T, 194Mr, 198Ml/Mr, 213Bl/Mr, 217Br/Tr, 223Tl/Tr, 238Br/Ml, 249Ml, 250Ml, 253Ll, 256Br, 278Br, 307Mr, 316Lr; ***Robin Gwen Agarwal:*** 74Tl, 160Br, 224U, 237Mr; ***Robert Aguilar (Smithsonian):*** 10Bl/Br, 23Br; ***Phil Allman:*** 169Ml; ***Alvaro Altenkirch:*** 244Ul; ***Leon Altoff:*** 24Br; ***Nestor Ardila:*** 58Br, 78Tl, 109Mr, 110Ll, 160Tr; ***Hanan Atia:*** 247Tl; ***Gwenaelle Barbotteau:*** 60Bl, 139Bl, 209Bl; ***Lucas Pescador Barcelos:*** 47Br; ***Andrea Bernecker:*** 300Tr; ***Athila Bertoncini:*** 81Bl, 156Tl, 264Ml/Mr/Tl/Tr, 320M; ***Hans Bertsch:*** 159Ml; ***Kara Birkenmayer:*** 241U; ***Bob Blauvelt:*** 95Br, 101B, 240B/Ml, 263Bl; ***Pol Bosch:*** 130Tl, 134B, 166Mr, 301Mr, 303Ll; ***Joyce Burek:*** 155T; ***Dianna Burkholder:*** 62Mr, 134Ll, 136Tl/Tr/U, 150Ur, 151Tl, 174Tl, 178Ml/Mr, 181Bl, 189B/Mr, 191Tl, 201Ml/Mr, 202Br, 206Bl/Br, 228Br, 231Bl/Br, 233Ul/Ur, 241Br, 267Tl/Tr, 273Mr, 277Tl, 278Tl/Tr, 279Lr/Ml, 286Bl/Mr, 287Ml/Tr, 291Bl, 295Tl; ***Yan Buske:*** 19Bl/Mr, 50T, 51Bl, 57Bl, 59Tr, 75Br, 77B, 91Bl, 95Mr, 101Tr, 106Tr, 127Mr, 160Tl, 193Br, 198B/Tr, 199Tl, 239Tr, 241Ll, 259Br/Tr, 271Tl, 294Bl, 315Ml; ***Manuel Caballer:*** 22Br/Tl, 24Tr, 53M/T, 68Bl/Br, 70B, 187Tr; ***Senem Çağlar and Bilal Ozturk:*** 31Tl; ***Rodrigo Canalli:*** 153Bl/Br/Ml; ***Tracy Candish:*** 201Br; ***Evelyn Caporali:*** 61Ur, 62Bl, 66Tl, 89Br, 131Ml, 132Tl/Tr, 139Mr, 150Br, 155Mr, 189Ml, 195Bl/Br, 204Bl/Mr, 208Br, 242Ur, 244Ll, 267U, 277Tr, 285Br, 286Ml, 287Mr, 291Tr, 295Tr, 316Tr; ***Gaby Carias Tucker, Isla des Aves, Venezuela:*** 242B; ***Enrique Carnivali:*** 88Tr, 173Br, 219Tr, 243Lr; ***João Luís Carraro:*** 78Mr; ***Chuck Catlett:*** 136Br; ***Mélodie Caussat:*** 5Tm, 9Bl/Ml, 14Tl, 17Tl, 39Bl, 40B, 45Mr, 54Tr, 91Tr, 95Tl, 122Tr, 124Bl, 138Ml/Mr, 139Br, 144Bl/Br/L, 149Tl/Tr/Ur, 157Tl, 158Tr, 160Bl, 163Tr, 165Ml, 166Bl, 169Mr, 172Bl, 173Tr, 179Ml, 184Bl, 190Bl, 201T, 206Ml/Mr, 214Lr, 235Ll, 243Ll, 244B, 250B/Tl, 251Tr, 256Mr, 260Br/L, 268Tl, 272Br, 280Bl/Br/Tl/Tr, 281Ul/Ur, 282B, 284B/Tr, 292Br, 293Tl/Tr, 295Ll, 296Ml, 299Ml/Tr, 300Bl, 301Br/Ml, 305Mm, 306Bl, 308Tr, 309Mr/Tr, 315Br/Mr, 316Tl; ***Scott Chandler:*** 140Mr; ***Mickey Charteris:*** 3Bl, 4Tr/Ul, 8Mr, 18Tl, 20Tl, 37Bl/Mr, 38Ml, 40Mr, 41Ml, 43Ml, 44Tr, 45Bl, 46Bl/Mr/Tl, 50Br, 54Tl, 56Mr, 61Bl, 64Br/Mr, 79B, 85Bl, 87L/Ul, 89Bl/Tl/Tr, 90Br/Tl/Tr, 91Br, 93Bl, 94Tl/Tr, 97Bl/Br, 98Mr, 99Tl/Tr, 102Ml/Mr, 103B/Ml/Mr/Tr, 109Bl, 110Tl/Tr, 112Br/Tr, 114Bl/Br/Ml/Mr/Tl/Tr, 116Ml/Mr, 118Bl/Br/Ml/Mr, 119Bl, 120Ml, 122Ml, 124Tr, 127Br, 129Mr, 137Br, 142L/U, 143Bl/Br/Tr, 161Bl/Ll, 167Tl, 168Br/Tl/Tr, 169Tl, 173Tl, 179Tr, 180Tl/Tr, 183B/T, 192T, 194Br/Ml, 196Bl/Ml/Mr/Tl/Tr, 197T, 208Bl, 214Bl, 215Lr/Tl/Tr/U, 228Tl/Tr, 232Tr, 235Tl, 238Bl, 239Tl, 241Tl/Tr, 245Ml, 251Tl, 253Bl/Lr, 255Tl, 261B/Ll/Lr/Tr, 263Mr/Tr, 266B, 267Ll/Lr, 268Ml, 272Ml/Mr, 273B/Ml/Tl/Tr, 274Br, 276Ml/Mr, 279Tr/Ur, 280Ml/Mr, 284Ml/Mr, 288Ml/Mr, 291Tl, 292Tl, 293Ml, 295B/Lr, 296Br, 297Br, 299Bl, 302Bl/Tl, 303Br, 305Tl, 306Tl, 307Tr, 308Ml, 309Bl; ***Florida Fish & Wildlife Conservation Commission, photos by Janessa Cobb:*** 84B/Ml/Mr; ***Conquiliologistas do Brazil:*** 1Br; ***Lazaro Coto:*** 202Mr; ***Courtesy of Jesus Ortea:*** 19Tl, 24Bl/Ml/Tl, 25Mr, 53B, 187Tl, 192Bl/Br, 243T, 270B/Tr, 278M, 281T, 298U; ***Courtesy of Jesus Ortea et al:*** 24Mr, 106Ml/Mr, 111Bl/Br/Mr; ***Courtesy of Roberta Crescini:*** 209Tl/Tm/Tr; ***Carol Cox:*** 149Bl/Br, 160Mr, 167Ml, 199Tr; ***Robert Cox:*** 154Br; ***Phil Croy:*** 6Bl/Br, 310Ml; ***Carlo Magenta Cunha:*** 83Bl/Tl/Tr, 141Tl/Tr, 153Tr, 156Tr, 212Bl, 271Tr; ***Nicholas Curtis:*** 47M; ***Sharon Davies:*** 277Ul; ***Peter de Graaf:*** 320Tl; ***Maria De las Alas:*** 105Tr, 243Br; ***Armando de Luca Junior:*** 63Br; ***André de Molenaar:*** 7Br, 125Bl, 163Ml; ***Carole Assier de Pompignan:*** 16Bl, 41Tl, 49Tr; ***Felipe De Vasconcelos Silva:*** 252B/M/Tl/Tr; ***Ned DeLoach:*** 37Ml, 71Tl; ***Deborah Devers:*** 33Bl, 35Tl/Tr, 147Br, 246Bl, 282Tr; ***Ariane Dimitris:*** 7Mr, 39Ml/Tl, 40Ml, 46Br/Tr, 47Tl/Tr, 51Ml, 56Bl, 62Ml, 66Bl, 68Ml/Tm, 73Bl, 74Ml, 80T, 82B/Mr/Tl/Tr, 89Ml, 91Tl, 98Br, 116Tl, 120Bl/Br, 130Tr/U, 139Tl/Tr, 140B, 145Bl/Br/Tl/Tr, 146Tr, 152Bl, 159Mr/Tl/Tr, 162Bl/Mr/Tl, 163Br, 172Br, 175Tl, 190Mr, 192Mr, 209Br, 218Tl/Tr, 221Tl, 226Tl, 234B/M, 235Tr/Ur, 236Ml, 239Br, 240Mr, 251Br, 256Bl, 260Ul/Ur, 263Ml, 265Bl, 268Bl, 274Tr, 275Ml/Mr, 277Lr, 302Ml, 308Mr, 310Mr, 313T, 315Tl; ***Alasdair Dunlap-Smith:*** 57Tl/Tm; ***Anne DuPont:*** 3Tl, 4Tl, 5Tr, 7Bl, 9Ur, 13Lr, 16Ml/Mr, 17Bl, 20Bl/Br, 36Bl, 38Bl/Lr, 40Tl/Tr, 41Tr, 45Tr, 73Tm, 74Br, 77Tl, 87Bl, 98Tl, 99Ml, 100M, 131Mr, 179Bl, 189Tl/Tr, 239Bl, 258Tl/Tr, 262Ml; ***Ben Eddy:*** 3Ml, 23Mr/Tr, 71Ml, 94B/Ml/Mr, 126Mr, 131B, 212Tr, 255Bl/Tr, 291Mr, 292Mr/Tr, 296Bl, 302Tr, 303Tl/Tr, 308Mm, 310Bl, 318Tl; ***Sandra Edwards:*** 2Ml, 5Tl, 10Tl, 15Mr, 21Bl, 41Mr, 42Mr, 51T, 52Mr/Tl, 61Ml, 62Tl, 72Br, 104Tr, 135Tr, 161Tl/Ur, 162Ml, 165B, 167Tr, 168Ml, 169Bl, 175B, 178B, 180Br/M, 181Br, 182Ml, 203Bl/Br/Ml/Mr, 204Br, 214Tl, 216Bl, 225U, 228Bl, 233Tr, 236Tl, 251Bl, 277B, 285Tr, 287Tl, 290Bl, 294Ml, 304Ll, 308Br; ***Lyndsey Emery:*** 66Ml; ***Erik Daniel Erikson:*** 55Ml, 298L; ***Raul Fernandez:*** 66Mr; ***Luis Fernando Cassino:*** 141M; ***Lureen Ferretti:*** 61Tl; ***Luke Foster:*** 3Mr, 71Br, 126Bl, 271Ml/Mr, 285Mr, 300Ml, 318Tr; ***John Fowler:*** 8Ml; ***Bill Frank:*** 1Ml, 23Bl; ***Corinne Fuchs:*** 288Br; ***Francisco Garcia & Jesus Troncoso:*** 86U, 212Br; ***Jim Garin:*** 86Ll/Lr; ***Jeff Goddard:*** 85Br, 92Bl/Br/Ml/Mr, 93M/Tl/Tr, 104Ml, 112Lr, 126Br/Tl/Tr, 261Tl/Ul/Ur, 311Tl/Tr/U; ***Russ Goldenberg:*** 291Br, 303Bl; ***Brian Gortney:*** 68Tr; ***Terence Gosliner:*** 10Tr, 13Ur, 14Ml/Mr, 16Tr, 17Mr, 23Ml, 25Bl/Br/Ml/Tl/Tr, 44Ml, 52Tr, 70L/Ul/Ur, 71Bl, 91L/Ur, 98Bl, 108Ur, 110Lr, 112Ll, 127Bl, 152Ml, 157Ml, 238Tr, 239Ml, 255Mr, 269L, 270Tl, 272Bl, 283B, 294Mr, 305Ml, 313U, 317Bl/Br/M/Tl/Tr; ***Emanuel Goyco:*** 75Mr, 77Ml/Mr/Tr, 88Bl/Br/Tl, 100Bl, 140Ml, 312B; ***Alain Goyeau:*** 13Tr, 36Br, 42Ml, 43Bl, 45Tl, 57Ml, 59Tl, 63Ll, 75Tl, 84Tl/Tr, 92Tl, 110Ur, 119Lr, 126Ml, 129Tl, 130B, 151Bl, 158Ml, 167Bl, 170Mr, 173Bl, 182Br, 190Br, 200Ml/Mr, 202Ml, 203Tl, 207Bl, 208Tr, 250Mr, 254Tr, 262Mr, 269U, 284Tl, 290Br, 292Bl, 293Mr, 306Bm/Ml, 308Bl, 320Bl; ***Greg Graeber:*** 171Br; ***Ross Gundersen:*** 141B, 269Tl, 283Tl; ***Robert Gundy:*** 203Tr; ***Ray Haberman:*** 7Tl, 45Br, 48Tl/Tr, 54Ll/Ul, 56Tl, 65Ml/Mr, 75Bl, 85Ul, 91Ul, 92Tr, 97Tl/Tr, 98Ml, 107Br/Lr/Tr, 109Br, 111Tl, 112Tl, 116Bl/Br/Tr, 117Bl/Br/Ml/Mr/Tl/Tr, 119T, 120Mr/Tl, 121Br/Tl/Tr, 122Bl/Br/Mr/Tl, 123Ml/Tl/Tr, 124Ml, 127Ml, 140Tl/Tr, 151Ll, 177Tl, 199Bl/Br/Ml/Mr, 213Tl, 255Br, 262Bl, 263Mm/Tl, 265Mr/Tr, 269Tr, 275Bl/Br, 281Bl/Br/L, 316B; ***Jeff Hamann:*** 80B, 115T, 243Ul; ***Kennedy Hanson, Apalachicola NERR (Florida DEP):*** 178Tl/Tr; ***Leslie Harris (LACM):*** 69T; ***Larry Harris:*** 319Bl; ***Ben Hartley:*** 226Tr; ***Cherie-Anne Henderson Dam:*** 16Tl, 21Ml, 48Mr, 186Tl/Tr, 202T, 249Tr, 269Bl/Br, 304Br/Ur, 307Br; ***Alicia Hermosillo McKowen:*** 29Tl, 36Tr, 51Br, 54Bl/Br, 75Ml, 80L, 86Bl, 90Bl, 93Br, 104Br, 108Bl/Br/Ll, 115Ml/Mr, 123Br, 134U, 138B/Tl/Tr, 144U,

146Tl, 147M/Tl/Tr, 149Ll, 150Tr, 151Tr, 152Tl/Tr, 154Bl, 155Bl/Br, 156Bl/Br, 162Br/Tr, 167Mr, 184Mr/Tl/Tr, 190Tr, 192Ml, 198Tl, 216Br, 219Br, 231M/Tl/Tr, 253Br, 259M, 294Tl, 312Tl/Tr/U, 319Br/T; ***Brenda Hill:*** 56Ml, 100Tl, 236Bl; ***Dave Holladay:*** 64Ml, 254Ml, 309Tl; ***Craig Hoover:*** 81Mr, 98Tr, 99Bl, 102T, 214Ll, 233B; ***Patricia Horie:*** 6Ml; ***Stephane Hourdez:*** 253T; ***Paul Humann:*** 154Mr; ***Linda Ianniello:*** 2Tl, 4Bl/Ll, 9Tl, 15Ml, 27Bl/Bm/Br/Ml/Mm/Mr, 28Bl/Br/Tm/Tr, 29Br/Mr/Tr, 30Bl/Bm/Br/Tl, 31Ml/Tr, 32Bl/Br/Tl/Tr, 33Br/Ml/Mr/Tl/Tr, 34Bl/Br/Mr/Tl/Tr, 35Bm/Br/Tm, 36Tl, 42Br/Tl, 49Bl/Br/Ml, 52Ml, 60Tr, 62Br, 66Br, 67Bl, 73Tl, 74Bl/Tr, 79M, 83Br, 95Tr, 97Mr, 104Tl, 108Tr, 132B/Ll, 133U, 135Bm/Tl, 146Br/Ml, 150Ll, 155Ml, 158Tl, 161Tr, 163Mr, 165Tl/Tr, 171Tl, 174Tr, 175Tr, 177M, 179Br/Mr/Tl, 180Bl, 181Ml/Tl, 182Mr, 185Tr, 190Tl, 191Tr, 195Ml, 196Br, 197Bl/Br, 205B/Tl/Tr, 206Tl/Tr, 207Br/Tl, 208Tl, 211Tl, 214Br/Tr/Ul, 216Mr, 220Br, 221Br, 222Br, 223Ml, 224L/Tr, 229Bl/Ml, 233Lr, 235Lr/Ul, 236Br/Tr, 237Tr, 239Mr, 241Lr, 242Tl/Tr, 243Ur, 245Bl/Br/T, 246Br/Tl/Tm/Tr, 247Br/Tr, 253Ul, 254Bl/Br, 257T, 264B, 266Ml/Mr, 276Tl/Tr, 285Ml, 289U, 294Tr, 305Bl/Mr, 307Bl, 308Tl, 315Tr; ***Invert E Base (CC0 1.0):*** 318U; ***Jack Israel:*** 9Lr, 132U, 267B, 268Tr, 310Tr; ***Rino Jackson:*** 44Bl, 64Bl, 79Tl/Tr, 87Ur, 102Br, 110Bl, 112Bl, 124Br, 129Bl/Ml, 158Bl, 175Ll/Lr, 185Tl, 194Bl, 205Ml/Mr, 234T, 260Bl, 277Ll, 279Bl, 292Ml, 300Mr, 301Tr, 303Lr, 304Bl, 309Br, 310Br, 311Br, 316U; ***Gabriel Jensen:*** 73Mr; ***Claire Jeuffroy:*** 259Tl; ***Christian Jones:*** 279Tl; ***Douglas Kahle:*** 249Br; ***S. Kaicher:*** 78Bl; ***John Kalimnios:*** 207Tr, 248Bl/Mr; ***Tim Kary:*** 57Br; ***Stephen C. Kempf:*** 159B; ***Hart Kerbel:*** 158Mr; ***Leila Kim:*** 58Ml/Mr; ***David Kirsh:*** 22Bl/Tr, 270M, 283Tr; ***Douglas Klug:*** 148M; ***Dewalt Kotze:*** 61Lr, 66Tr, 68Mm, 76Ml, 139Ml, 151Ur, 186B, 214Ur, 230Tr, 232Br, 286Br; ***Steven Kovacs:*** 31Bl; ***Nick Kowalske:*** 87T; ***Mike Krampf:*** 65Bl/Tr; ***Frank Krasovec:*** 8Tl, 18Tr, 38Br, 80Ul, 137Bl/Tr, 186Ml/Mr, 193Mr, 195Mr/Tr, 217Tl, 232Ml/Tl; ***Marlo F. Krisberg:*** 1T, 11T, 12Br/Tr, 15Tr, 16Br, 17Br, 18Br/Ml; ***Laura Kushner:*** 118Tl/Tr, 150Lr, 295U; ***Jose H. Leal:*** 109Tl; ***Diane Leazenby:*** 229Br; ***Patric Lengacher:*** 249Bl; ***Don Lewis:*** 164Tl; ***Kerry Lewis:*** 19Br, 21Tr, 104Bl/Mr, 105Tl, 106Tl, 233Tl, 249Mr, 298Tl/Tr, 313Ll/Lr; ***Phil Liff-Grieff:*** 23Tl; ***Jessica Link:*** 153Mr; ***Jorge Lopez:*** 37Br; ***Jack Mackay:*** 161Br/Lr; ***Leah Marcus:*** 55Bl; ***Magali Marquez-Ramos:*** 188Ml/Mr, 200Br/T, 208Mr, 314B; ***Andy Martinez:*** 38Tl/Tr; ***Angelica Matos:*** 226Ml; ***Jorge Mayorga:*** 52Bl/Br, 78Ul, 123Mr, 260T, 314U; ***April McCormack:*** 207Ml, 243Bl; ***Scott McDuff:*** 68Tl; ***Greg McFall (NOAA):*** 86Br; ***Andrew McKinlay:*** 55Tr; ***Melissa Meadows:*** 135Ml/Mr; ***Susan Mears:*** 28Bm, 29Ml, 31Mr, 35Bl, 233Ll; ***Suzan Meldonian:*** 28Tl, 29Bl, 30Tm/Tr, 34Ml, 41Br, 108Tl, 247Bl/Ml; ***Julie Mellinger:*** 13Bl/Tl, 36Mr, 41Bl, 44Br/Tl, 46Ml, 49Tl, 59Ml, 99Mr, 129Tr, 130Ll/Lr, 157Tr, 160Ml, 168Bl, 169Br, 210Ll, 220Bl, 244Tl/Tr, 251M, 255Ml, 257B, 258Bl/Mr, 268Br, 271Bl/Br, 279Br, 282Mr, 286Tr, 288Bl, 289Ll/Lr, 290Tr, 294Br, 296Mr, 299Mr, 306Br, 310Tl, 313B, 314T, 315Bl; ***Flavio Mendes:*** 36Ml, 78Ll, 81Br, 101Ml/Mr, 115Bl, 123Bl, 128Tl, 167Br, 318B/Ll/Lr; ***Michael Middlebrooks:*** 275T, 296T; ***Alvaro Migotto:*** 133Tl/Tr, 187B, 230B; ***Cristobal Molina:*** 311Bl/L; ***Leopoldo Moro Abad:*** 86T; ***Ellen Muller:*** 2Tr, 13Ll, 75Tr, 85Lr, 125Ml/Mr, 136Ll/Lr, 164B/Ml/Mr, 170B/Ml/Tr, 174Ml, 210Ur, 219Bl, 221Tr, 222Tr, 256Ml, 259Bl, 274Ml, 300Br, 301Tl, 320Br/Tr; ***Ryan Nelson:*** 248T; ***Tommy Nezik:*** 143Mr, 220Ml, 244Lr; ***Markus Niemeyer:*** 3Tr, 17Tr, 51Mr, 58Ml (inset)/Tr, 100Br, 135Br, 166Ml, 202Bl, 204Tr, 207Mr, 212Tl, 223Br/Mr, 240Tr, 285Tl; ***Martin Nobs:*** 301Bl; ***Jose Nunez:*** 131T; ***Giulia Oliveira:*** 226B; ***Karen Oliver:*** 58Bl; ***Tricia O'Malley:*** 50Bl, 125Br, 225B; ***Deneb Ortigosa:*** 14Bl, 59Bl/Br; ***Karen Osborn:*** 31Br; ***Alev Ozten:*** 4Mr, 170Tl; ***Vinicius Padula:*** 78Tr; ***Alex Parry:*** 171Tr; ***Jessica Pate:*** 42Tr; ***Cedric Pau:*** 158Br; ***Gustav Paulay (FMNH):*** 18Bl, 157B, 287Bl/Br; ***Elizabeth Pelkey:*** 193Ml, 241Bl; ***Jose Juan Calderin Peñate:*** 133B; ***Anderson Rabello Pereira:*** 38Ur; ***Jason Phillip:*** 60Lr, 67Br, 72Tl, 88Lr/Ur; ***Bernard Picton:*** 65Br, 69Ml/Mr, 113Bl/Br, 222Mr; ***Seth Pikal:*** 80Ur; ***Daniel Pinelli:*** 59Mr; ***Kadu Pinheiro:*** 63Bl, 69Bl/Br, 128Tr, 153Tl, 154Tl/Tr, 156M; ***Julie Piron:*** 44Mr; ***Tom Poff:*** 9Br, 95Bl, 146Mr, 147Bl, 211Ul, 220Tr; ***Denisse Pohls:*** 218M, 316Ll; ***Daniel Prairie:*** 61Tr, 72Bl/Mr, 76Mr/Tr, 82Ml, 99Br, 132Lr, 191Br, 209Mr, 220Mr, 266Tr, 289T, 307Ml; ***Betsey Putnam Hansen:*** 47Bl; ***Lee Richter:*** 57Tr, 297Bl; ***Brian Ricker:*** 148Tl/Tr; ***Jessica Riederer:*** 71Mm/Mr/Tr, 238Mr, 254Mr; ***Francisco Riquelme Tapia:*** 37Tl/Tr; ***Mauricio Riquelme:*** 6Mr, 63Lr, 89Mr, 96Tl, 101Tl, 102Bl, 103Tl, 105B, 182Bl, 277Ur, 302Br, 306Tr; ***Nelsy Rivero:*** 81Ml; ***Laetitia Rossignol:*** 13Ul, 42Bl, 43Mr, 49Mr, 176Bl/Br, 210B/Lr, 244Ur, 304Lr, 309Ml; ***Lazaro Ruda:*** 288Tr; ***Mike G. Rutherford:*** 70T; ***Xavier Salvador:*** 15Bl/Br; ***Pablo Samper Mendez:*** 133Lr; ***David Sanchez:*** 60Ll/Ul, 73Br, 76Tl, 87Br, 95Ml, 120Tr, 134Tl, 135Bl, 148B, 151Ul, 176T, 184Br, 216Ml/Tl/Tr, 222Bl, 229Tr, 285Bl, 286Tl, 289B; ***Dan Schofield:*** 212M; ***Matthew J. Scripter:*** 300Tl, 314L; ***Paul Selden:*** 258Br/Ml; ***Nancy Sheridan (Florida Fish and Wildlife):*** 76Bl/Br, 78Br, 113Ml; ***Prof. Carlos Sigmaringa & PhD Alexandre Pimenta:*** 11M; ***Paula Sills:*** 142Bl/Br; ***Audrey Smith:*** 298B; ***Smithsonian Marine Station:*** 19Tr, 188B; ***Vance Solseth:*** 194Tl/Tr; ***Randall Spangler:*** 109Tr; ***Svetlana Speransky:*** 62Tr, 67Ur, 113Tl/Tr, 134Lr, 150Tl, 152Br, 163Bl, 177Tr, 200Bl, 223Bl, 229Mr/Tl, 235B, 237Tl, 282Tl, 288Tl, 302Mr; ***Robert Stansfield:*** 174Br, 181Mr/Tr, 215B, 249Tl, 272Tr, 276B, 299Br; ***Joshua Stone:*** 149Lr; ***Robert Stuke Jr.:*** 225T, 237Bl; ***Matt Sullivan:*** 56Tr, 175Ul; ***Stephanie Taylor:*** 171Bl; ***Arturo Telle:*** 133Ll; ***Joe Tomoleoni:*** 58Tl; ***Anna Torres:*** 226Mr; ***Judy Townsend:*** 107Tl, 109Ml, 157Mr, 242Ul, 262Tl; ***Amy Tripp:*** 128Bl/Br/Ml/Mr; ***Daphne Tsapalas:*** 2Br; ***Heather Turek:*** 146Bl; ***Shawn Turek:*** 191Bl, 193Bl, 204Ml, 237Ml; ***Everett Turner:*** 2Mr, 4Br, 17Ml, 18Mr, 39Br/Tr, 43Tl, 50M, 73Tr, 96Tr/Ul, 110Br/Ul, 119Br/Ur, 137Ml/Mr, 144T, 166Tl, 172Mr, 183Ml/Mr, 204Tl, 210Tr, 217Bl, 242Lr, 253Ur, 256Tr, 278Bl, 293Br, 299Tl, 305Br/Tr; ***Nuch Uthairat:*** 8Br, 129Br, 218B, 304Ul; ***Angel Valdes:*** 15Tl, 106B, 125T, 127T; ***Chris Venus:*** 55Br; ***Peter Vermeul:*** 81T; ***Alfe Villan Bolos:*** 201Bl, 250Tr, 282Ml; ***Oliver Vincent:*** 105Ml/Mr; ***Marli Wakeling:*** 217M; ***Pauline Walsh Jacobson:*** 5Bl, 6Tr, 13Br, 60Ur, 68Mr, 100Tr, 107Ul, 166Br, 168Mr, 174Bl, 185Bl/Br, 188Tl/Tr, 190Ml, 195Tl, 197Ml/Mr, 211Bl/Br/Mr, 224Bl/Tl, 230Ml/Mr/Tl, 237Br, 240Tl, 263Br, 268Mr, 291Ml, 297Tl, 306Mr; ***Jacob Weston:*** 11B; ***Keri Wilk:*** 5Ml, 6Tl, 7Ml, 8Bl, 21Br/Mr, 43Tr, 48Ml, 55Tl, 57Mr, 60Br/Tl, 63Tr/Ur, 64Tl/Tr, 67Lr/Tl/Ul, 72Tr, 85Tr, 113Mr, 137Tl, 142Tl/Tr, 163Tl, 164Tr, 169Tr, 173Ml, 174Mr, 210Tl, 213Br, 232Mr, 238Tl, 254Tl, 256Tl, 262Br, 265Br, 272Tl, 290Tl, 297Ll/Lr/Ul/Ur, 303Ul, 304Tr; ***Kris Wilk:*** 9Tr, 112Ul, 248Ml; ***Benjamin Wilson:*** 96B/Ll/Lr; ***Tracey Winholt:*** 172Ml, 208Ml, 215Ll, 232Bl, 293Bl; ***Paul Young:*** 61Br; ***Terence Zahner:*** 55Mr, 63Tl, 65Tl, 67Ll, 134Tr, 165Mr, 177Bl/Br, 209Ml, 219Tl, 222Ml, 225Ll/Lr, 303Ur, 312L; ***Kierstin Zetterberg:*** 21Tl; ***Don Zoecklein:*** 115Br. Uncredited images were taken by the author.